D1451534

Biomolecular Stereodynamics

Biomolecular Stereodynamics, Volume IV

Proceedings of the Fourth Conversation in the Discipline Biomolecular Stereodynamics held at the State University of New York at Albany, June 4-8, 1985

Edited by

R. H. Sarma & M. H. Sarma

Institute of Biomolecular Stereodynamics
State University of New York at Albany and
Naitonal Foundation for Cancer Research

Adenine Press, P.O. Box 355
Guilderland, New York 12084

Adenine Press
Post Office Box 355
Guilderland, New York 12084

Cover illustration: Nucleosome structure: 3Å resolution view of the histone octamer with 165 b.p. DNA model built around it. From *Biomolecular Stereodynamics III*, "The Structure of the Histone Octamer and Its Dynamics in Chromatin Function" by R. W. Burlingame, W. E. Love, T. H. Eickbush and E. N. Mourdrianakis. Copyright © E. N. Moudrianakis, 1986.

Library of Congress Cataloging-in-Publication Data
(Revised for vol. 3 & 4)

Biomolecular stereodynamics.

Vol. 3— edited by R.H. Sarma & M.H. Sarma.
Includes bibliographical references and indexes.
Contents: v. 1-2. Proceedings of the Second SUNYA Conversation in the Discipline Biomolecular Stereodynamics held at the State University of New York at Albany April 26-29, 1981 under the auspices of the Department of Chemistry and organized by the University's Institute of Biomolecular Stereodynamics—v. 3-4. Proceedings of the Fourth Conversation in the Discipline Biomolecular Stereodynamics held at the State University of New York at Albany, June 4-8, 1985.
1. Biomolecules—Congresses. 2. Stereology—Congresses. I. Sarma, Ramaswamy H., 1939- . II. Sarma, M.H. (Mukti H.), 1940- . III. SUNYA Conversation in the Discipline Biomolecular Stereodynamics (2nd : 1981 : State University of New York at Albany) IV. State University of New York at Albany. Dept. of Chemistry. V. State University of New York at Albany. Institute of Biomolecular Stereodynamics. VI. Conversation in Biomolecular Stereodynamics (4th : 1985 : State University of New York at Albany)
QH506.B554 574.8'8 81-14867
ISBN 0-940030-00-4 (v. 1)
ISBN 0-940030-01-2 (v. 2)
ISBN 0-940030-14-4 (v. 3)
ISBN 0-940030-18-7 (v. 4)

Made in New York, USA

Preface

These are the proceedings of the Fourth Conversation in Biomolecular Stereodynamics held at the State University of New York at Albany June 04-08, 1985 under the auspices of the Department of Chemistry and organized by the University's Center for Biological Macromolecules and the Institute of Biomolecular Stereodynamics. Over 500 scientists from 15 countries were gathered at Albany in early June 1985 for the Fourth Conversation. These volumes essentially constitute the invited presentations. The papers contributed at the poster sessions of the congress have already appeared in the various issues of the *Journal of Biomolecular Structure & Dynamics* released after the conference.

The conference and these volumes were made possible by generous support from public and private institutions. We are immensely grateful to the following institutions for supporting the congress:

State University of New York
State University of New York at Albany: Office of the President, Office of Vice-President for Research, Dean, College of Science and Mathematics, Department of Biological Sciences, Chemistry and Physics
Adenine Press
Domtar, Inc
General Electric Company
Hoffmann-La Roche Inc
IBM Instruments Inc
Lederle Laboratories
Merck & Co, Inc
Naitonal Foundation for Cancer Research
National Institutes of Health
Shell Development Company
Smith, Kline and French Laboratories
Upjohn Company
Wilmad Glass Company

We thank our friends and colleagues who have either served in the Organizing Committee or have provided valuable suggestions or have physically helped us in several ways. We must particularly mention: M. M. Dhingra, G. Gupta, C. W. Hilbers, N. R. Kallenbach, S. Manrao, W. D. Phillips, A. G. Redfield, A. Rich, N. C. Seeman and A. H.-J. Wang. We acknowledge with gratitude the help from Virginia Dollar and Charles Heller of Chemistry, John Elliot of Biology, Al Dasher of the College of Science and Mathematics and Don Bielecki of the conference department of SUNYA.

We thank David L. Beveridge for editing two of the articles in these volumes *viz.*, that by E. W. Prohofsky and the one by F. Vovelle and J. M. Goodfellow.

Most of all we thank, congratulate and applaud the participants of the Fourth Conversation for thier notable contributions and their continuing and unabated dedication to the discipline of biological structure, dynamics, interactions and expression.

We are looking forward to the pleasure of being your hosts for the Fifth Conversation June2-6, 1987.

Ramaswamy & Mukti H. Sarma
Albany, New York
March 20, 1986

CONTENTS

Volume IV

CONTENTS

Volume III

Biomolecular Stereodynamics IV, Proceedings of the Fourth Conversation in the Discipline Biomolecular Stereodynamics, State University of New York, Albany, NY, June 04-09, 1985, Eds., Ramaswamy H. Sarma & Mukti H. Sarma, ISBN 0-940030-18-7, Adenine Press, ©Adenine Press 1986.

Intramolecular Collaps of DNA: Structure and Dynamics

Dietmar Porschke
Max-Planck-Institut für biophysikalische Chemie,
3400 Göttingen, FRG

Abstract

The structure and dynamics of collapsed DNA has been investigated by electro-optical and stopped flow measurements. The limit dichroism of DNA collapsed by spermine is about −0.4 close to the value expected for a toroidal organisation. In the absence of spermine the limit dichroism of the same DNA is −1.4, close to the value expected for usual straight B-form DNA. Addition of spermine to free DNA leads to a "cooperative" increase of the light scattering intensity at a threshold concentration indicating the transition to the collapsed form, whereas the electric dichroism increases gradually with the spermine concentration and approaches a plateau value at a concentration, which is sufficient to induce a complete collaps according to the light scattering intensity. Stopped flow measurements of the DNA collaps detected by light scattering at low DNA and spermine concentrations around 1 μM show a characteristic induction period, which is attributed to the time required for spermine binding, until the threshold degree of binding is exceeded. As soon as the threshold is exceeded, the DNA collapses in a relatively fast reaction. Simulation of the spermine binding according to an excluded site model and comparison with the stopped flow data demonstrate that spermine molecules have a relatively high mobility along the double helix corresponding to a rate constant of approximately 200 s^{-1} for movement by one nucleotide residue. At high spermine concentrations, where the rate of DNA collaps is not determined by spermine binding anymore, the dynamics of DNA folding—analysed by electric field jump experiments with scattered light detection—is reflected by a spectrum of time constants ranging from 25 μs to 2 ms. The high rate of DNA folding demonstrates a remarkable flexibility of the double helix with very fast bending motions. The DNA may be compared with a spring, which is kept under tension by electrostatic repulsion; upon release of this tension by binding of positively charged ions the spring collapses almost immediately. The gradual decrease of the DNA dichroism with increasing degree of spermine binding below the threshold value suggests that the helix is bent by spermine binding to a superhelix structure before the transition to the collapsed state is observed.

Introduction

Since DNA molecules are often exposed to conditions leading to degradation and thus to loss of the genetic information, long DNA molecules are usually protected by some packing procedure. A particularly simple form of packing is observed, when DNA is exposed to spermine or spermidine (1,2). Binding of these ions to DNA reduces the electrostatic repulsion between DNA strands and induces the formation of a very compact form of DNA. Gosule and Schellman (1,2) demonstrated that in the limit of very low concentrations ($\sim 1\ \mu M$ of both reactants) intermolecular association of DNA molecules can be avoided and long DNA strands are "condensed" in an intramolecular reaction to a tightly packed toroidal form (3). According to electron microscopy the DNA is organised in the torus by "circumferential wrapping" (4). The collaps of DNA is not only observed in the presence of ions like spermine or spermidine, which are known to participate in natural packing reactions in phage heads, but also by ions like $(Co(NH_3)_6^{3+}$ (5). Apparently the ions merely serve to reduce the charge density of the double helix and do not provide any substantial specific contribution to the interactions in the collapsed state (6). This interpretation is also supported by measurements of the degree of ligand binding at the midpoint of the collaps reaction (7). These measurements indicate that for different ions associated with different affinities to the helix a collaps is always observed at the same degree of binding, corresponding to the same degree of compensation of phosphate charges. All these results provide more insight into the collaps reaction; however, the nature of the interactions leading to torus formation remains unclear. Probably tight packing of DNA helices is supported by salt bridges between phosphate residues and spermine molecules (8).

The ligand induced collaps of DNA is a remarkable reaction not only with respect to the nature of interactions in the toroidal form but also with respect to its dynamics. A very long thread of DNA extended randomly over a large space with properties analogous to a wormlike coil is wrapped into a well ordered and compact torus. This particular form of a transition from a disordered to an ordered form has been studied by electric field jump and stopped flow measurements. The results provide new information on the dynamics of DNA helices and the structures involved in the collaps reaction. This information obtained for the packing of DNA by simple ligands should also be useful for an understanding of more complex DNA packing reactions as for example in chromatin.

Materials and Methods

λ-DNA (Boehringer, Mannheim, FRG) and T4-DNA (Miles Laboratories, Elkhart, IN) were dialysed extensively against a standard buffer containing 1mM NaCl, 1mM Na-cacodylate pH 6.5 and 50 μM EDTA. Spermine*4HCl and spermidine*3HCl were obtained from Serva, Heidelberg, FRG.

The electric dichroism was measured by a pulse generator developed by Grunhagen (9) and an optical system composed of a 600 W mercury-xenon lamp, a double monochromator (2 Schoeffel GM250), a Glan-air polariser and a fast photomultiplier head. The optical pathlength of the measuring cell was 20 mm with Pt-electrodes at a distance of 7.4 mm. Illumination of the samples was restricted by an automatic shutter to a few seconds for each pulse experiment, in order to avoid photoreactions. The "total light intensity" was measured during this time interval (before the pulse) and was stored by a sample and hold circuit. The light intensity and the electric field strength was recorded as a function of time by a transient digitizer 'tektronix 7612D'. The data were transferred to a LSI 11/23 (digital equipment) for an evaluation of absorbance changes and of the electric field strength (10). The time constants were evaluated by an efficient deconvolution procedure on the Univac 1108 of the Gesellschaft für wissenschaftliche Datenverarbeitung, Göttingen (11). Field jump and stopped flow measurements with scattered light detection were performed as described in references (10) and (12) respectively.

Results

The structure of collapsed DNA in solution

The spermine induced collaps of DNA is a remarkable reaction for various reasons. The most interesting aspect appears to be the observation that the "collaps" leads to a well ordered toroidal form. DNA toroids have been observed by electron microscopy in different laboratories (3,4,13,14). However, it is difficult to exclude artifacts resulting from the preparation of specimens for electron microscopy and thus it should be useful to obtain independent evidence for the existence of toroids in aqueous solution. A method for the analysis of structures in solution, which is particularly sensitive and proved to be useful for the characterisation of nucleic acids, is the electric dichroism (15-17). By this method it is possible to obtain information on the structure of collapsed DNA at concentrations around 2 μM nucleotide residues, where intermolecular association can be avoided. As shown in Fig. 1, the dichroism of collapsed DNA observed at low field strengths is higher than that of "free" DNA, but approaches saturation already at relatively low field strengths providing a limit dichroism of about -0.4. The dichroism of "free" DNA continues to increase at higher field strengths and finally extrapolates to a limit value of -1.4. The measurements of the dichroism for the collapsed DNA cannot be extended to higher electric fields, because high field pulses convert collapsed DNA to the free form (cf. below and ref. 20).

A toroidal organisation of the collapsed DNA implies that the DNA helices are wrapped in circles. Electric field pulses induce orientation of these circles in the direction of their largest polarisability, which is in the plane of the circle. Thus the circles should be aligned with their diameters parallel to the electric field vector. For this case the theoretical limit dichroism is -0.375 (21). The close agreement of the experimental dichroism with the value expected for a torus structure corroborates the evidence for a toroidal organisation of DNA helices in the spermine induced collapsed form. Additional information on the structure in solution can be obtained from dichroism experiments by an analysis of the rotation time constants. Compared to the rather broad spectrum of time constants observed for free DNA in the range from 5 μs to 20 ms the corresponding spectrum for toroidal DNA is relatively narrow. The time constants found for DNA collapsed by spermine in the range around 80 μs are consistent with a toroidal structure.

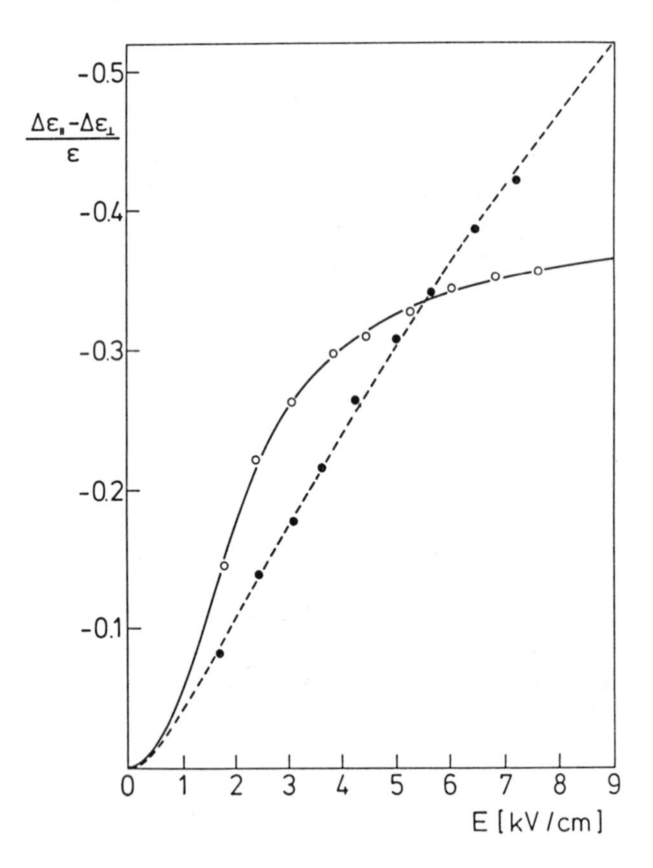

Figure 1. Electric dichroism $(\Delta\epsilon_{\parallel}-\Delta\epsilon_{\perp})/\epsilon$ of 2 μM λ-DNA as a function of the electric field strength E without spermine (●) and in the presence of 1 μM spermine (○). Least squares fitting by a saturating induced dipole model (with a transition from the induced to the saturated dipole according to the "square root" function, cf. ref. 18, see also ref. 17 and 19) provides values for the limit electric dichroism of −1.4 for the free DNA and −0.4 for the collapsed DNA. The saturated dipole moments are $1.4*10^{-26}$ Cm for the free and $1.1*10^{-25}$ Cm for the collapsed DNA (½ standard buffer, 20°C).

"Gradual approach" or "instantaneous transition" to the collapsed state

According to the measurements of the light scattering intensity the transition from the "free" to the collapsed form of DNA occurs in a narrow range of the spermine concentration and thus resembles a cooperative transition (7,12). This cooperativity results from the fact that the degree of ligand binding must exceed a threshold value to induce the DNA collaps. The process of spermine binding to the double helix is not reflected by any large change of the light scattering intensity, as long as the degree of binding remains below the threshold. This does not mean, however, that the DNA structure is not affected by spermine association at low degrees of binding. Detection of effects below the threshold has been attempted by measurements of the electric dichroism. These measurements illustrate a completely different aspect of the spermine induced DNA collaps. As shown in Fig. 2 the limit dichroism shows a continuous decrease with increasing spermine concentration and arrives at a plateau value of about −0.4 at a spermine concentration, which is just sufficient to induce complete DNA collaps according to measurements of the light scattering intensity performed under corresponding conditions. These data demonstrate that the "cooperative" transition indicated by a large change of the light scattering intensity is preceded by a gradual change of the DNA structure. The decrease of the dichroism suggests that the gradual change involves bending of the DNA helix and the formation of a superhelix structure.

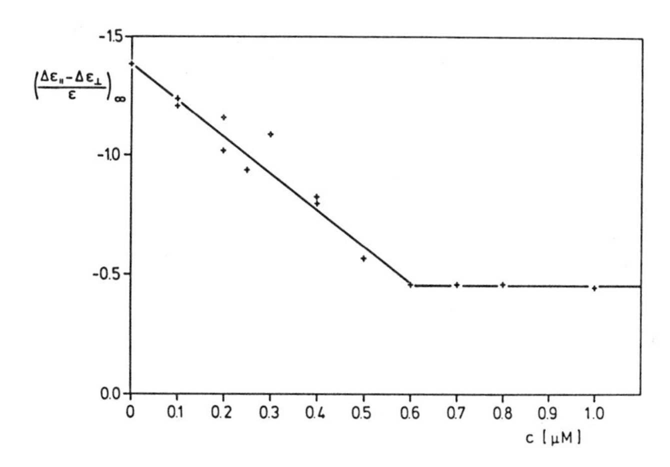

Figure 2. Limit electric dichroism $[(\Delta\epsilon_{\parallel}-\Delta\epsilon_{\perp})/\epsilon]_{\infty}$ of 2 μM λ-DNA as a function of the spermine concentration (standard buffer, 20°C).

Dynamics of the intramolecular collaps

At a first glance an investigation of the DNA collaps dynamics seems to be relatively simple and apparently does not require more than simple mixing of DNA with a condensing agent like spermine and follow the reaction by established techniques. However, the investigation is not simple at all, since the data have to be obtained at very low concentrations, in order to avoid aggregation between DNA strands. Moreover, the intramolecular collaps of DNA proves to be too fast for investigation by simple mixing, whereas fast mixing by conventional stopped flow techniques shears long DNA molecules into small pieces. The difficulty has been settled by using a stopped flow apparatus with a reduced flow rate to avoid shearing at the expense of a high time resolution (12). The reaction followed by measurements of the light scattering intensity (cf. Fig. 3) shows an induction period, which is clearly visible at low reactant concentrations and is reduced with increasing spermine concentration c_s. When the spermine concentration is increased beyond 5 μM the main part of the reaction is too fast for characterisation by the modified stopped flow apparatus. These results indicate that the rate of DNA collaps under the conditions of the stopped flow measurements is determined by the rate of spermine binding.

The spermine induced intramolecular collaps of DNA involves a large number of reaction steps. The sequence of reaction steps is expected to correspond to the following general mechanism:

$$\begin{aligned} DNA_{free} + \text{spermine} &\Leftrightarrow DNA_{free}{*}\text{spermine} \\ DNA_{free}{*}\text{spermine}_n + \text{spermine} &\Leftrightarrow DNA_{free}{*}\text{spermine}_{n+1} \qquad (1) \\ DNA_{free}{*}\text{spermine}_m &\Leftrightarrow DNA_{collapsed}{*}\text{spermine}_m \end{aligned}$$

Upon mixing of the reactants spermine molecules will bind to the double helix at many different sites. This part of the reaction is not reflected by any measurable change of the light scattering intensity and corresponds to the induction period found in the stopped flow experiments. When the degree of spermine binding

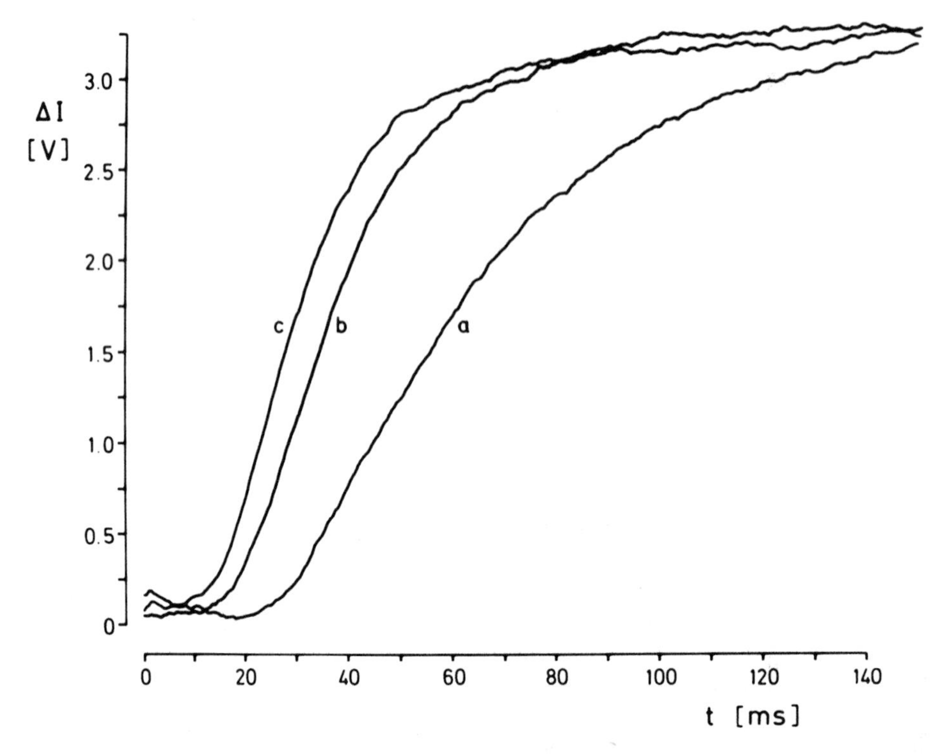

Figure 3. Change of scattered light intensity ΔI as a function of time t after mixing of 1 μM T4 DNA with (a) 0.7, (b) 1.0 and (c) 1.5 μM spermine in a "slow" stopped flow apparatus (standard buffer, 20°C, from ref. 12).

arrives at a threshold value, the DNA molecule is converted from the "free" wormlike coil to a compact toroidal state. This conversion is accompanied by a large change of the light scattering intensity.

The set of equations (1) is already relatively complex; however the reaction is still more complex due to the fact that each ligand covers more than a single nucleotide residue resulting in "exclusion of binding sites" (22). The excluded site phenomenon has a particularly strong influence on the reaction at high degrees of binding and leads to a very complex coupling of reaction steps, which cannot be described by analytical procedures but requires numerical methods (23). According to equilibrium titrations the DNA collaps is induced, when more than 86% of the phosphate charges are compensated by spermine molecules. Numerical simulation of the ligand association up to this degree of binding demonstrate that the stopped flow data can only be explained, when it is assumed that spermine molecules are mobile along the double helix (12). By comparison of measured and simulated data it is concluded that the rate constant for moving spermine by one nucleotide residue is about 200 s^{-1}.

The stopped flow data provide information about the DNA collaps in the limit of low spermine concentration, where the rate of DNA collaps is determined by the dynamics of ligand binding. The rate of the DNA collaps itself can only be studied in the limit of high spermine concentrations, where the reaction is too fast to be followed by stopped flow techniques. This part of the reaction could be studied by the electric field jump technique. Electric field pulses are used to induce dissociation of spermine molecules from collapsed DNA by a "dissociation field effect" (20). At high pulse amplitudes the degree of spermine binding is pushed below the threshold and thus the DNA is converted from the toroidal to the free form. The reverse reaction can be observed after pulse termination. Owing to the high time resolution of the field pulse technique, the dynamics of DNA collaps can be followed up to high spermine concentration. The reaction curves obtained under these conditions cannot be described by a single exponential, but requires at least three exponentials extending from 25 μs to about 2 ms (cf. ref. 12). A very similar distribution of time constants is observed at high spermidine concentrations (cf. Fig. 4). Apparently these reaction curves reflect a continuous spectrum of helical folding motions. Analogous bending motions have been observed previously for short restriction fragments by measurements of the linear dichroism (24). The time constants of these motions are about 700 ns at a chain length of 250 base pairs and strongly increase with increasing chain length. Thus the results on bending of short restriction fragments and on the collaps of long DNA molecules appear to be compatible.

Pathway of the field induced transition from the "toroidal" to the "free" DNA

As described above electric field pulses induce dissociation of spermine molecules from toroidal DNA and thus lead to a field induced transition to the free DNA.

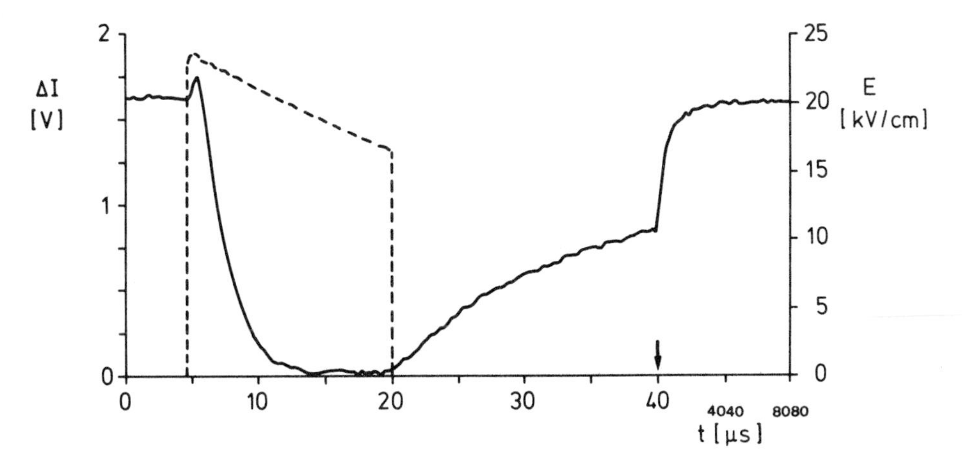

Figure 4. Change of scattered light intensity ΔI as a function of time t induced by an electric field pulse (broken line, field strength E given on left scale) on a solution of 1 μM T4-DNA + 1 mM spermidine in the standard buffer. The relaxation after pulse termination can be fitted by 3 exponentials (convolution due to limited response time of the detector considered) τ_1=45 μs (21%), τ_2=280 μs (57%) and τ_3=1.3 ms (22%).

The mechanism of this field induced reaction has been analysed by measurements of the transition rate at different field strengths. The observed linear increase of the time constants with the electric field strength suggests that the spermine molecules are pushed from the DNA by a dissociation field effect (20). This result provides information about the mechanism of ligand dissociation. Some information on the subsequent step of toroid unfolding may be obtained by measurements of the electric dichroism. Fig. 5 shows the electric dichroism of toroidal DNA induced by an electric field exceeding the threshold required for the transition from the collapsed to the free form. Under the conditions of this experiment the time constant of the field induced transition, which has been characterised independently by measurements of the light scattering intensity, is around 10 μs, whereas the orientation risetime of the toroids is about 1μs. Thus the DNA toroids are first oriented into the electric field. However, at the same time spermine molecules are dissociated from the collapsed DNA, and the toroids are unfolded. This reaction is apparently reflected by a decrease of the dichroism resulting in a minimum value, before ultimately the dichroism increases again to the level characteristic of free DNA. The experiment demonstrates that the collapsed DNA is unfolded by electric field pulses via an intermediate of relatively low dichroism. Probably the intermediate is a disorganised state resulting from the collapsed DNA directly after dissociation of spermine molecules by repulsion between negatively charged phosphate residues.

According to this assignment of the effects shown in Fig. 5 the slow increase of the dichroism observed after the minimum is due to orientation of free DNA. A comparison with risetime constants of the same DNA in the absence of spermine demonstrates that the risetime found in Fig. 5 is unusually large. This observation may be explained by a topological problem of disentangling the DNA strand from the toroidal organisation.

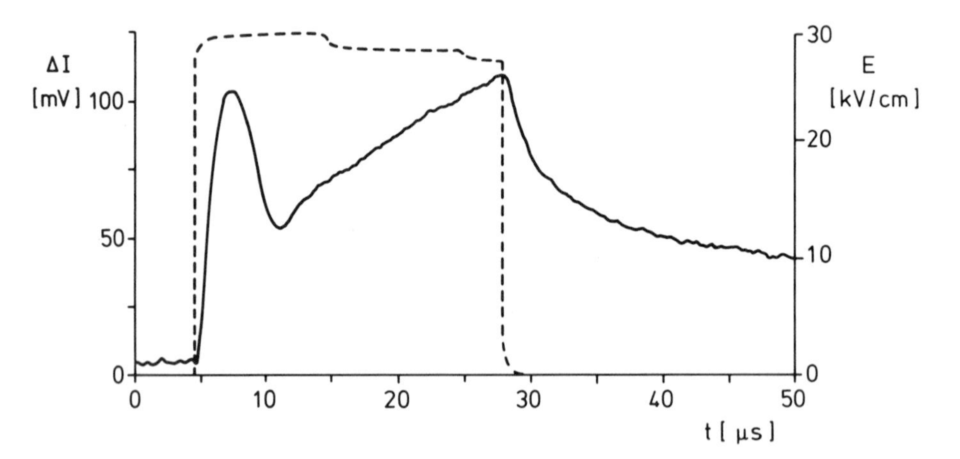

Figure 5. Change of the light intensity ΔI at a parallel orientation of polarisation plane and field vector as a function of time t induced by a field pulse (broken line, left scale) for 4.3 μM λ DNA + 1.5 μM spermine (standard buffer, 20°C; the optical signal is convoluted due to a limited response time of the detector of about 0.7 μs).

Discussion

Since the genetic information contained in long DNA molecules may be damaged or destroyed by many different reagents and also by physical effects like shearing, these molecules have to be protected against degradation. At the same time, however, the genetic information should remain easily accessible. The demands for efficient data protection and fast data processing appear to be in conflict with each other. This antithesis demonstrates that an understanding of the protection procedures used in nature requires knowledge both of the structures and their dynamics.

In the present investigation it is attempted to analyse both the structures and the dynamics involved in a particularly simple example of a DNA packing reaction. The electro-optical measurements of DNA collapsed by addition of spermine provide evidence that the toroidal structure of the collapsed DNA observed by electron microscopy (3,4,13,14) also exists in dilute aqueous solution. The transition to the collapsed state as reflected for example by measurements of the light scattering intensity (7,12) is usually regarded as a "cooperative" transition, which is induced, when the degree of ligand binding exceeds a threshold value. A complementary picture of the collaps process is provided by the electro-optical measurements. According to these measurements the cooperative collaps is preceded by a gradual change of the DNA structure, which leads in an apparently smooth transition to the collapsed state. The continuous decrease of the limit dichroism upon addition of spermine suggests that the structure of DNA, which is expected to correspond to a wormlike coil in the absence of spermine, gradually changes towards a torus structure, which can be envisaged by intermediate states with an increasing degree of superhelicity. Bending of the double helix to a superhelix structure may be supported by an asymmetry of the charge distribution along the double helix resulting from spermine binding.

A superhelix structure appears to be a reasonable starting point for the formation of a toroid and also provides a simple explanation for the observed high rate of DNA collaps. In this connection the analogy of a spring, which is kept under tension by electrostatic repulsion, seems to be pertinent: upon release of the tension by compensation of the phosphate charges the spring collapses almost immediately. By this simple mechanism the DNA can be protected at a high rate by a small change of ligand binding and the reverse process can also be completed in a relatively short time. The characterisation of the ligand binding reactions preceding the collaps reaction demonstrates that the mobility of ligands along the double helix is crucial for a short response time. A similar mechanism is expected to be even more important for packing of DNA by large and rigid ligands as in the case of chromatin.

Acknowledgement

The expert technical assistance of Mr. J. Ronnenberg is gratefully acknowledged.

References and Footnotes

1. L.C. Gosule and J.A. Schellman, *Nature (London) 259,* 333 (1976).
2. L.C. Gosule and J.A. Schellman, *J. Mol. Biol. 121,* 311 (1978).
3. D.K. Chattoraj, L.C. Gosule and J.A. Schellman, *J. Mol. Biol. 121,* 327 (1978).
4. K.A. Marx and T.C. Reynolds, *Proc. Nat. Acad. Sci. USA 79,* 6484 (1982).
5. J. Widom and R.L. Baldwin, *J. Mol. Biol. 144,* 431 (1980).
6. G.S. Manning, *Quarterly Reviews Biophys. 11,* 179 (1978).
7. R.W. Wilson and V.A. Bloomfield, *Biochemistry 18,* 2192 (1979).
8. J.A. Schellman and N. Parthasarathy, *J. Mol. Biol. 175,* 313 (1984).
9. H.H. Grünhagen, *Messtechnik 1974,* p. 19.
10. D. Porschke, H.J. Meier and J. Ronnenberg, *Biophys. Chemie 20,* 225 (1984).
11. D. Porschke and M. Jung, *J. Biomol. Struct. Dynamics 2* (1985) *2,* 1173 (1985).
12. D. Porschke, *Biochemistry 23,* 4821 (1984).
13. S.A. Allison, J.C. Herr and J.M. Schurr, *Biopolymers 20,* 469 (1981).
14. K.A. Marx and G.C. Ruben, *Nucleic Acids Res. 11,* 1839 (1983).
15. E. Fredericq and C. Houssier, *Electric Dichroism and Electric Birefringence,* Clarendon Press, Oxford (1973).
16. C.T. O'Konski, Ed., *Molecular Electro-optics,* Marcel Dekker, New York (1978).
17. S. Diekmann, W. Hillen, M. Jung, R.D. Wells and D. Porschke, *Biophys. Chem. 15,* 157 (1982).
18. S. Diekmann, M. Jung and M. Teubner, *J. Chem. Phys. 80,* 1259 (1984).
19. K. Yoshioka, *J. Chem. Phys. 79,* 3482 (1983).
20. D. Porschke, *Biopolymers, 24,* 1981 (1985).
21. D.M. Crothers, N. Dattagupta, M. Hogan, L. Klevan and K.S. Lee, *Biochemistry 17,* 4525 (1978).
22. J.D. McGhee and P.H. von Hippel, *J. Mol. Biol. 86,* 469 (1974).
23. I.R. Epstein, *Biopolymers 18,* 2037 (1979).
24. S. Diekmann, W. Hillen, B. Morgeneyer, R.D. Wells and D. Porschke, *Biophys. Chem. 15,* 263 (1982).

Biomolecular Stereodynamics IV, Proceedings of the Fourth Conversation in the Discipline Biomolecular Stereodynamics, State University of New York, Albany, NY, June 04-09, 1985, Eds., Ramaswamy H. Sarma & Mukti H. Sarma, ISBN 0-940030-18-7, Adenine Press, ©Adenine Press 1986.

Motional Dynamics of the DNA Double Helix

E.W. Prohofsky
Department of Physics
Purdue University
West Lafayette, Indiana 47907

Abstract

We discuss a number of recent experiments which indicate that the vibrational modes of the double helix have long lifetimes and long coherence lengths. This situation indicates that lattice dynamics is the correct approach to a theoretical study of the vibrational modes of the helix for helices longer than tens of basepairs. We discuss the predictions of such a theoretical approach and the experimental verification of many effects and quantitative mode frequency predictions. We also discuss a number of applications of the basic theory. One application deals with methods for calculations of local motional dynamics at unique regions such as termini, replicating forks, or unique sequence regions. Another application is the development of a microscopic mean field theory of helix melting. A third application is a soft mode approach to a theory of conformation change.

Introduction

The methods described in this paper are based on methods devised for the study of crystaline solids. It may at first appear strange to many that such methods can be successfully applied to the study of a biological entity such as the DNA double helix. Experimental observations during the last year have indicated however that the approach works very well indeed and that it is likely the best approach to the study of thermal motions and directed motions in long strands of DNA. The DNA double helix does act like a very thin but long mini solid. It seems to have vibrational modes which are modes of the helix proper and above $\approx$ 2.5 GHz they are to a large extent decoupled from the helix surroundings. They have lifetimes above hundreds of picoseconds and mean free paths or coherence lengths of thousands of base pairs. These modes can be observed in DNA in many stages of aggregation and seem to be relatively independent of the state of aggregation.

In the second section we discuss the experimental evidence that leads to these conclusions. Many of the observations can be interpreted in terms of well known physical principles and don't depend on any detailed theory. Other points do depend on specific theories and the consistency between the theory and a large number of observations will be described. The lattice dynamics approach has made numerous prior predictions of specific mode frequencies and characters and all such predictions are in very good agreement with subsequent experimental observation. The apparent correct view of excitations in long strands of double helical DNA is that of propagating quantized phonons with relatively long lifetimes and coherence lengths. The connectivity of the helix dominates the nonlinear and stochastic factors to bring this about.

For those who are unfamiliar with lattice dynamics Section III contains a brief description of how lattice dynamics relates to calculations of normal vibrations used in smaller molecules. It develops the way in which such calculations can be extended to large (infinite) systems if a certain symmetry exists. It shows that wavelike solutions arise in these systems.

In Section IV we discuss the application of lattice methods to the problem of conformation change. Both theoretical and experimental observations indicate that at least the dynamics of the A to B change is soft mode mediated. Analysis along these lines can greatly enhance the ability to understand the details which drive conformation change.

In Section V we discuss an approach to incorporating nonlinear or anharmonic effects into lattice dynamics. The method leads to a theory of helix melting in which vibrational modes dominate the dynamics of melting in a way that resembles the way in which such modes dominate the dynamics of conformation change in soft mode transitions. The method is applied to a simple model system and results from calculations on poly(dG)•poly(dC) are presented.

Because the coherence length in DNA is large one has to be careful in calculating local effects. For example one may be interested in how the modes of one particular region, say particular codon, differ from that of another region, say a different codon. The codon is imbedded in a helix with strong connective effects and coherent interaction over thousands of base pairs and one shouldn't just calculate the behavior of an isolated codon and expect it to be an appropriate solution. One must either include many thousands of base pairs in the calculation cell or use specialized methods developed for such calculations. In Section VI we discuss such a special method using Green functions to calculate the behavior of a unique region of helix. There is an experimental observation which is in agreement with predictions of an end mode based on such calculations.

Lifetime and Coherence Length of Modes of the DNA Double Helix

The observation of resonant microwave absorption (1) by DNA in dilute solution is the most straightforward indicator that excitations exist in the single macromolecule of the double helix and that they have surprisingly long lifetimes. The simplest modes in matter are those of compressional acoustic waves. In a long thin system of finite length like the double helix standing acoustic waves can occur. These waves are familiar as the standing waves on a violin string, or a closer analogy as the standing sound waves of an organ pipe. Such standing waves must fit into the length of the system so as to match the boundary conditions at the ends. The standing waves occur when a number of half wavelengths fit the distance between ends and allow for either node or antinode boundary conditions. Such acoustic waves can exist and be sharply tuned on the helix if these waves are localized to the helix and don't couple strongly to the water etc. of the helix surroundings. Such waves are expected to be at the microwave frequencies (2) and could be observed as microwave absorption peaks. Of these standing waves only those with odd half wavelengths would have net coupling to a microwave field. The even half wavelength modes would cancel the net absorption. The wavelengths for a strand of length L would be given by

$$(n + \tfrac{1}{2})\lambda = L \tag{1}$$

and the frequency ν by

$$\nu = \frac{C}{L}(n + \tfrac{1}{2}) \tag{2}$$

where C is the velocity of sound on the helix. The frequencies have the ratios of 1:3:5:7---. This pattern is observed when care is taken to have DNA strands or identical length (1). The frequencies do depend on the length of the DNA strand and have an acoustic velocity for the linear pieces the same as that observed by Brillouin scattering (3,4). In the experiment described the DNA is in very dilute solution and the individual DNA molecules are separated and the sound wave must be associated with the single DNA molecules. Similar resonances are observed in closed natural plasmid DNA but in that case the standing waves are integral whole wavelength modes as needed for that topology.

The intact plasmids are supercoiled. The observation of the standing waves predicted for these entities implies that the acoustic wave is able to propagate around all the bends and twists resulting from the supercoiling. The velocity is greater than that for the linear DNA but this is consistent with observations in fibres that the velocity increases as the packing of the fibers is tighter.

The fundamental mode, i.e., the one with $n = 0$ in Eq. 1 is seen for a segment of DNA $\approx$ 1000 b.p. in length. This necessarily implies a coherence length of that magnitude as the wave must be coherent over the length of the sample to express the standing wave resonance. The authors report lifetimes $>$ 300 p.s. and coherence lengths $>$ 1000's of b.p. for DNA in dilute solution. From a comparison of the shape of the higher harmonics as compared to theoretical calculations of absorption the lifetime of these modes are increasing with frequency. The damping of the modes presumably by the aqueous surroundings seems to fall off above $\approx$ 2.5 GHz.

The discussion till now has centered on acoustic modes. Most of the vibrational modes of a complex system like DNA are optical modes, that is they are not characterized by a simple inverse relationship between frequency and wavelength. One may expect that the long lifetime and large coherence lengths are more likely for acoustic modes than optical modes. This assumption is likely wrong as the acoustic modes should couple more strongly to the aqueous surroundings as they tend to drag water along in their motions. The optical modes are such that the center of mass of each unit cell remains fixed in space. There is no gross motion for the surrounding water to couple to. In any case there is experimental data which allows us to estimate coherence lengths for the low lying optical modes as well.

It is more difficult to determine lifetime and coherence length for the optical modes because the experimental material is different. In the microwave experiments referred to, the DNA was identical single molecules made from identical plasmids. The data from Raman and Brillouin measurements comes from DNA fibres and films. The individual molecules are of different length and base sequence. In addition the fibres are considerably disordered in their relative positions in the macroscopic state of aggregation. The observed lines are heterogeneously broadened and one is unlikely to be measuring intrinsic linewidths. Nonetheless reasonable lines are detected. It is observed that the lines do show k-conservation scattering (5,6). Because of the polarization selection (7) of the observed lines one knows that these lines correspond in the DNA helix to modes of wavelength of 10-11 base pairs. For the k-conservation to be expressed the coherence length must be many wavelengths or at least of the order of hundreds of b.p. This lower limit does not depend on any detailed theoretical model.

By using a second observation one can get a longer estimate of the coherence length but it does depend on a particular theoretical model. Our calculations (8,9) of the spectrum A poly(dG)•poly(dC) show a number of modes which satisfy the selection rules for Brillouin and Raman scattering (see Table I). One mode, the 12 cm^{-1} mode is a zone center mode which corresponds to infinite wavelength. That it is of such wavelength is verified by its being seen in (z,z) polarization scattering. We predict that it is the conformational soft mode of the A to B conformation change (see Sect. III). Another mode or group of modes at $\approx$ 22 cm^{-1} are at wavelength of eleven base pairs. This wavelength is verified as this mode is seen in (x,z) or (y,z) scattering polarization. This mode is predicted to be closely related in character to the 12 cm^{-1} mode. This relationship is born out by the experimental observation that it also softens as predicted at the A to B conformation change. For the shift in frequency from 22 cm^{-1} at eleven base pairs to 12 cm^{-1} to be expressed the coherence length must be of the order of thousands of base pairs. It appears that this low lying optical mode has a coherence length as large or larger than the acoustic modes above 2.5 GHz. The optical modes are expected to have less coupling to water around the helix and therefore longer lifetimes and coherence lengths.

In addition to the particular modes discussed above, the lattice calculations predict a number of modes which could be observed by Raman scattering the comparison of these predictions with observation is listed in Table I. An overall pattern of quantitive agreement exists in the cases where particular characteristics can be assigned, such as soft mode behavior, or H-bond character (85 cm^{-1}) these characteristics seem to be correct. In addition to the low frequency spectrum discussed here we have used lattice methods to calculate vibrational modes at much higher frequencies and relatively accurate refinements compared to Raman observations have been achieved.

The existence of large coherence lengths does dictate the kinds of theoretical analysis that can be applied. The lattice approach is the only feasible approach for such large coherent excitations. For example, a molecular dynamics simulation of these modes would require a calculation cell of thousands to tens of thousands of base pairs. To see modes in the gigahertz range one would have to simulate for times of nanoseconds after equilibrium. The use of smaller simulation cells would miss completely the dispersion, i.e. change in frequency with wavelength, that is so important at the lower frequencies. One would also not be able to calculate the correct selection rules.

Lattice Dynamics: A Method for Large Systems

Lattice dynamics has evolved specifically to deal with systems so large that one could not possibly deal with all the essential degrees of freedom in a straightforward way. It relies heavily on simple group theory for its validity. It applies to any large system which is uniform in the sense that one region looks very much the same as some other region. The essential result is that for such a system the vibrational excitations are wavelike and that these wavelike solutions form bands. There is always one or more low lying bands which have a frequency which approaches zero when the wavelength becomes infinite. There are a number of other bands called optical bands which do not have a frequency of zero for any wavelength. In this section we will describe features of these vibrational modes that can be related to the size of the system and how the various bands relate to modes familiar in smaller molecules.

Because we will apply this to the DNA helix we need only consider a system extended in one dimension and in that case we can say that a symmetry operator A exists which translates whatever wave function exists at one region to another region a distance "a" away. If the material is the same over distances "a" then this is a symmetry operation. Consider the quantum equation

$$H\,\psi(r) = E\psi(r) \tag{3}$$

where H is the Hamiltonian and ψ is the wave function of a normal mode of energy E. Then

$$\begin{aligned} AH\psi &= AE\psi \\ AHA^{-1}A\psi &= AE\psi \\ HA\psi &= EA\psi \end{aligned} \tag{4}$$

Table I
Predicted Frequencies and Experimental Observation

Theory (cm^{-1})	Experiment (cm^{-1})
A-DNA	
11.5*	12.5*
14.1	15
20*	22*
34	33
64	65
100	100
B-DNA	
11.5	12.
14.8	16
83†	85†

*Signifies predicted soft modes which are seen to soften experimentally.
†Signifies H-bond modes which is seen experimentally to behave anomalously at melting.

as $AHA^{-1} = H$ is a result of the symmetry and $AE = EA$ as E is simply a number.

$$A\psi(r) = \psi(r + a) \tag{5}$$

For a large system A can be applied many times

$$A^n\psi(r) = \psi(r + na) \tag{6}$$

and with periodic boundary conditions the operations A form a cyclic group and n is an integer. Therefore

$$\psi(r + na) = e^{in\theta_k}\psi(r) \tag{7}$$

where θ is a discrete set of phase shifts with separation $\Delta\theta = \theta_{k+1} - \theta_k = \frac{2\pi}{N}$ where N is the number of cells of length "a" in the cyclic system. This solution which is dictated by a simple symmetry condition guarantees that all the solutions are wavelike. The wavelength λ is $\lambda = ma$ where $m = \frac{2\pi}{\theta}$. The range of θ is taken to be the discrete values $-\pi < \theta \leq \pi$ by convention. The values $\theta < 0$ are degenerate with values $\theta > 0$ by time reversal invariance. The parameter θ_k plays the role of a quantum number specifying a particular wavelength. In a very large system $M \Rightarrow \infty$ and θ becomes a continuous variable i.e. all wavelengths are possible on an infinite system but only discreet waves are possible on a finite system. The minimum value of λ is $\lambda = 2a$ and corresponds to $\theta = \pi$. Even for an infinite system only a few values of θ need be calculated to determine the band shapes.

If we consider a real system that fits the formalism above the distance "a" defines a unit cell and translations by "na" shift these unit cells over by n units. Assume that there are M atoms in this unit cell of length "a". And assume there are N such unit cells in a large one dimensional chain. Then there are NM total atoms and 3NM total degrees of freedom in the system. The number of unique translations "n" is limited to $n = N$. Therefore there are N unique values of θ and therefore N different wavelengths. The total number of degrees of freedom 3MN then determines that there are 3M degrees of freedom associated with each λ and this implies 3M independent vibrations at each of the N wavelengths. All of the modes at different wavelength are orthogonal due to wave incommensurability and the lattice problem reduces to solving for each θ the 3M modes associated with that θ. It is as if the unit cell were a separate molecule of M atoms having 3M vibrational modes. The larger system has 3M modes per allowable wavelength and the frequencies depend on the wavelength.

For a large system i.e. large N the allowed values of θ are close together as $\Delta\theta = \frac{2\pi}{N}$. The shift in each frequency of the 3M solutions from one θ to a nearby θ is small. The 3M solutions therefore form an almost continuous band on a two dimensional plot of ω vs. θ. This is displayed in Fig. 1. There are 3M such bands. The system that is large due to N being large can be solved by solving a secular equation of dimension 3M for many values of θ and plotting them as shown in Fig. 1. The mathematical details can be found in any text on lattice dynamics.

Choosing the appropriate unit cell or equivalently the size of "a" depends on the material being studied. If one analyzes homopolymer double helical DNA the unit cell is one base pair with the associated backbone parts. The symmetry operation A from Eq. 4 is not just a translation but is a screw axis operator. It advances along the axis and rotates, all in one operation, thereby generating a helix. If the substance is alternating polymer DNA the appropriate cell contains two base pairs and two sets of associated backbone etc. The transformations lead to a picture like that of Fig. 1, and details of DNA lattice calculations can be found in the literature (10,5).

The smaller the unit cell the smaller the matrices to be diagonalized and the easier it is to do the calculations. A first attempt would be to do calculations for homopolymer DNA and to see how well the results apply to general double helical DNA systems. This should not be too bad an approximation as the base pairs are very much alike in their gross characteristics. One might expect this approximation to work fairly well at low frequencies and long wavelengths as for these modes, the base pairs move as coherent units. The approximation should be bad at higher frequencies and short wavelengths. Experimental observations at low frequencies which at present can only be measured at long wavelengths bear out these assumptions (8).

In small molecules there are a number of degrees of freedom not contributing to vibrational modes rather they are associated with center of mass motion and rotations of the entire molecule. Those degrees of freedom in the unit cell of a long system contribute to the acoustic modes. These have vanishingly small internal displacements

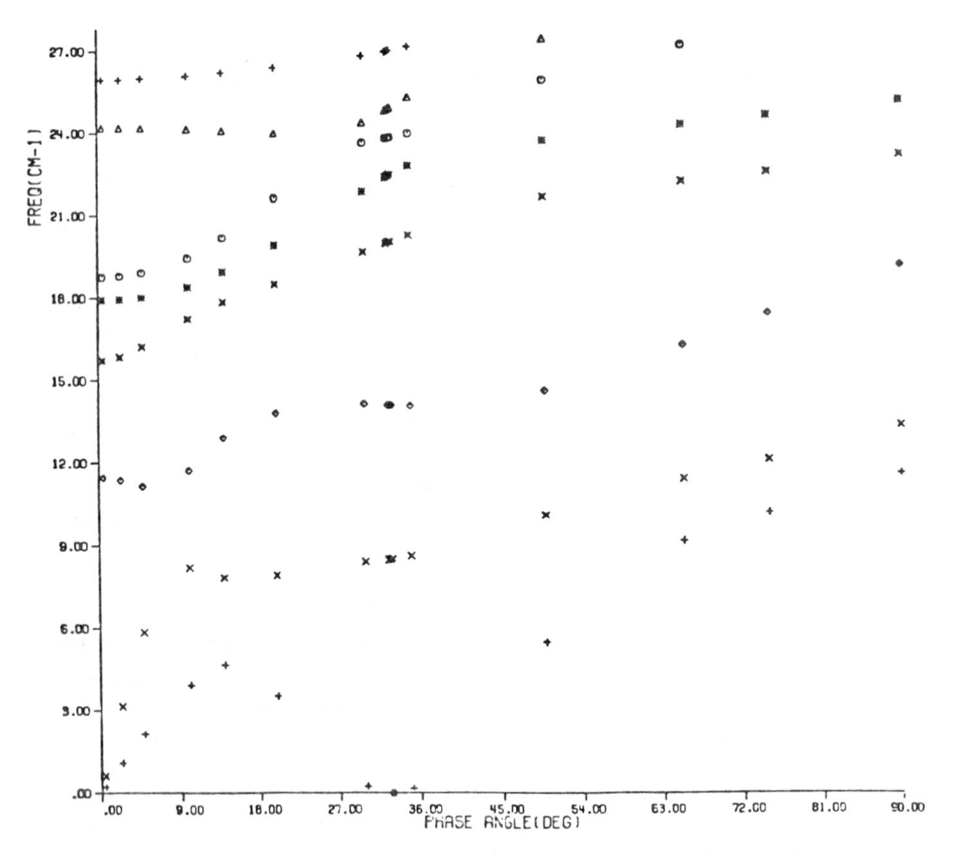

Figure 1. The frequencies of the normal modes of A conformation poly(dG)•poly(dC). A number of modes are solved for each value of phase angle. The grouping of the solutions into bands is obvious. The two bands linear in θ near $\theta = 0$ are the compressional and torsional acoustic modes. The band touching zero at $\theta = 32.7°$ along with the touch at $\theta = -32.7°$ forms the two transverse acoustic modes.

within the unit cell but the center-of-mass of a unit cell can move relative to the center-of-mass of another unit cell causing, for example, a compressional wave. In the DNA helix there is such a compressional acoustic mode and two polarizations of transverse acoustic modes. Since the cells in DNA are connected and form a long chain small rotational oscillations can only occur about an axis parallel to the helix axis and this forms the torsional acoustic mode. All other vibrational modes do have relative displacements within a unit cell and are called optical modes. Acoustic modes have the property that as the wavelength gets long no compressions etc. are occurring at all. In that case the frequency goes to zero at $\lambda = \infty$ which is equivalent to $\theta = 0$ for the compressional and torsional modes and $\theta = \psi$, where ψ is the helix pitch angle, for the transverse displacement modes (13).

Conformation Change: Soft Modes

Conformation change is an area in which knowledge of vibrational modes can be important. The connection between conformation change and vibrational motion comes about because such changes can be soft mode mediated. In the simplest description of a soft mode conformation change one vibrational mode becomes very soft i.e. its restoring force against displacement drops and eventually the material moves along the vibrational mode displacement and doesn't return. That is if the mode is a vector displacement X_o with time dependence ω s.t.

$$X = X_o e^{-i\omega t} \tag{8}$$

and if ω^2 decreases and becomes negative then $\omega \Rightarrow i\alpha$ and

$$X = X_o e^{\alpha t} \tag{9}$$

i.e. exponential departure from the former equilibrium position. The reason ω^2 decreases rather than ω is that ω^2 is the proper eigenvalue for the solution of vibrational mode problems. In a soft mode mediated change the frequency of the soft mode is observed to decrease on approaching the transition falling to zero at the transition. The atom displacements in going from one phase to the other are along the vibrational eigenvector.

A much deeper insight into the workings of the soft mode mechanism can be gotten by analysing a more sophisticated approach to the theory based on the Landau theory of displacive phase change (11). The free energy of a system determines its equilibrium state and for transitions to occur the free energy must be a function of the parameter that describes the transition. That is the free energy can be written as

$$F = F_o + \tfrac{1}{2} A\eta^2 + B\eta^3 + \ldots \tag{10}$$

where F_o is a constant part of the free energy, and η is the order parameter describing the transition, and A and B etc. are coefficients which can depend on temperature and other control parameters of the system. For a magnetic transition η could be the magnetization or spin order parameter and hence the name order parameter. For a liquid gas transition η could be the density. For a displacive change or conformation change η would have to be vector displacements of the atoms which describe the changed conformation.

As long as the quadratic coefficient is positive and the higher order contributions small the minimum in F is at $\eta = 0$ and this is the state the system occupies. If a lower minimum occurred for some value of η the system would transform to the new conformation with displacement specified by the value of η at the minimum. Second order phase transitions occur when A goes to zero and the minimum then shifts smoothly to some other value of η. In all cases large A helps maintain the minimum at $\eta = 0$ and decreases in A help in the occurence of other minima and transitions.

For a system with a well defined structure F must have the full symmetry of the system and from group theory analysis η must be a member of an irreducible representation of that symmetry group. This also indicates that the linear term in η not be present in the expansion.

The connection between order parameters for displacive changes and vibrational modes comes from the fact that vibrational mode eigenvectors are also members of displacement vector irreducible representations (12). The vibrational modes derive from a Hamiltonian which has the full symmetry of the system. The eigenvectors form a complete set for the vector displacement of the system that are also irreducible representations.

To make this connection more specific it has been shown (13) that the free energy of a system can be expressed as

$$F = \tfrac{1}{2} \sum_{qj} \omega_j^2(q) |Q_j(q)|^2 + \frac{1}{n!} \sum_{n>2} \sum_{q_n j_n} V^n_{j_1 \ldots j_n}(q_1 \ldots q_n) \times Q_{j_1}(q_1) \ldots Q_{j_n}(q_n) \tag{11}$$

In this equation the atomic displacements are described in terms of the amplitudes of normal vibrational coordinates $Q_j(q)$ where q and j specify the specific vibrational mode and branch, ω is the frequency of the vibration and the V^n are the nth order derivatives of the molecular potential function. The vibrational modes in Eq. (11) are quasi-harmonic or thermodynamic phonons, and their frequency ω is the true frequency at the system temperature and in the system's true environment. The force constants related to these modes are not the simple second derivatives of the molecular potential function but a renormalized set of force constants which are derived from the true potential weighted over the thermodynamic range of the atomic displacements.

If the system is initially in a stable conformation defined by all Q's having a zero time average, F must be a minimum in this conformation. A conformation change will occur as the minimum in F shifts to some other point in Q space. A smooth change in conformation can occur if one or more $\omega_j^2(q)$'s go through zero to negative values. The minimum flattens out at $\omega_j^2(q) = 0$ and becomes a local maximum for $\omega_j^2(q) < 0$. The minimum moves away from the origin in Q space along the particular $Q_j(q)$ multidimensional vector. This type of conformation change fits the Landau definition of a thermodynamic second order phase change, and is the simplest version of a soft mode phase change.

Although it is possible for many ω's to decrease simultaneously in a noninteracting system in which all modes remain irreducible representations such behavior would be the equivalent of an accidental degeneracy and very unlikely. It is more likely that one mode alone was the soft mode. However when modes move in energy they cause crossovers and interactions become more likely. Interactions of this type can often be understood if one knows enough about all the modes in the frequency region. When modes interact the higher order terms become large and first order transitions can result.

Thermodynamic first order transitions occur when a second minimum drops in free energy and becomes of equal or lower free energy than the initial minimum. In this case the equilibrium conformation jumps to the new minimum conformation as distinct from having the minimum move smoothly out to a new conformation. The appearance of a new minimum necessarily involves negative contributions from higher order terms in the free energy expression, Eq. (11). The most likely way in which such a first order transition could occur still involves the onset of a mode softening. The downward shift in free energy of the second minima can be brought about either by a decrease in the positive contribution of the quadratic terms or an increase in the higher order potential derivatives which are the coefficients of those higher order terms which contribute negatively to F. The increase in higher order derivatives would however also increase the interaction between the quasi-harmonic vibrational states and would necessarily also change the quasi-harmonic frequencies. A decrease in the positive contributions from the quadratic term in Eq. (11) is a likely part of even first order transitions.

One can even analyze as complex a transition as the liquid vapor phase change by soft mode analysis. In this case the appropriate soft mode is not a vibration but rather a density fluctuation as changes in density characterize this transition. Density fluctuations become easier and easier to make as one approaches the critical temperature i.e. the mode is softening. Large density fluctuations characterize the approach to transition and can be observed as the phenomena of critical opalescence.

A likely sequence of events then leading to a conformation change is that a particular mode begins to soften as a result of changes in the environment of the system, the most common such change being temperature change. This mode may proceed to soften completely giving rise to a second order transition. It may before softening completely, interact with other modes which cause a rapid change in free energy leading to a first order transition.

In our calculations of the vibrational modes of the double helix we found one (the lowest zone center optical mode) whose eigenvector was almost exactly that needed to be the soft mode of the A $\rightleftarrows$ B conformation change (9). We suggested that this mode was the conformational soft mode and that the conformation change was a soft mode mediated transition (14). These modes have been experimentally observed to soften as predicted (8,9). The significance for understanding the conformational dynamics is that in such a transition one can experimentally observe the important term for the transition in the free energy expansion in Eq. (11). One can get very detailed information about the causes of the onset of the conformation change.

The question may arise as to whether it is necessary or sensible to complicate an apparently simple event such as a conformation change with additional concepts such as soft modes. Cannot more straightforward approaches such as potential calculations or molecular mechanics calculations provide a simpler theoretical approach to understanding conformation change? The problem with all potential model calculations, including molecular mechanics calculations, is that they are attempts to find small energy differences between a large number of large potential terms where the potential terms themselves are only crude approximations. The error in the potential formulation can be of the order of the differences in net energy for different conformations. the advantage in a soft mode formulation is that one can determine important energy difference terms directly in terms of experimentally calibrated parameters. The possibility of quantitative theoretical predictions exists for this approach.

As an example of the sensitivity of the soft mode approach I will describe work in progress on the A to B conformation change. Since it is experimentally verified to be a soft mode mediated transition the factors that cause the mode to soften are those that bring about the conformation change. Using our mode calculation methods we can explore which changes in interatomic interactions soften the mode. Electrostatic force changes such as those that arise from changes in dielectric shielding (i.e. changes in effective dielectric constant) do soften the mode but not enough to be the principal cause of the transition. Changes internal to the furanose ring such as associated with soliton theories (15) seem to have no tendency to bring about the transition. Factors which would arise from altering the filiments of water structure in the vicinity of the helix such as those suggested by x-ray analysis (16) do seem to be just what is needed to soften the appropriate mode and are the likely cause of the conformation change. When we project out the interactions which would soften the important mode we find these interactions must exist between phosphate groups and furanose oxygen atoms along the helix and across grooves. Such interactions would be the result of water and ion bridges (17) between these positions. The details of these calculations will be published elsewhere (29).

Anharmonic Effects: Mean Field Theory of Helix Melting

Our Lattice calculations on the double helix indicate that a number of vibrational bands exist which can be described as being modes that stretch the hydrogen bonds between base pairs. We have examined a number of different repeating polymers, all those that can be made where as unit cell is no larger than two base pairs, and find that for all of them the frequencies of these modes are in the neighborhood of 85 cm^{-1}. We show the actual frequencies of these bands at zone center (or $\theta = 0$) for these cases in Fig. 2. The height of the lines in the figure correspond roughly to the degree to which these modes are pure H-bond stretch modes. The various bands are H-bond modes that are split by motion of varying degree of other atoms and regions of the helix. We expect that in native DNA with varying sequence the resulting heterogeneously broadened modes of H-bond stretch character would form a broad band with center and width corresponding to the center and width of the modes in Fig. 2.

Urabe et al. (18) have reported a band centered at 85 cm^{-1} in B-DNA which they assign H-bond character to as it softens and disappears on approaching the melting

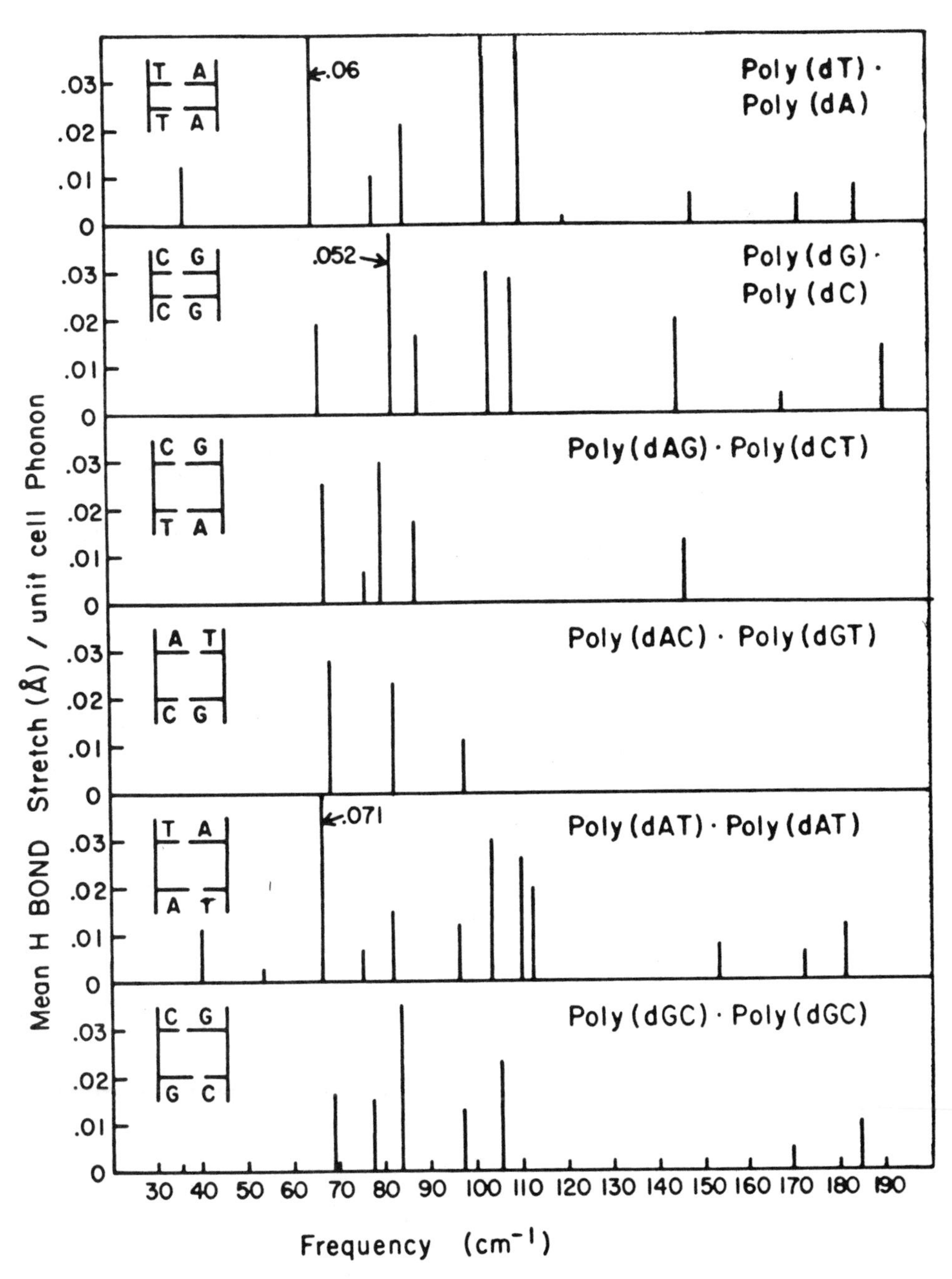

Figure 2. The H-bond coherent stretch modes. The figure shows the amplitude of the average H-bond stretch per phonon at the indicated frequency for the indicated DNA polymer. Only those modes that had in phase stretch for all H-bonds of a given base pair are shown. The normalization of the amplitudes is one quantum per unit cell. The extended phonon amplitudes should be multiplied by $N^{-1/2}$ where N is the number of base pairs.

temperature. The width and location of the line is in good agreement with the distribution shown in Fig. 2.

We had predicted (19) that these H-bond modes would play a role in the dynamics of melting somewhat similar to the role soft modes play in soft mode conformation change. The reasoning is as follows. Melting results from thermal excitations which increase the thermal motion of atoms. This motion both weakens and eventually overcomes the bonding energy which opposes melting. Since the melting we are discussing involves the breaking of a specific set of bonds, the H-bonds between bases and that the motion of atoms along these bonds is due to a few vibrational modes; then one can assume that these few vibrational modes dominate the particular bond melting. Another complimentary view would be to realize that there are many bonds of varying strength in DNA. One expects that many bond separations could occur that could be described as a series of meltings which occur at different temperatures. That melting associated with a particular temperature probably involves a particular set of excitations which occur at a particular set of frequencies. The lower temperature melting is associated with lower frequency excitations and again one can then assume some set of modes dominates a particular melting such as helix strand separation. Of course melting is a cooperative effect. The excitations both weaken the bonds and overcome them. Thus one must follow the excitations involved as they change going toward the melting transition. This is equivalent to following the mode softening in the conformation change problem.

In the case of displacive change (conformation change) the problem could be fit into an existing framework. Soft mode displacive change has been studied extensively in solids. In the case of melting dominated by a few modes, no framework existed for calculations. We developed such a framework by developing a modified version of selfconsistent phonon theory (MSPA) based on the selfconsistent phonon approximation (SCPA). A brief description of MSPA follows and the working of the method are demonstrated by application to a particularly simple system.

Lattice dynamics is usually presented in elementary texts as using only the harmonic terms in an expansion of a true potential. At nonzero temperature thermal motion causes displacements which introduce anharmonic effects which cause deviations from the simple harmonic approximation which can be large. In a more sophisticated approach it is realized that vibrational modes occur and are observed experimentally and that the correct quasiparticle solutions of even the true potential must reflect this fact. This means that the sophisticated renormalized solutions must resemble the known vibrational modes. One particularly useful method for determining excitations that are phonon like is the method of selfconsistent phonon theory (20) (SCPA). In this method one writes a true Hamiltonian

$$H_T = \text{K.E.} + V(r_1 \ldots r_n) \tag{12}$$

where K.E. is the kinetic energy parts and $V(r_1 \ldots r_n)$ is the true many body potential. This potential has harmonic and all anharmonic contributions. Because one knows that the final solutions are phonon-like one can write an effective Hamiltonian which has a pseudopotential part which explicitly displays the expected quadratic form

$$H_E = K.E. + \tfrac{1}{2}\,\phi\mu^2 \tag{13}$$

where μ is a relative displacement between atoms and ϕ is an effective force constant. All the subscript notation which specifies the various atoms and the various components of displacement has been suppressed but can be found in the references (20,21). This effective force constant ϕ is not the second derivatives of V at the equilibrium positions but rather that renormalized or pseudopotential value which gives the best fits to the observed phonon spectrum. SCPA gives a prescription for determining ϕ from $V(r_1 \ldots r_n)$ in a selfconsistent manner.

Since such a theory must be applicable at elevated temperatures the thermodynamic effects on the system are incorporated by working with the free energy rather than the Hamiltonian. One defines a true free energy F_T and a quadratic form effective free energy F_E by substituting H_T and H_E into

$$F = -kT \log \mathrm{Tr}\left\{\exp\left(-\frac{H}{kT}\right)\right\}. \tag{14}$$

The value of ϕ is determined then by picking that ϕ which minimized the difference between F_T and F_E, i.e.

$$\frac{d(F_T - F_E)}{d\phi} = 0. \tag{15}$$

The value ϕ is different for each temperature and is the renormalized force constant which best fits the true free energy in a selfconsistent phonon picture. Again the details of such a procedure are in published literature (20,21).

Carrying out the analysis described above by SCPA gives very intuitively simple forms for ϕ. Ignoring component breakdown and subscripts one finds (20,21)

$$\phi = \frac{\int d^3\mu \, e^{-\frac{\mu^2}{2D}} \frac{d^2V}{d\mu^2}}{\int d^3\mu \, e^{-\frac{\mu^2}{2D}}} \tag{16}$$

where μ is a relative displacement between atoms, D is the mean square correlated displacement of μ i.e.

$$D = \langle \mu \cdot \mu \rangle \tag{17}$$

and $\frac{d^2V}{d\mu^2}$ is the local force constant for various amounts of displacement. The effective force constant is then the weighted average of the force constant, weighted over the statistical range of displacements for a given temperature or level of excitation.

The value of D assuming that the vibrational modes are in thermal equilibrium at temperature T becomes

$$D = \sum_{k\lambda} \frac{1}{2\omega_\lambda(k)} \text{ctanh}\left(\frac{\hbar\omega}{KT}\right) \quad N_\lambda^2(k) \tag{18}$$

where k and λ refer to all the specific phonon modes, $\omega_\lambda(k)$ is the frequency of such modes and $N_\lambda(k)$ is the projection of the particular mode eigenvector onto the displacement of the particular atoms involved in the potential under consideration. D is the thermal mean square displacement of the appropriate interatomic distance.

We have modified the usual SCPA formalism by explicitly including thermal expansion into the determination of the effective force constant. We call this approach the modified selfconsistant phonon approximation (MSPA). This is accomplished by introducing a parameter for the mean distance between atoms which can vary with temperature and is not forced to coincide with the true potential minimum. The effective phonon oscillations are then centered about this mean position R_T. This parameter is determined from the condition

$$V_T(R_T + \mu_o) = V_T(R_T - \mu_o) \tag{19}$$

where V_T is the true potential and μ_o is the amplitude of oscillation of the thermal phonons. That is R_T is centered between the oscillatory turn around points. Since μ_o is selfconsistently determined in the procedure (it is linearly related to the square root of D) R_T is also selfconsistently determined. In our scheme the following set of parameters which depend on other members of the set are selfconsistently calculated. The parameters are

$$\phi = \phi(R_T, D) \tag{20}$$

$$R_T = R_T(D) \tag{21}$$

$$\omega = \omega(\phi) \tag{22}$$

$$D = D(\omega, T) \tag{23}$$

where ϕ are the effective force constants which are determined from V_T weighted by a spread determined by D centered in the potential by R_T, R_T is determined only by being the midpoint of the excursion in V_T characterized by D, ω is the set of vibrational mode frequencies determined from the geometry and the value of the effective force constants ϕ, and D is determined by the frequencies ω and the temperature T using a Bose-Einstein distribution for the phonon amplitudes (the phonon eigenvectors are also important as they determine the actual contribution to the D of a particular phonon).

In principal every force constant changes with temperature and all force constants used to solve for the phonon spectrum should be these effective force constants. In practice most strong valence force constants change very little over moderate changes in temperature. In our calculations of the vibrational modes for frequencies above 200 cm^{-1} we fitted force constants (22,23) to experimental observations. These fitted force constants would be the effective force constants at room temperature and most force constants would not change appreciably over the entire range from the freezing point of water to the melting point of the DNA. This is the practical temperature range over which DNA is studied. The strong valence force constants don't change as their potential is deep enough to keep displacements fairly localized in the potential. Nonbonded force constants such as the electrostatic force constants don't change drastically even if the displacements become large because these forces are long range and have fairly weak dependence on displacement. One believes however that some force constants particularly those of the hydrogen bonds should change rapidly as one approaches the DNA melting point, as these bonds break on melting. In the rest of this section we will concentrate on the temperature dependence of the H-bond effective force constants.

The MSPA requires a true potential between atoms. Baird (24) has carried out an ab initio calculation in which he calculated the bond energy for an N-H---N hydrogen bonded system as a function of the distance between the outer atoms assuming the H atom occupied a position which minimized the overall bond energy. In our calculations (19) the main H-bond vibrational modes are predicted to be centered about 85 cm^{-1} and indeed modes of this character are observed (18) at this frequency. Since the H atom oscillates at several thousands of cm^{-1} it can be expected to be able to relax to its equilibrium position and a potential like that of Baird is a reasonable one to describe the interaction between the heavy atoms of a H-bonded system for frequencies $\lesssim$ 100 cm^{-1}. The numerical potential was shown to fit very well to a Morse potential.

$$V = V_o \left| 1 - e^{-a(r-R_o)} \right|^2 - V_o \tag{24}$$

where r is the distance between the two nitrogen atoms. The potential constants are

$$V_o = 3.48 \text{ kcal/mole}$$

$$R_o = 3.37 \text{ A}$$

$$a = 1.22 \text{ A}^{-1}$$

The analytical potential is used in this calculation because it considerably simplifies the programming without reducing the physical significance.

The potential used is essentially a one dimensional potential and the model problem is two ammonia masses interacting via the potential of Eq. (16). Since all the theoretical analysis pertains to solids the problem examined should be thought of as a one dimensional solid ammonia chain. Since we solve only for one vibrational frequency the solid ammonia chain is treated in the Einstein approximation where the compressional mode oscillates only at the zone edge frequency. In one dimension the equations of MSPA simplify considerably. Eq. (23) becomes

$$D = \frac{1}{M\omega} \text{ctanh} (\tfrac{1}{2} \beta\omega) \tag{25}$$

where M is the mass of an ammonia molecule $\hbar = 1$ and ω is determined from Eq. (22) which reduces to

$$\tfrac{1}{2} M\omega^2 = \Phi \tag{26}$$

The equation for ϕ in MSPA is reduced to

$$\Phi = \int_{R_m}^{\infty} du\, e^{-\frac{u^2}{2D}} \frac{d^2}{d\mu^2} V(R_T + u) \Big/ \int_{R_m}^{\infty} du\, e^{-\frac{u^2}{2D}} \tag{27}$$

in which the displacement μ has origin at R_T the thermal mean position rather than at R_o the potential minimum. The integration can be done in closed form and

$$\Phi = 2a^2 V_o \frac{A-B}{C} \tag{28}$$

where

$$A = 2e^{2a^2D - 2a(R_T - R_o)} \text{erfc} \left(\frac{2aD - R_T + R_m}{\sqrt{2D}}\right) \tag{29}$$

$$B = e^{\frac{a^2D}{2} - a(R_T - R_o)} \text{erfc} \left(\frac{aD - R_T + R_m}{\sqrt{2D}}\right) \tag{30}$$

$$C = \text{erfc} \left(\frac{R_m - R_T}{\sqrt{2D}}\right) \tag{31}$$

The lower limit of the integral R_m is the point where the inner value of the potential is equal to the energy at infinity to prevent the non-physical overweighting of the hard core region. This necessity to restrict the hard core contribution has been discussed in detail by others (25). We note that this is still necessary even when we have introduced the shift of weighting function to R_T. The thermal expansion is certainly not excessive in light of this behavior. The factor R_m is given by

$$R_m = R_o - \frac{1}{a} \log 2 \tag{32}$$

The thermal expansion estimation can also be solved from the expression of V in Eq. 24 and is

$$R_T = R_o + \frac{1}{a}\log[\cosh(a \cdot u(T))] \tag{33}$$

This can be expressed in a more useful form which shows the selfconsistent coupling of the equations by expressing u(T) as

$$u(T) = 2(2\log 2)^{1/2} D^{1/2} \tag{34}$$

The selfconsistency is achieved by simultaneous solution in Eqs. 25, 26 and 27. These solutions were found for temperature ranging from 1K to 500K. A sense of the working of this calculation can be gotten by examination of Fig. 3. In that figure the thermal spread of the N-N distance for three temperatures is shown in the potential well. The change in width of this spread as well as the shift in mean position with temperature can be seen. The second derivative is also displayed and the variation of the weighting of this term by the spread in position can be roughly observed.

Fig. 4 shows how the thermal expansion R_T depends on the temperature. Below $T = 300°K$, R_T is almost linear with T and $\frac{dR_T}{dT} \sim 0.001A/°K$. At about $T = 325°K$,

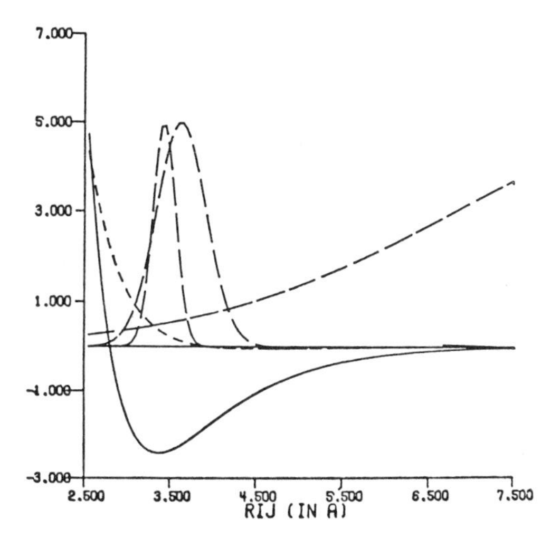

Figure 3. The solid line is the interatomic potential as a function of N-N distance in Angstroms. Each unit is 10^{-13} ergs. The short dash line is the second derivative of the potential. Each unit is 2×10^4 dyne/cm. The long dashes are the probability densities for N-N distances for three different temperatures. The furthest left is T = 1K the middle peak is for T = 250K and the right most curve for T = 500K. These curves are arbitrarily fitted to a height of five. It can be seen that the probability density broadens due to thermal motion and is forced to larger mean position as well due to the asymmetry of the potential. The shift in mean position is the thermal expansion.

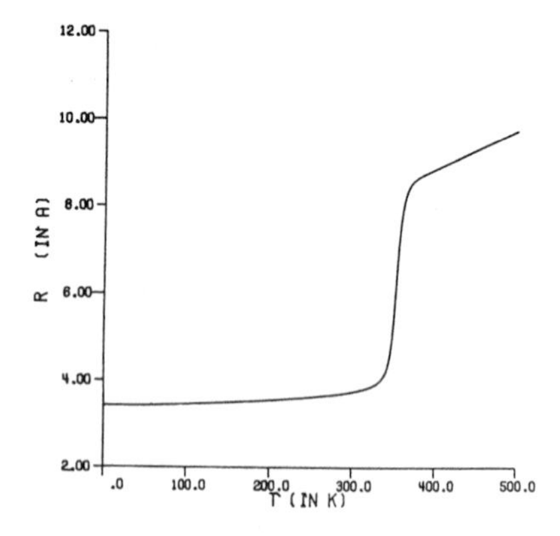

Figure 4. The mean instrinsic N-N distance in Angstroms as a function of temperature. The sudden increase is what we associate with bond melting.

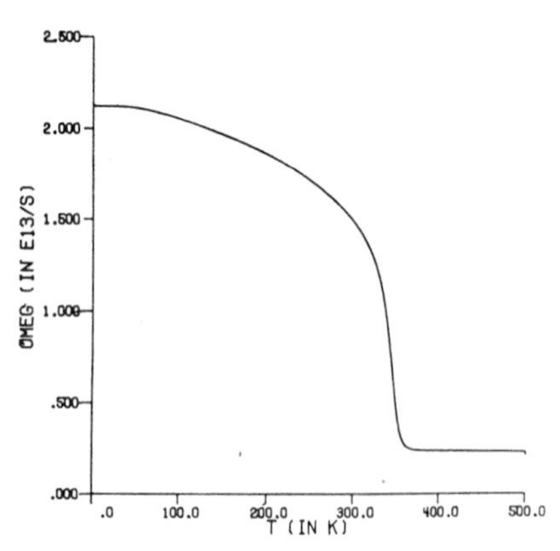

Figure 5. The frequency of the N-N vibrational mode in units of 10^{13}/s as a function of temperature.

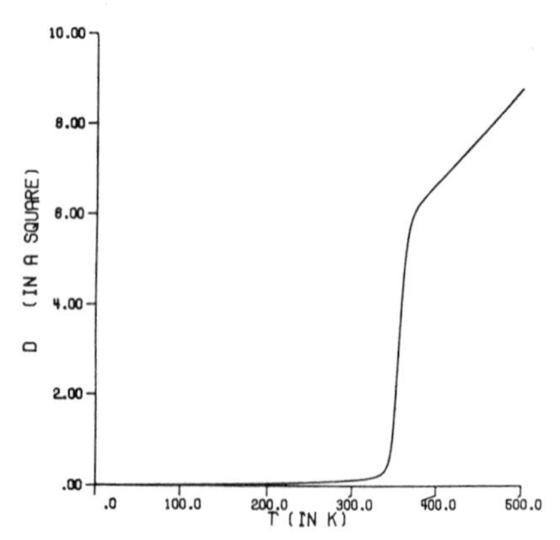

Figure 6. The correlation function D in Angstroms squared as a function of temperature.

R_T starts increasing rapidly from 3.9A to 7.0A in 30°K. The frequency ω and the correlation D also show the similar drastic behavior in the same temperature region, as could be seen in Fig. 5 and Fig. 6. We would associate the instability with the melting of the hydrogen bond. A hydrogen bonded molecule would certainly be expected to have melted in undergoing such a change in bond length. Similarly from Fig. 5 one can see that ω softens which implies ϕ softens and the loss of particular frequency modes and the loss of Hookian response against particular displacement is also associated with melting. Finally in Fig. 6 one sees that the displacement is small for all temperatures below the region of rapid change. This indicates that the bond is a reasonable solid bond up to the point at which it ceases to be a reasonable bond. The association of this instability with melting seems reasonable. In the region after the onset of the instability D becomes large and all the approximations involved in the development of the formalism become suspect. However, the calculation should be an indicator of the onset of melting when approached from the low temperature side.

From Fig. 3 it is clear how the calculated features arise. As the temperature increases the vibrational mean displacement increases and in the asymmetric potential the far turnaround point is pushed to regions of even weaker restoring forces. These small contributions to the force constant reduce the effective force constant and the mean thermal displacement increases even further. This coupled behavior gives rise to a positive feedback effect causing the instability at high enough temperature. In reality when the interatomic distances become large enough so that the return force is very weak the return motion would be greatly reduced. The resulting vibrational motion would become very anharmonic. In tracing the frequency verses T for a real system one would expect some initial softening followed by considerable broadening possibly to the point of becoming a central feature. This expected behavior of the vibrational mode is quite close to what is observed in DNA upon approach to the melting temperature (18).

This selfconsistent approach to the determination of force constant becomes a mean field approach to melting. The force constant ϕ plays the role of the mean field and the combination thermal motion (characterized by D) and thermal expansion ($R_T - R_o$) are determined by thermal excitation. As T increases R_T and D increase and that increase weakens ϕ which causes further increase in R_T and D. The process has positive feedback and can lead to dramatic change.

Application to poly(dG)•poly(dC)

A calculation such as that described above can be carried out for a large system like the DNA double helix. We have carried out such a calculation (19) for poly dG•poly dC. The main differences from the simple case is in the large number of vibrational modes which contribute to the appropriate D parameters and the fact that there are three H-bonds which imply 3 D's, 3 R_T's, and 3 ϕ's.

In our calculations the D factors were calculated by summing over all vibrational modes. The large H-bond stretch modes did contribute greatly to the result. The true potentials used in the calculation of ϕ's was again assumed to be Morse like. The values of the parameters V_o, a, and R_o were fitted so as to give reasonable force constants and R_T's at room temperature, which agrees with the known atom distances and observed vibrational spectrum. The values of V_o, a and R_o for the three H-bond potentials are listed in Table II. The selfconsistent solutions had to be solved iteratively on computer. The results of the calculations are shown in Figs. 7-10. It can be seen that well defined R_T's, ϕ's, D's and frequencies are found for temperatures below 340K. All the approximations appear valid in this range. At 340K singularity occurs in R_T and D and this is associated with melting. This model predicts that melting in poly(dG)•poly(dC) begins with the separation of the H-bond near the major groove.

Table II
The Morse Potential Parameters for the Three Hydrogen Bonds

Bond	N(1)HN(3)	N(2)HO(2)	N(4)HO(6)
V_o(mdyn-A)	0.015	0.0282	0.0284
a(1/A)	1.84	2.38	2.42
R_o(A)	2.805	2.698	2.691

From Fig. 7 it can be seen that those modes whose character is mostly H-bond stretch soften and disappear on approaching the melting temperture. This is in good qualitative agreement with the observed behavior of the H-bond mode for increasing temperature (18).

This analysis is only a mean field theory of melting. Like all mean field theories it does not deal properly with fluctuations. The fluctuations expected to be important for a one dimensional system such as DNA are the formation of melting nucleation sites. With the formation of such nucleation sites one expects adjacent base pairs to melt more easily and the actual melting of DNA will proceed to grow from such nucleation sites. The incorporation of such details is outside the scope of a mean field theory. We are developing an approach which deals with the propagation of melt from a nucleation site using a local version of the mean field theory discussed above.

The significance of this work lies in the fact that a method for incorporating nonlinearities into the basic lattice approach is developed. It seems to work over the temperature range, from room temperature up to the melting temperature, which is the useful range for DNA analysis. Nonlinearities are important but their importance does not invalidate lattice methods. Sophisticated lattice methods can be used at various temperatures if proper concern for modifying effective force constants is exercised. Whether the extension of the theory into a theory of melting is of importance is yet to be demonstrated.

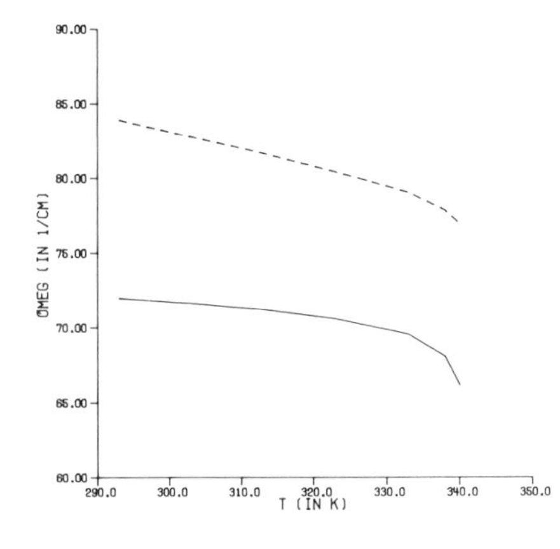

Figure 7. The frequencies for two zone center phonons as a function of temperature. The upper frequency was used to fit V_0 and a to the frequency of an observed Raman line. The two chosen have large components of H-bond motion.

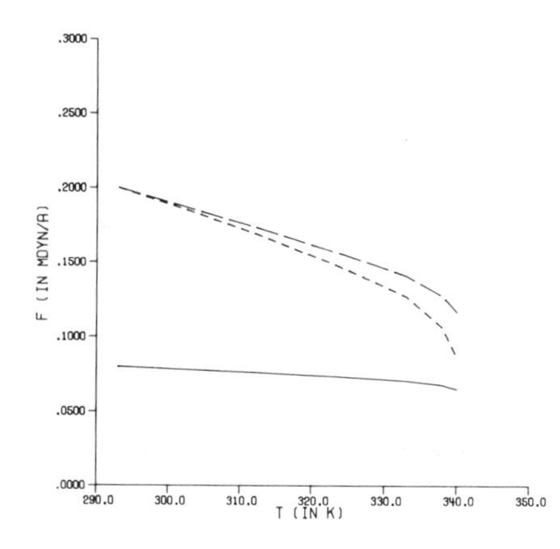

Figure 8. The mean distance between end heavy atoms (thermal expansion) for the three H-bonds as a function of temperature. The solid line is for the central or N(1)-H-N(3) bond, the short dash line is for the O(6)-H-N(4) bond adjacent the major groove and the long dash line is the N(2)-H-O(2) bond adjacent the minor groove.

Local Effects in Systems with Long Coherence Lengths

Biological processes occur at specific sites along the helix. Bulk vibrational modes of long coherence length are not good probes of local phenomena. However, it has long been recognized that where there are local differences vibrational modes become localized. By a local difference we mean for example a terminus of the helix, or a replicating fork, or a melted bubble, or a region of unique conformation or even a region of unique base sequence. These unique regions do have vibrational modes which are different from those in nonunique regions but the nature of these differences are complicated by the fact that the unique regions are imbedded in a large system which itself has modes of large coherence length.

The proper theoretical approach is to use methods such as the methods developed for dealing with defect modes in condensed matter physics. The methods we have used make use of Green functions where the differences in a region are the Green function source terms. The best developed example of such an approach is our

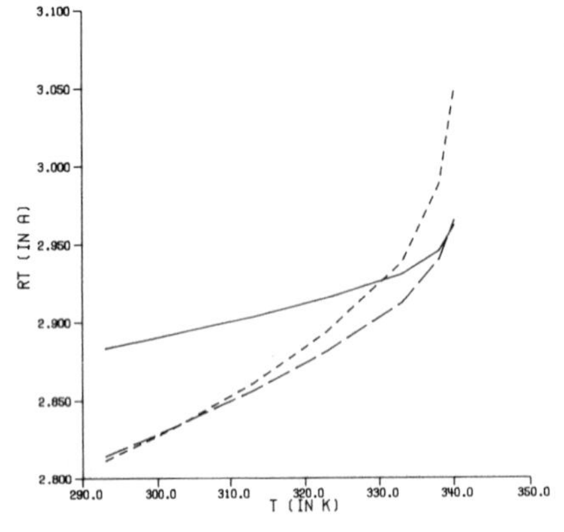

Figure 9. The effective force constants for the three H-bonds as a function of temperature. The lines refer to the same bonds as in Figure 8.

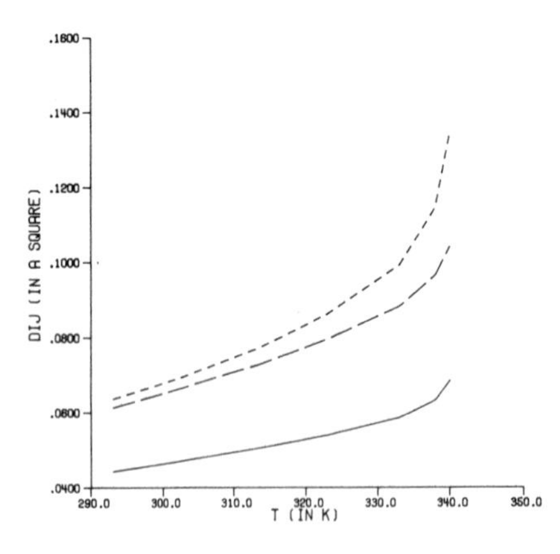

Figure 10. The correlation function D for each H-bond as a function of temperature. The lines refer to the same bonds as in Figure 8.

calculations of the local mode at a DNA terminus (26). In that calculation the connections between two parts of a double helix were cut theoretically by putting in the negatives of all the links between two adjacent base pairs. These negative values were the source terms for a Green function solution of the new modes that arise due to the perturbation. The negative links combined with the positive links that were already in the original bulk solution add up to give an effective severed helix.

The calculation predicted a strong narrow mode below one GHz. This is a region of the spectrum where no other source of a sharp resonance has been predicted. Such a strong narrow mode was observed by Lindsay et al. (27) at 600 MHz. That this mode is a terminus local mode is indicated by microwave absorption measurements where a shoulder of this mode is observed in linear DNA which has ends but is not observed in intact plasmid DNA which does not have ends (1).

The point to be made is that this mode is a creature of the end of a large section of helix. No mode anywhere near this frequency could raise from an analysis of a single or even a group of base pairs if they were not attached to a semi-infinite section with strong coherent connectivity. The mode is characterized as an enhanced rotation of the end base pair relative to the rest of the helix. Coupled with this rotation is enhanced stretching of the base joining H-bonds. When one adds up the enhanced H-bond stretch at the terminal base pair for all the modes affected by the perturbation one finds that the mean square stretch is twice as big at the end base pair as it is in a base pair in the middle of a long stretch of helix. This enhanced H-bond stretch is related to the enhanced probability of melting at the ends of the helix.

Another unique region which has been explored is that of the replicating fork (28). This calculation uses a Green function method to calculate modes in a region where the double helix changes to two single strands not H-bonded to each other. The last H bonded base pair had large H-bond stretch motion compared to base pairs far from this region. This is also related to the ease of propagating a melted region into the double helix.

Acknowledgements

This work was supported in part by NIH grant GM 244443.

References and Footnotes

1. G.S. Edwards, C.C. Davis, J.D. Saffer and M.L. Swicord, *Phys. Rev. Lett. 53,* 1284 (1984).
2. M. Kohli, W.N. Mei, E.W. Prohofsky and L.L. Van Zandt, *Biopolymers 20,* 853 (1981).
3. G. Maret, R. Oldenbourg, G. Winterling, K. Dransfeld and A. Rupprecht, *Colloid. Polym. Sci. 257,* 1017 (1979).
4. M.B. Hakim, S.M. Lindsay and J. Powell, *Biopolymers 23,* 1185, (1984).
5. H. Urabe, Y. Tominaga and K. Kubata, *J. Chem. Phys. 78,* 5937 (1983).
6. S.M. Lindsay and J. Powell, *Structure and Dynamics: Nucleic Acids and Proteins,* Eds. E. Clementi and R.H. Sarma, Adenine, N.Y. 241 (1983).
7. S.M. Lindsay and J.W. Powell, *Phys. Rev. Lett. 53,* 1853 (1984).
8. K.V. Devi-Prasad and E.W. Prohofsky, *Biopolymers 23,* 1795 (1984).
9. S.M. Lindsay, J.W. Powell, E.W. Prohofsky and K.V. Devi-Prasad, *Structure and Motion: Nucleic Acids, Proteins and Membranes,* Eds. E. Clementi, G. Corongin, M.H. Sarma and R.H. Sarma, Adenine, N.Y. 531 (1985).
10. J.M. Eyster and E.W. Prohofsky, *Biopolymers 13,* 2505-2526 (1974).
11. L.D. Landau, *Physik Z. Sowjet Union 11,* 26 (1937).
12. W. Chochran, *Proceedings of NATO Advanced Study Institute on Structural Phase Transitions and Soft Modes,* Geilo Norway, April (1971) p. 1-13.
13. P.B. Miller and P.C. Kwok, *Solid State Commun. 5,* 57 (1967).
14. J.M. Eyster and E.W. Prohofsky, *Biopolymers 16,* 965 (1977).
15. J.A. Krumhansl and D.M. Alexander, *Structure and Dynamics: Nucleic Acids and Proteins,* Eds. E. Clementi and R.H. Sarma, Adenine, N.Y. 61 (1983).
16. R.E. Dickerson, H.R. Drew, B.N. Conner, R.M. Wing, A.V. Fratini and M.L. Kapka, *Science 216,*
17. W.K. Lee, Y. Gao and E.W. Prohofsky, *Biopolymers 23,* 257 (1984).
18. H. Urabe and Y. Tominaga, *J. Phys. Soc. Japan 50,* 3543 (1981).
19. E.W. Prohofsky, *Comments Mol. Cell. Biophys. 2,* 65 (1983).
20. N.R. Werthamer, *Am. J. Phys. 37,* 763 (1969).

21. Y. Gao, K.V. Devi-Prasad and E.W. Prohofsky, *J. Chem. Phys. 80,* 6291 (1984).
22. W.N. Mei, M. Kohli, E.W. Prohofsky and L.L. Van Zandt, *Biopolymers 20,* 833 (1981).
23. B.F. Putnam and L.L. Van Zandt, *J. Comp. Chem. 3,* 297 (1982).
24. N.C. Baird, Int. *J. Quant. Chem. Symp. 1,* 49 (1974).
25. P.F. Choquard, *The Anharmonic Crystal,* W.A. Benjamin, N.Y. (1967).
26. B.F. Putnam, L.L. Van Zandt, E.W. Prohofsky, and W.N. Mei, *Biophys. J. 35,* 271 (1981).
27. S.M. Lindsay and J. Powell, *Biopolymers 22,* 2045 (1983).
28. B.F. Putnam and E.W. Prohofsky, *Biopolymers 22,* 1759 (1984).
29. K.V. Devi Prasad and E.W. Prohofsky, *J. Biomol. Struct. & Dyn. 3,* 551 (1985).

Biomolecular Stereodynamics IV, Proceedings of the Fourth Conversation in the Discipline Biomolecular Stereodynamics, State University of New York, Albany, NY, June 04-09, 1985, Eds., Ramaswamy H. Sarma & Mukti H. Sarma, ISBN 0-940030-18-7, Adenine Press,

Ionic Stabilization and Modulation of Nucleic Acid Conformations in Solution

Dikeos Mario Soumpasis
Institute for Theoretical Physics, Freie Universität Berlin,
Arnimalle 14, D-1000 Berlin 33, FRG

Many mickles make a muckle
(J.M. Ziman, Elements of Advanced Quantum Theory)

Abstract

All levels of the structural organization of nucleic acids in solution depend rather strongly on both the types and the concentrations of ionic species present. I briefly review some of the most striking experimental findings and discuss the general modes of interaction of alkali, alkaline earth and transition metal cations with DNA conformations. Ionic effects on the B-Z transition are summarized and discussed in somewhat greater detail.

Introduction

The structural organization of nucleic acids in solution is the result of a subtle interplay between intramolecular interactions (e.g. covalent and H-bonds, atomic core repulsions, van der Waals and Coulomb interactions) and interactions of the molecules with other components of the cellular environment (e.g. water, ions, specific ligands and proteins). It is by means of this complex system of physicochemical interactions that the genetically transmitted primary structure information is translated into specific, hierarchically ordered three dimensional structures (i.e. the secondary, tertiary, etc. levels of organization), directly affecting the diverse biological functions of DNA and the RNA's. (Comprehensive discussions of the structural morphology of nucleic acids and chromatin may be found in several places, (e.g. see (1-4) for DNA, (1,5-7) for RNA and (2,8-10) for chromatin).

By virtue of their primary structures these biopolymers, are giant polyanions (backbone phosphates) containing in addition many solvent accessible electronegative atoms of the bases and backbone sugars. Therefore, hydration and interactions with charged species (salt ions, basic polypeptides, histones, etc.) play a central role in maintaining specific structures and inducing transitions between them.

In this work I will first recall some of the experimentally documented highlights of ionic structural effects in vitro, then analyze the basic modes of ion-nucleic acid interactions from a physicochemical point of view and finally discuss recent theoretical developments in describing the salt induced right-left transition of DNA. The main emphasis is on cationic effects on DNA in vitro. Hydration and effects in vivo are not understood well enough yet (at least by this author!) and a fairly complete discussion of ion-RNA interactions (see (11,12) for reviews) requires a full length paper by itself.

Ions affect all levels of structural organization in solution. A brief review of experimental findings.

Effects on secondary DNA structure

Helix-coil transition studies have clearly shown that the thermodynamic stability of ordered DNA conformations is strongly dependent on the types and concentrations of the ionic species present (13-16). However, more recently interest shifted to helix-helix transitions and their possible roles in gene expression and regulation. Perhaps, the most dramatic example of such a structural change, is the salt induced right-left (B-Z) transition first observed with d(C-G)•d(C-G) helices in NaCl and $MgCl_2$ (17) and currently studied in many laboratories. The structural morphology of both the Z (18-22,1-4) and B (23-26,1-4) families of conformations has been analyzed in great detail and several comprehensive accounts of many physicochemical and biological aspects of the right-left isomerization including a plethora of original references may be found in Refs. (2,27,28). Ionic aspects of particular interest to us will be discussed in greater detail further below.

In addition to the B-Z transition of deoxyribonucleotide sequences ions can also cause A-Z transitions, as recently observed with poly[r(G-C)•r(G-C)] (29) and the synthetic hexanucleotide $(dCdGFl)_3$ (30). Pronounced, albeit less cooperative, structural changes have been also reported for d(A-T)•d(A-T) sequences (31-36). In this case the high salt conformations are not left handed but probably of the alternating B type (35,36).

In addition to transitions between distinct families of structures, ions also mediate gradual changes of the helical parameters within one family (e.g. B) as suggested by many lines of spectroscopic evidence (34,37-41).

Effects on tertiary structure

Several polyamines and complex cations such as Cobalt hexamine cause intramolecular collapse of many linear, B form DNA's into toroidal forms (42-49). Similar condensations of tertiary DNA structure are also observed with simple salt cations in alcohol-water solvents (50-53) or in the presence of neutral polymers (54-56). Z-DNA forms many types of compact structures (57-59) and has great propensity for intermolecular aggregation (129) especially in the presence of Mg(II) and alcohol. Particularly interesting phenomena occur with covalently closed supercoiled DNA. In this case the secondary and tertiary levels are strongly coupled due to the topological constraints present and ions may induce changes in both levels at the same time, or induce secondary structure transitions much more effectively (as compared to their action on linear DNA) due to the synergism of the tertiary structure.

For instance, potentially Z forming sequences inserted in plasmids, undergo the B-Z transition at physiological monovalent salt concentrations (60-62,27) which also have been found to induce cooperative structural changes at the tertiary level as well (63-65).

High order structures

Ionic effects at higher levels of DNA structural organization are most clearly seen in studies of 'dressed' DNA (i.e. DNA in nucleoprotein complexes such as chromatin and chromosomes) Na^+ and Mg^{2+} are known to cause condensation of the extended 100 Å thick chromatin fiber into a compact 250-300 Å thick solenoid (66-69) the second step (the first being nucleosome formation) in the chain of compactions leading to the enormous levels of DNA packing present in chromosomes.

Increasing ionic strengths cause progressive dissociation of the non-histone and histone proteins associated with DNA in chromatin and eventually lead to destruction of the nucleosome structure (70,71). This is a typical example of how simple salts can affect DNA-charged ligand binding equilibria also observable with other functional DNA-protein complexes. Salt induced chromatin condensation and decondensation (via protein dissociation) are intimately related to so called chromosome puffing phenomena directly observable with giant polytene chromosomes under the microscope (72-75). An impressive example is shown in Figure 1, kindly provided by Michel Robert-Nicoud, MPI Göttingen, which shows an isolated polytene chromosome of Chironomus thummi initially in the extended (decondensed) state (b-d) condensing (e-i) and decondensing again (j-l) upon increasing the concentration of supporting electrolyte (NaCl).

Effects on RNA structure

In the case of RNA, ion mediated structural changes may occur not only at the secondary and tertiary levels but most strikingly at the primary level as well. Many divalent and trivalent cations can cause depolymerization of polyribo nucleotide chains (76,77,16). Pb(II) catalyzes backbone cleavage of t-RNA (78-80) and Mg(II) seems to be an essential factor in the recently discovered autocatalytic cleavage of RNA (81,82) and RNA induced splicing of precursor t-RNA's (83,84).

This brief and certainly not complete review of ionic effects on nucleic acid structure nevertheless suffices to demonstrate how important and complex these effects can be. Our understanding of the underlying mechanisms is still poor and quantitative theories describing even the simplest phenomena are either non-existent or still in their infancy. The rest of this work is an effort to draw a first order picture

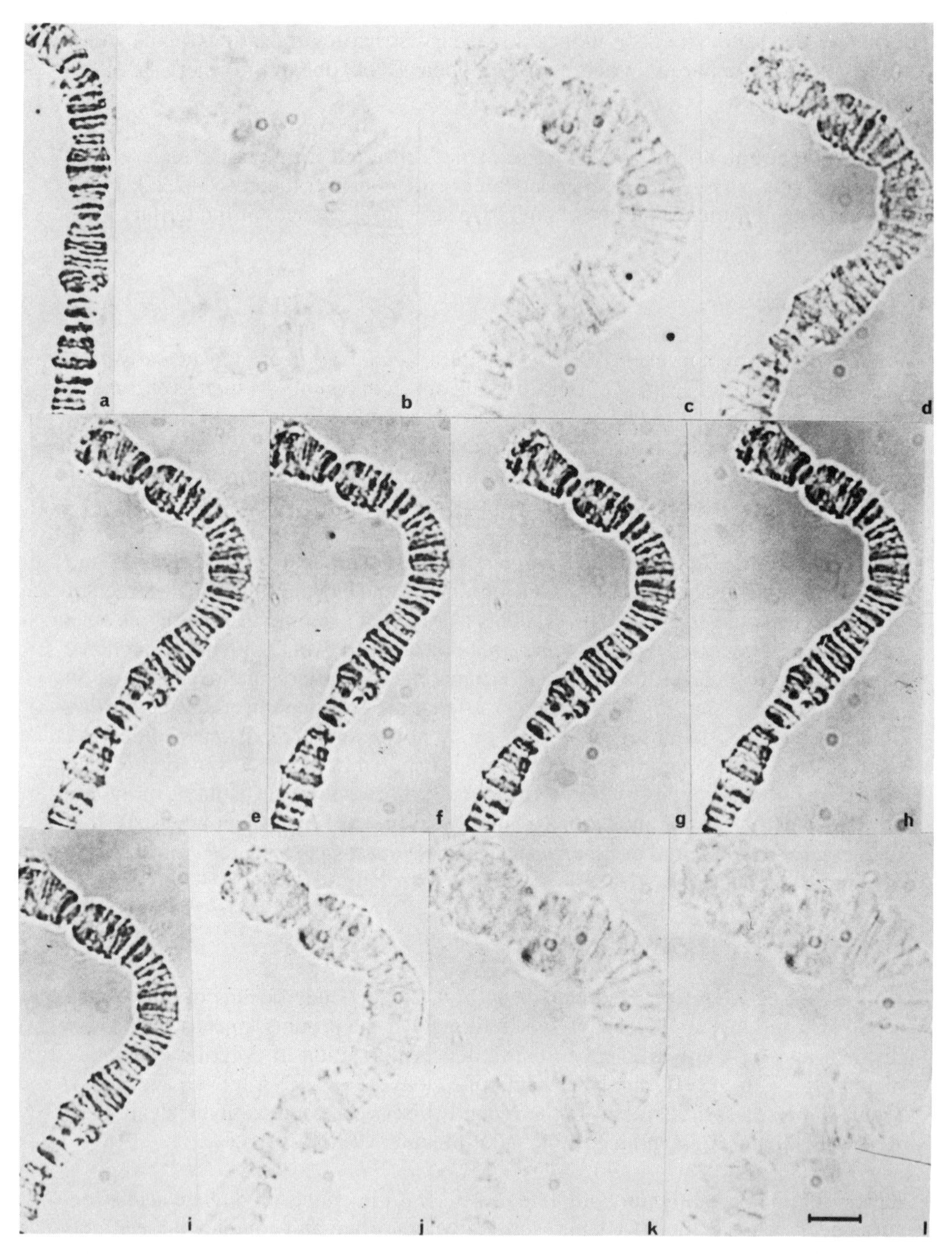

Figure 1. Ion-induced condensation-decondensation phenomenon in a polytene chromosome of Chironomus Thumii. From c to 1 NaCl concentration increases in steps of 50 mM. b: no salt added (Courtesy of M. Robert-Nicoud)

of the underlying molecular aspects and prepare the ground for more quantitative future work.

Modes of ion-DNA interaction in solution

Analysis of a large body of experimental evidence extensively reviewed in many places (16,85-89) indicates that ions may affect secondary DNA structure in two distinct ways.

(i) *they either form a diffuse cloud or atmosphere statistically screening the electrostatic phosphate repulsions and transferring impulse to the molecule via collisions* or

(ii) *they bind to specific DNA sites thereby changing the charge distribution of the polyanion.*

In the former case, both the ions and the DNA retain an intact hydration sheath. The ions are mobile but the high DNA electric field confines many of them in the immediate neighborhood of the molecule (hence the term territorial binding (90) also used to describe the situation). By and large, this type of ion-DNA interaction provides gross thermodynamic structural stabilization and is *sequence independent.* As a rule, structural transitions driven by diffuse ionic cloud interactions occur at *high critical salt concentrations (91) showing no definite stochiometric relation to the number of DNA monomers present.* In the case of specific site binding, both the DNA sites and the ions binding to them undergo changes in their state of hydration (*inner sphere complexation*). Sometimes, the amount of electrostricted water released from the hydration spheres is enough to produce measurable volume changes. The ions become immobilized for times much longer than the characteristic times for molecular motions in fluids. They become parts of the DNA structure and may lead to detectable spectroscopic changes. Especially with multivalent site bound cations, the overall DNA charge distribution is drastically altered and this may lead to structural transitions at *very low critical salt concentrations, with definite stochiometric relations between the number of binding ions and the number of DNA sites or monomers.* By and large, this mode of interaction is DNA *sequence dependent* and provides fine stabilization and modulation of DNA secondary structures. Due to both the limitations of current experimental methodologies and the primitiveness of the theories used to analyze the data, it is not always possible to decide whether particular ionic species are site bound or simply confined near DNA due to the high local electric field (diffuse atmosphere, territorial 'binding').

However, conceptually, things are somewhat clearer. (a) Whenever DNA is in solution, formation of a diffuse ionic atmosphere is the more common phenomenon with small amounts of site binding only in special (albeit very important!) cases. In fact the diffuse ionic atmosphere is a consequence of the dissolution process energetically driven mainly by hydration. (b) Substantial charge neutralization via site binding is tantamount to substantial dehydration and this means that the molecule has to leave the aqueous phase. Whether or not ions of a given species will be appreciably site bound to a given DNA species in a given environment depends crucially on

(i) *the availability and geometry of suitable, accessible DNA sites (sequence and conformation dependence)*

(ii) *the electronic structure of the ions,* and

(iii) *the environmental conditions, i.e. temperature, pressure, P_H (protons often compete for the same sites), organic cosolvents, water activity, the concentrations and types of all charged species present and the state of the sample (e.g. gel, crystal or dilute DNA solution).*

(i) Possible DNA binding sites

X-rays structural studies of DNA constituents (nucleotides, nucleosides) and DNA fragments cocrystallized with diverse salts (92,93,1) provide direct information about possible sites and the geometry of the ion-site complexes whenever these are stable and the resolution high enough to detect them. Quantum chemical calculations and accessibility algorithms (94-96) are also very helpful in a priori estimations of the binding capacity of particular sites. The most useful spectroscopic tools are usually NMR, IR, and Raman and UV spectroscopy. In the case of DNA conformations, the strongest candidates for inner sphere complexation with cations turn out to be the *anionic oxygens of the phosphates, base ring nitrogens (N7 of Guanine, and Adenine, N3 of Adenine) and base oxygens (O6 of Guanine, O2 of Thymine and Cytosine).*

(ii) The role of electronic ion structure

The ability of inorganic cations to form inner sphere complexes with atoms having lone electron pairs (such as nitrogen and oxygen) can be fairly well understood by means of classification schemes based on simple quantum chemical considerations, proved to be quite useful in ionic solution chemistry (97,98). According to such a scheme proposed by Schwarzenbach a long time ago (99), cations are grouped into three main classes:

1. Cations with a *noble gas electronic configuration* e.g. alkali and alkaline earth metal cations, Al(III), La(III), etc.).

2. Cations with electronic configuration; *noble gas plus a completely filled d-subshell* (e.g. Zn(II), Cu(I), Ag(I), etc.).

3. Cations with electronic configuration; *noble gas plus an incompletely filled d-subshell* (e.g. transition metal cations).

By and large cations of the first class do not hydrolyze and are comparatively *weak complex formers* (100). Their interactions with nitrogen and oxygen are expected to be *predominantly electrostatic,* increasing with increasing charge and decreasing size (i.e. the bare ionic radius in inner sphere complexes but the hydrated radius in outer sphere complexes). Complexes involving cations of the second class, show more covalent character, the extent of inner sphere coordination being related to the *difference in electronegativities* of the cation and the electron donor atom of the ligand. Thus, their affinity for the typical electron donors decreases in the order

$$S > Br > Cl > N > O > F$$

In addition to the above characteristics, cations of the third class exhibit *special ligand field stabilization effects* due to their only partially occupied degenerate d-orbitals. In general, transition metal ions form stable inner sphere complexes with many ligands and have more affinity for nitrogens than for oxygens. Studies of many different complexes (101-104,98) have shown that given a particular ligand, stability increases in the order Mn(II) < Fe(II) < Co(II) < Ni(II) < Cu(II) > Zn(II) (i.e. the so called Irving-Williams series). Most divalent complexes have coordination number 6 and octahedral ligand coordination (e.g. the typical aquo complex $M(H_2O)_6^{2+}$, M:Metal). However, 4 coordinations with tetrahedral or square geometry are also quite common (e.g. with Co(II), Cu(II) and particularly Ni(II) complexes).

(iii) Environmental conditions

The influence of environmental factors on the ionic site binding equilibrium can be very complex and cannot be discussed in a simplified manner. However, a few general remarks are in place. Site occupation will in general increase in the presence of dehydrating agents (e.g. alcohols) or condensed DNA phases (i.e. fibers, crystals, etc.) It is favored by very low or very high binding ion concentrations. It will decrease upon increasing the monovalent salt concentration although most monovalent salt cations do not really compete for sites (as protons often do) since they do not bind to them but rather form a diffuse atmosphere around DNA. At very high monovalent salt concentrations site binding can increase again due to dehydration effects similar to those of organic cosolvents. Effects of temperature can be very complex. Depending on the specific composition of the system (cosolvents, types and concentrations of ions) and the characteristics of the binding-sites lattice (i.e. DNA sequence and conformation) one may have both monotonic and nonmonotonic changes of site occupation with increasing temperature.

This brief discussion of the general aspects expected to influence ionic binding to DNA suffices to rationalize the main experimental findings qualitatively. In typical solution environments close to room temperature, and low ratios of DNA monomers to salt concentrations *alkali cations (except possibly Li(I) do not bind to DNA but rather form a diffuse atmosphere around it retaining intact hydration sheaths (105-107).* This is in accord with an earlier discussion containing additional references (110). However, the apparent affinity of DNA for alkali cations decreases in the order

$$Cs(I) > K(I) > Li(I) > Na(I)$$

as judged from a competition study using ^{23}Na MMR (108). We do not understand the position of Li(I) in this series but the rest of it is consistent with a diffuse atmosphere picture where the *smallest hydrated species* (i.e. Cs) come closer to DNA and therefore interact with it somewhat stronger that the larger ones. A recent NMR study (109) of the Na-DNA association using both right and left conformations clearly shows appreciable conformation dependence. Both this dependence and the ionic size effect above, cannot be explained by currently used polyelectrolyte theories such as Manning's counterion condensation—Debye Hückel (CC) approach (110) or the so called Poisson-Boltzmann (PB) eq. approach (111-113) since ions are treated as *point particles and DNA is modeled too primitively* (as an infinite charged line with CC and infinitely long conducting cylinder with PB). Both theories predict that the counter-ion concentration near DNA is essentially independent of bulk salt concentration. This is consistent with experimental evidence for NaCl up to 1.3 M (108). However, it must be understood that any reasonable statistical mechanical theory would yield the same prediction, simply because the spatially fixed DNA phosphates configuration creates a high local concentration of negative charge (of the order molar) and sufficiently many counterions of the diffuse atmosphere have to remain close to DNA in order to preserve local electroneutrality. Bulk concentrations lower than this DNA structure dependent critical (phosphate) concentration cannot appreciably change the counterion concentration near the DNA (i.e. the so called 'condensation' layer). At bulk concentrations higher than that, condensation looses all its significance because then the system of ions is dense everywhere, not only close to the DNA.

In typical solution environments the *interaction of Mg(II) with DNA is also of the diffuse atmosphere type with negligible inner sphere complexation to specific sites (114-121).* However, at very low or very high Mg(II)/phosphate ratios or low water contents this cation interacts predominantly with anionic phosphate oxygens as judged from volumetric (119) P^{31} NMR (116) and IR (124) studies. It is not seen in the B-dodecamer structure (25) but one Mg(II) has been seen forming an inner sphere complex with N7 of Guanine and an outer sphere complex (via its water of hydration) to the end phosphate in the Z structure of $(CpG)_3$ (22). In a very recent study (139) four Mg(II) have been localized in the same structure but the majority of them forms only outer sphere complexes bridging different helices in the crystal. It is almost certain that all these complexes are not stable enough to survive the dissolution process. Ca(II) has also been found to coordinate directly to the phosphates and indirectly to base atoms in crystals of CpG dinucleotide mini helices (122). Different lines of experimental evidence clearly show that transition metal cations (e.g. Mn(II), Co(II), Ni(II), Pt(II), etc.) can form strong inner sphere complexes particularly with N7 of purines both in DNA solutions (16,85-89,125,126) and in the crystals of DNA constituents (1,92,123). Inner sphere complexation of transition metal cations to Z-DNA has been recently demonstrated (124). Especially the N7 of Guanine proves to be perhaps the most important binding site for transition metals, which can form octahedral inner sphere complexes with it further stabilized by coordination—water mediated hydrogen bridges to nearby phosphate or base oxygens (139).

Computer generated graphs, kindly provided by Reinhard Klement, MPI Göttingen, and shown in Figure 2 vividly illustrate the high accessibility of this site both in right handed and left handed DNA and its geometric relation to nearby phosphates. This accessibility and its high electrostatic potential (94-96) make it a very attractive site for ions which due to their electronic structure like to form complexes in aqueous solution, such as transition metal cations.

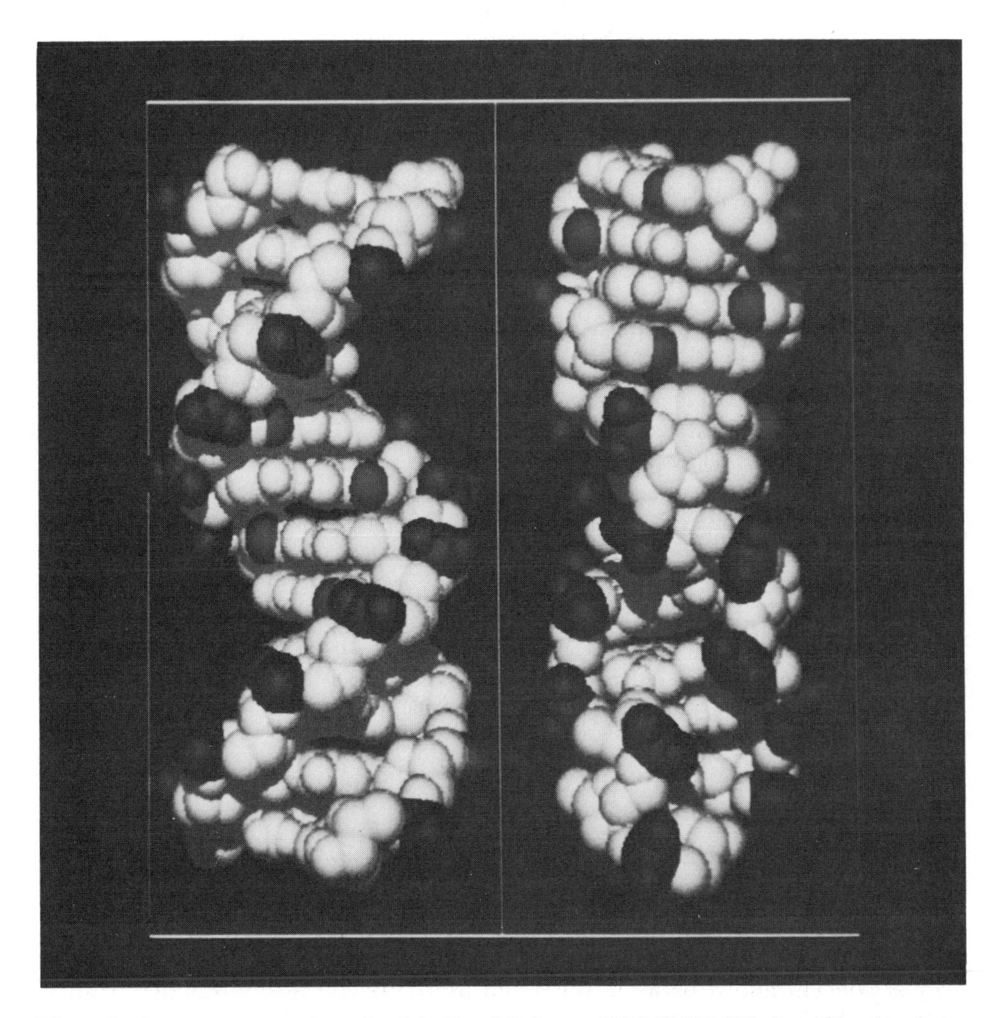

Figure 2. Computer generated graphs of the B and Z_1 forms of d(G-C)•d(G-C) helices. The phosphates and N7 of Guanine are shown in darker shade (Courtesy of R. Klement)

A specific example of ionic structural modulation. The B-Z transition revisited.

Various types of cations and solvents affect the B-Z transition of linear poly[d(C-G)•d(C-G)] in solution in drastically different ways (17,127-138) which, however, can be fairly well understood in the light of the discussion presented above. In the case of other sequences and chemical modifications thereof (e.g. methylation, halogenation, sulfur substitutions, etc.) the situation is even more complex (27,28,140) and will be discussed in the near future, along with the case of the transition in supercoiled DNA.

Table I
Parameters of the B-Z transition of poly [d(C-G)•d(C-G)] induced by various simple cations and organic cosolvents†

cations	cosolvent	$\overset{*}{C}_1$/mM	$\overset{*}{C}_2$/mM	$\overset{*}{r}$	Ref.
Na^+	—	2300	—	—	(17)
"	48% ETOH	3	—	—	(129,127)
"	62% TFE	1	—	—	(128)
Li^+	—	5000	—	—	(137)
Mg^{2+}/Na^+	—	3	660	—	(17,129)
"	10% ETOH	3	4.00	—	(129)
"	20% ETOH	3	0.40	9	(129)
Mn^{2+}/Na^+	—	10	0.15	4	(131)
"	—	25	0.50	—	(131)
"	—	75	1.80	—	(131)
"	25% EG or 15% ETOH	7.5	0.25	—	(132)
Co^{2+}/Na^+	—	10	0.60	9	(131)
Co^{2+}/Na^+	25% EG or 15% ETOH	7.5	0.20	—	(132)
Ni^{2+}/Na^+	—	10	0.60	9	(131)
Zn^{2+}/Na^+	—	2	0.15	2	(136)

†All data for room temperature, except (Li)I (30°C) and Zn(III) (35°C). The anion is chloride in all cases. In the case of Co(II) and Ni(II) transitions are biphasic. Data here refer to the main transition at the higher salt concentration. $\overset{*}{C}_1$, $\overset{*}{C}_2$; concentrations of mono and- divalent salt at transition mid point. $\overset{*}{r}$; approximate divalent ion/phosphate ratios. Abbreviations: ETOH, ethanol, TFE trifluoroethanol, EG, ethyleneglycol.

Table I summarizes the transition data in various solvents and types of simple cations. It is immediately evident that *Na(I) and Mg(II) in water induce the transition at high salt concentrations while transition metals are much more efficient in stabilizing the electrostatically unfavorable left handed form.* The difference in critical concentrations between Na(I) and Mg(II) is almost entirely due to their different charges as can be easily seen when critical *ionic strengths* (2300 mM for Na^+, 2000 mM for Mg^{2+}/Na^+) instead of *concentrations* are compared. But the mode of interaction is the same. Both ions induce the transition via diffuse cloud statistical interactions with the phosphates. The same is also true for K, Rb and Cs (138) where differences in critical concentrations (not shown) are due to the different hydrated radii of the ions. Li(I) does not induce the transition at room temperature (134,138) but it has been shown recently (137) that one can find ranges of higher temperature and high salt concentrations where a transition is possible. This anomalous behavior may be due to stabilization of the B form via specific binding. Ca(II) and Ba(II) at concentrations similar to Mg(II) also cause abrupt changes in the CD spectra (130) but these high salt spectra are anomalous. Addition of alcohols greatly enhances the effects of Na(I) and Mg(II). In the case of Na(I) ethanol system the bulk alcohol concentration is high enough to cause appreciable lowering of water activity, but in the case of Mg(II)-ethanol we think that the effect is more subtle. It is possible that adsorption of ethanol on DNA causes substantial Mg(II) binding to the phosphates. The resultant dehydration and partial neutralization would favor Mg(II) mediated crosslinking of different helices, i.e. the observed aggregations (129). In the case of transition metal cations the *extremely low critical concentrations and stochiometric ratios* observed clearly show that Z family forms are stabilized *via site specific binding* almost certainly to N7 of Guanine. Increasing Na(I) concentrations destabilize the complex (e.g. see data for Mn(II)/Na(I) in Table I). The biphasic transitions with Co(II) and Ni(II) (131,132) could result from the presence of an additional population of sites (phosphates? O2 cytosine?)

Table II

Transition Parameters for some complex cations in aqueous solution†

cations	$\overset{*}{C}_1$/mM	$\overset{*}{C}_2$/mM	$\overset{*}{r}$	Ref.
$[Co(NH_3)_6]^{3+}/Na^+$	50	0.02	—	(130)
Zn^{2+}-A/Na^+	2	0.09	1	(136)
Zn^{2+}-Cysteine/Na^+	2	0.08	1	(136)
Zn^{2+}-B/Na^+	2	0.04	1/2	(136)
Zn^{2+}-C/Na^+	2	0.003	1/24	(136)
dien-Pt^{2+}/Na^+	200	—	1/7	(135)
dien-Pt^{2t}/Na^+	6	—	1/7	(135)

†Transitions at room temperature except with the Zn complexes (35°C) A: ethylene diamine, B: diethylene triamine, C: tris (2 aminoethyl) amine, Dien-Pt; chlorodiethylene triamine Pt(II).

Table II summarizes transition data for complex cations. Cobalt (III) hexamine is very effective and the structural basis for its effectiveness is clear considering recent crystallographic work (139) which shows that in the Z form the trivalent ion interacts with G N7 and in addition forms five H-bridges to nearby oxygens, which is not possible in the B geometry. In solution, increasing NaCl concentrations again destabilize the complex $\overset{*}{c}_2$ and $\overset{*}{c}_1$ being related by $\overset{*}{c}_2 \sim \overset{*}{c}_1^{2.8}$ (130). A recent study (136) shows that Zn(II) complexes are also very effective, one of them (Zn(II)-C of Table II) being able to induce the transition at a ratio of 1 ion to 12 bp's albeit in the presence of very low Na^+ concentration in water. Zn(II) is biologically very interesting because it is found complexed to several enzymes working on DNA.

An interesting phenomenon is exhibited by dien-Pt(II) (135, Table II) which when *bound to DNA at a ratio of 1 ion to 7 phosphates* induces a Z-B transition at 6mM•NaCl and a second B-Z at 200 mM NaCl.

Theoretical developments

The first quantitative, statistical theory describing the salt dependence of the B-Z transition in the absence of specific binding has appeared last summer (141). Briefly, the free energy difference of two given DNA conformations (say B and Z_I)ΔG is decomposed into three terms ΔG_{in}, ΔG_h and ΔF_I containing the contributions of intramolecular interactions, hydration and phosphate-ion cloud interactions, respectively. The latter is then approximated in the form:

$$\beta \Delta F_I(Z_I,B) = \sum_{ij}^{M} W(r_{ij}(Z_I)) - W(r_{ij}(B)) \qquad (1)$$

where $r_{ij}(x)$ is the distance of the charge centers of gravity of phosphates i and j in conformation X,$\beta = (K_BT)^{-1}$ and W is the pair potential of mean force in a homogeneous salt solution. Modeling the hydrated ions as equal size charged hard spheres of effective diameter σ in a dielectric continuum (water) and using known techniques of the modern theory of liquids (in order to calculate potentials of mean force) in conjunction with the atomic DNA coordinates, one obtains quantitative agreement with experiment for σ = 4.90 A (141,146). Furthermore, the theory describes correctly the reduction of critical concentration with increasing charge of the cation (e.g. going from Na(I) to Mg(II) and predicts that the critical transition concentrations of equally charged non-binding cations should increase with decreasing hydration radius. For example, in the order Na < K < Rb < Cs for alkali cations, as observed experimentally (138). As will be shown elsewhere, formula (1) is not an ad hoc approximation but the *first order term in the rigorous expansion of the free energy difference ΔF_I*. Higher order terms turn out to be negligible due to cancellations. The success of the theory is due to:

(i) *inclusion of the short range ion repulsions which at high salt concentrations are as important as the long range Coulomb interaction*

(ii) *the realistic description of DNA charge configurations* and

(iii) *the use of advanced potentials of mean force which describe the many body correlations in the system in a satisfactory way.*

After this theory appeared, two papers using discretized versions of the PB approach have been published in an effort to describe the B-Z transition (142,143). They both fail to find the high salt transition experimentally observed, which is no wonder because only *average energies* not *free energies* are calculated and in addition the PB approach, which is a mean field theory, is totally *incapable of describing concentrated solutions.* Even more recently, an abstract appeared (144) using once again the good old infinitely long uniformly charged cylinder model for DNA, along with the PB approach. It is claimed that this kind of treatment is able to predict a Z-B-Z, i.e. two transitions, one at low and one at high concentration due to the fact that Z-DNA is thinner than B-DNA! First of all in the absence of special effects (such as methylation, halogenation, specific ligand binding or torsional stress) *a low salt transition has never been observed,* and *second the PB approach is not applicable at 2.0 M NaCl where the real diffuse cloud induced transition occurs.* Furthermore, it can be rigorously shown (145) that within the framework of traditional mean field theory the electrostatic free energy difference of two fixed charged configurations *never changes sign, i.e. the one which is unfavorable at one salt concentration remains unfavorable at other concentrations too.* In the meantime, we have extended the calculations in several ways. (i) Using refined potentials of mean force we have estimated the salt contributions to the relative stabilities of the A,B,C, alternating B, Z_I andZ_{II} conformations of DNA (146). (ii) Using the AMBER force field of P. Kollman and collaborators, USF, in order to estimate the term ΔG_{in}, and our formalism for ΔF_1 we have again obtained a high salt B-Z transition for DNA (147) corroborating the earlier conjecture (141) that *intramolecular interactions are approximately equal in the two forms for sufficiently long DNA's and therefore cancel.* (Using the force field alone it is not possible to obtain a B-Z transition (148) because the important statistically averaged ionic interactions driving the transition are not included). (iii) A generalization of the theory suitable for treating the effects of specific ion binding on polyionic structural transitions has been recently derived and first results pertaining to the B-Z transition will be published in the near future.

Acknowledgements

I wish to thank T.M. Jovin, M. Robert-Nicoud, R. Clement, J. Wiechen and E. von Kitzing for many stimulating discussions and collaborations, R.H. Sarma for warmest hospitality and the Max Planck Institute for Biophysical Chemistry, Göttingen, for travelling funds.

References and Footnotes

1. W. Saenger, *Principles of Nucleic Acid Structure,* Springer Verlag, N.Y. (1984).
2. *Cold Spring Harbor Symp. Quant. Biol. 47, Parts 1 and 2,* Cold Spring Harbor Laboratory, (1983).
3. S.B. Zimmermann, *Ann. Rev. Biochem. 51,* 395 (1982).
4. *Biomolecular Stereodynamics,* R.H. Sarma Ed. Adenine Press, Guilderland, N.Y. (1981).
5. S.H. Kim in *Topics in Molecular and Structural Biology,* S. Neidle and W. Fuller, Eds., Vol. 1, 83 (1981).
6. *Transfer DNA: Structure, properties and recognition,* D. Söll, I.N. Abelson and R.P. Schimmel Eds., Cold Spring Harbor Laboratory (1980).
7. *Transfer RNA,* S. Altmann, Ed., MIT Press, Cambridge (1978).
8. T.J. Richmond, J.T. Finch, B. Rushton, D. Rhodes and A. Klug, *Nature 311,* 532 (1984).
9. J.D. McGhee and G. Felsenfeld, *Ann. Rev. Biochem. 49,* 1115 (1980).
10. *Cold Spring Harbor Symp. Quant. Biol. 42,* Cold Spring Harbor Laboratory (1977).
11. M.M. Teeter, G.J. Quigley and A. Rich in *Nucleic Acid-Metal Ion Interactions,* T.G. Spiro, Ed., 147, Wiley, N.Y. (1980).
12. R.R. Schimmel and A.G. Redfield, *Ann. Rev. Biophys. Bioeng. 9,* 181 (1980).
13. W.F. Dove and N. Davidson, *J. Mol. Biol. 5,* 467 (1982).
14. Y.A. Shin and G.L. Eichorn, *Biochemistry 7,* 1026 (1968).
15. D.W. Gruenwedel, Chi-Hsia Hsu and D.S. Lu, *Biopolymers 10,* 47 (1971).
16. G.L. Eichorn in *Inorganic Biochemistry,* G.L. Eichorn Ed., 1210, Elsevier Amsterdam (1975).
17. F.M. Pohl and T.M. Jovin, *J. Mol. Biol. 67,* 375 (1972).

18. A.H.J. Wang, G.J. Quigley, F.J. Kolpak, J.L. Crawford, I.H. van Boom, G. van der Marel and A. Rich, *Nature 282,* 680 (1979).
19. J.L. Crawford, F.J. Kolpak, A.H. Wang, G.J. Quigley, J.H. van Boom, G. van der Marel and A. Rich, *Proc. Natl. Acad. Sci. USA 77,* 4016 (1980).
20. H.R. Drew, T. Takano, S. Tanaka, K. Itakura and R.E. Dickerson, *Nature 286,* 567 (1980).
21. S. Arnott, R. Chandrasekaran, D.L. Birdsall, A.G. Leslie and R.L. Ratliff, *Nature 283,* 743 (1980).
22. A.H.J. Wang, G.J. Quigley, F.J. Kolpak, G. van der Marel, J.H. van Boom and A. Rich, *Science 211,* 171 (1981).
23. R.M. Wing, H.R. Drew, T. Takano, C. Broka, S. Tanaka, K. Itakura and R.E. Dickerson, *Nature 287,* 755 (1980).
24. R.E Dickerson and H.R. Drew, *J. Mol. Biol. 149,* 761 (1981).
25. H.R. Drew and R.E. Dickerson, *J. Mol. Biol. 151,* 535 (1981).
26. A.G. Leslie, S. Arnott, R. Chandrasekaran and R.L. Ratliff, *J. Mol. Biol. 143,* 49 (1980).
27. A. Rich, A. Nordheim and A.H.J. Wang, *Ann. Rev. Biochem. 53,* 791 (1984).
28. T.M. Jovin, L.P. McIntosh, D.J. Arndt-Jovin, D.A. Zarling, M. Robert-Nicoud, J.H. van de Sande, K.F. Jorgenson and F. Eckstein, *J. Biom. Struct. and Dynamics 1,* 21 (1983).
29. K. Hall, P. Cruz, I. Tinoco Jr., T.M. Jovin and J.H. van de Sande, *Nature 311,* 584 (1984).
30. G.V. Fazakerley, S. Uesugi, A. Izumi, M. Ikehara and W. Guschlbauer, *FEBS Lett. 182,* 365 (1985).
31. M. Vorlickova, J. Kypr and V. Sklenar, *J. Mol. Biol. 166,* 85 (1983).
32. M. Vorlickova, J. Kypr, V. Kleinwächter and E. Palecek, *Nucleic Acids Res. 8,* 3965 (1982).
33. D. Patel, S.A. Kozlowski, J.W. Suggs and S.D. Cox, *Proc. Natl. Acad. Sci. USA 78,* 4063 (1981).
34. Chi-Wan Chen and J.S. Cohen, *Biopolymers 22,* 879 (1983).
35. M.H. Sarma, G. Gupta and R.H. Sarma, *J. Biom. Struct. and Dynamics 1,* 1423 (1984).
36. H. Shindo and S.B. Zimmerman, *Nature 283,* 690 (1980).
37. W.A. Baase and W.C. Johnson Jr., *Nucleic Acids Res. 6,* 797 (1979).
38. M.J.B. Tunis-Schneider and M.T. Maestre, *J. Mol. Biol. 52,* 521 (1970).
39. V.I. Ivanov, L.E. Minchenkova, A.K. Shyolkina and A.I. Poletayev, *Biopolymers 12,* 89 (1973).
40. B.B. Johnson, K.S. Dahl, I. Tinoco Jr., V.I. Ivanov and V.B. Zhurkin, *Biochemistry 20,* 73 (1981).
41. A. Chan, R. Kilkuskie and S. Hanlon, *Biochemistry 18,* 84 (1979).
42. L.C. Gosule and J.A. Schellman, *J. Mol. Biol. 121,* 311 (1976).
43. L.C. Gosule and J.A. Schellman, *J. Mol. Biol. 121,* 311 (1978).
44. D.K. Chattoraj, L.C. Gosule and J.A. Schellman, *J. Mol. Biol. 121,* 327 (1978).
45. J. Widom and R.L. Baldwin, *Biopolymers 22,* 431 (1980).
46. J. Widom and R.L. Baldwin, *Biopolymers 22,* 1595 (1983).
47. K.A. Marx and G.C. Ruben, *Nucleic Acids Res. 11,* 1839 (1983).
48. K.A. Marx and T.C. Reynolds, *Biochim. Biophys. Acta 741,* 279 (1983).
49. K.A. Marx and G.C. Ruben, *J. Biom. Struct. and Dynamics 1,* 1109 (1984).
50. R.W. Wilson and V.A. Bloomfield, *Biochemistry 18,* 2192 (1979).
51. T.H. Eickbush and A.N. Moudrianakis, *Cell 13,* 295 (1978).
52. D. Lang, T.N. Taylor, D.C. Dobyan and D.M. Gray, *J. Mol. Biol. 106,* 97 (1976).
53. R. Huey and S.C. Mohr, *Biopolymers 20,* 2533 (1981).
54. L.S. Lerman, *Cold Spring Harbor Symp. Quant. Biol. 38,* 59 (1973).
55. T. Maniatis, J.H. Venable jr. and L.S. Lerman, *J. Mol. Biol. 84,* 37 (1974).
56. Yu.M. Evdokimov, A.L. Platonov, A.S. Tikhonenko and Ya.M. Varshavsky, *FEBS Lett. 23,* 180 (1972).
57. H. Castleman and B.F. Erlanger in *Ref. 2,* 133 (1983).
58. H. Castleman, L. Specthrie, S. Makowski and B.F. Erlanger, *J. Biom. Struct. and Dynamics 2,* 271 (1984).
59. G.L. Eichorn, Y.A. Shin and J.J. Butzow in *Ref. 2,* 125 (1983).
60. C.K. Singleton, C.J. Klysik, S.M. Stirdivant and R. Wells, *Nature 299,* 312 (1982).
61. S.M. Stirdivant, J. Klysik and R.D. Wells, *J. Biol. Chem. 257,* 10159 (1982).
62. L.J. Peck, A. Nordheim, A. Rich and J.V. Wang, *Proc. Natl. Acad. Sci. USA 79,* 4560 (1982).
63. H.J. Vollenweider, T. Koller, S. Parello and J.M. Sogo, *Proc. Natl. Acad. Sci. USA 73,* 4125 (1976).
64. A.M. Campbell, *Biochem. J. 171,* 281 (1978).
65. G.W. Brady in *Book of Abstracts Fourth Conversation in Biomolecular Stereodynamics, SUNYA,* R.H. Sarma Ed., 112 (1985).

66. J.T. Finch and A. Klug, *Proc. Natl. Acad. USA 73,* 1897 (1976).
67. F. Thoma, T. Koller and A. Klug, *J. Cell Biol. 83,* 403 (1979).
68. P. Suau, E.M. Bradbury and J.P. Bradbury, *Eur. J. Biochem. 97,* 593 (1979).
69. A. Ruiz-Carrillo, P. Puigdomenech, G. Eder and R. Lurz, *Biochemistry 19,* 2544 (1980).
70. H.H. Ohlenbush, B.M. Olivera, D. Tuan and N. Davidson, *J. Mol. Biol. 25,* 299 (1967).
71. D.R. Burton, I.J. Butler, I.E. Hyde, D. Phillips, C.J. Skidmore and I.O. Walker, *Nucleic Acids Res. 5,* 3643 (1978).
72. H. Kroeger and G. Müller, *Exptl. Cell Res. 82,* 89 (1973).
73. M. Lezzi and L.I. Gilbert, J. Cell. *Science 6,* 615 (1970).
74. M. Robert, *Chromosoma (Berlin) 36,* 1 (1971).
75. M. Robert and E.N. Moundrianakis, *J. Cell Biol. 70,* 62a (1976).
76. J.J. Butzow and G.L. Eichorn, *Biopolymers 3,* 95 (1965).
77. G.L. Eichorn in *Adv. in Inorg. Biochemistry,* G.L. Eichorn and L.G. Marzilli Eds., Vol 3, 1 (1981).
78. W.R. Farkas, *Chemico-Biol. Interactions 11,* 253 (1978).
79. R.S. Brown, B.E. Hingerty, J.C. Dewan and A. Klug, *Nature 303,* 543 (1983).
80. J.R. Rubin and M. Sundaralingam, *J. Biom. Struct. and Dynamics 1,* (1983).
81. K. Kruger, P.J. Grabowski, A.J. Zaug, J. Sands, D.E. Gottschling and T.R. Cech, *Cell 31,* 147 (1982).
82. A. Zaug, J.R. Kent and T.R. Cech, *Science 224,* 574 (1984).
83. C. Guerrier-Takada, K. Gardiner, T. Marsh, N. Pace and S. Altman, *Cell 35,* 849 (1983).
84. C. Guerrier-Takada and S. Altman, *Science 223,* 288 (1984).
85. R.M. Izatt, J.J. Christensen, J.H. Rytting, *Chem. Rev. 71,* 439 (1971).
86. Ch. Zimmer, *Zeitschrift f. Chemie 11,* 441 (1971).
87. M. Daune in *Metal Ions in Biological Systems,* H. Sigel Ed., Vol. 3, 1, M. Dekker, N.Y., Basel (1974).
88. H. Pezzano and F. Podo, *Chem. Rev. 80,* 365 (1980).
89. *Nucleic Acid Metal Ion Interactions,* T. G. Spiro Ed., Wiley, N.Y., etc. (1980).
90. G. Manning, *Acc. of Chem. Res. 12,* 443 (1979).
91. True for linear DNA but not for supercoiled DNA.
92. V. Swaminathan and M. Sundaralingam, *CRC Crit. Rev. Biochem. 6,* 245 (1979).
93. R.W. Gellert and R. Bau, in *Metal Ions in Biological Systems,* H. Sigel Ed., Vol. 8, 1, M. Dekker, N.Y. (1979).
94. A. Pullman and B. Pullman, *Prog. Nucleic Acid Res. and Mol. Biol. 9,* 327 (1969).
95. A. Pullman and B. Pullman, *Quart. Rev. of Biophysics 14,* 3 (1981).
96. K. Zakrewska, R. Lavery, A. Pullman and B. Pullman, *Nucleic Acids Res. 8,* 3917 (1980).
97. F. Basolo and R.G. Pearson, *Mechanisms of Inorganic Reactions* 2nd Ed., Wiley, N.Y. (1967).
98. G.H. Nancollas, *Interactions in Electrolyte Solutions,* Elsevier Amsterdam (1966).
99. G. Schwarzenbach, *Experientia Suppl. 5,* 162 (1956).
100. Exceptions are the smallest ones, i.e. Li(I) and Be(II) and of course the case of chelates with multidentate ligands such as EDTA.
101. D.P. Mellor and L. Maley, *Nature 159,* 370 (1947).
102. D.P. Mellor and L. Maley, *Nature 161,* 436 (1948).
103. H. Irving and R.J.P. Williams, *Nature 162,* 746 (1948).
104. H. Irving and R.J.P. Williams, *J. Chem. Soc. 31,* 92 (1953).
105. J. Reuben, M. Shporer and E.J. Gabbay, *Proc. Natl. Acad. Sci. USA 72,* 245 (1975).
106. J.T. Shapiro, B.S. Stannard and G. Felsenfeld, *Biochemistry 8,* 3233 (1969).
107. C.F. Anderson, M.T. Record, Jr. and P.A. Hart, *Biophys. Chem. 7,* 301 (1978).
108. M.L. Bleam, C.F. Anderson and T.M. Record, Jr., *Proc. Natl. Acad. Sci. USA 77,* 3085 (1980).
109. L. Nordenskiöld, D.K. Chang, C.F. Anderson and M.T. Record, Jr., *Biochemistry 23,* 4309 (1984).
110. G.S. Manning, *Q. Rev. Biophys. 11,* 179 (1978).
111. A. Katchalsky, Z. Alexandrowicz and O. Kedem in *Chemical Physics of Ionic Solutions,* B.E. Conway and R.G. Baradas Eds., 321, Wiley, N.Y. (1966).
112. M. Gueron and G. Weisbuch, *Biopolymers 19,* 353 (1980).
113. G.F. Anderson and M.T. Record, Jr., *Ann. Rev. Phys. Chem. 33,* 191 (1982).
114. S. Skerjanc and Y.P. Strauss, *J. Am. Chem. Soc. 90,* 3081 (1972).
115. G. Luck and C. Zimmer, *Eur. J. Biochem. 29,* 528 (1972).
116. J. Granot, J. Feigon and D.R. Kearns, *Biopolymers 21,* 181 (1982).

117. D. Pörschke, *Nucleic Acids Res. 6,* 883 (1979).
118. D. Pörschke, *Biophys. Chem. 4,* 383 (1976).
119. R.M. Clement, J. Sturm and M.P. Daune, *Biopolymers 12,* 405 (1973).
120. J.W. Lyons and L. Kotin, *J. Am. Chem. Soc. 87,* 1781 (1965).
121. D. Mùrk Rose, M.L. Bleam, M.T. Record, Jr. and R.G. Bryant, *Proc. Natl. Acad. Sci. USA 77,* 6289 (1980).
122. B. Hingerty, E. Subramanian, S.P. Stellman, T. Sato, S.B. Broyde and R. Langridge, *Acta Cryst. B32,* 2998 (1976).
123. H.A. Tajmir-Riahi and T. Theophanides, *Can. J. Chem. 61,* 1813 (1982).
124. J.A. Taboury, P. Bourtayre, J. Liquier and E. Taillandier, *Nucleic Acids Res. 12,* 4247 (1984).
125. J. Reuben and E.J. Gabbay, *Biochemistry 14,* 1230 (1975).
126. J. Granot and D.R. Kearns, *Biopolymers 21,* 203 (1982).
127. M. Pohl, *Nature 260,* 365 (1976).
128. V.I. Ivanov and E. Minyat, *Nucleic Acids Res. 9,* 4783 (1981).
129. J.H. van de Sande and T.M. Jovin, *EMBO J. 1,* 115 (1982).
130. M. Behe and G. Felsenfeld, *Proc. Natl. Acad. Sci. USA 78,* 1619 (1981).
131. J.H. van de Sande, L.M. McIntosh and T.M. Jovin, *EMBO J. 1,* 777 (1982).
132. W. Zacharias, J.E. Larson, J. Klysik, S.M. Stirdivant and R.D. Wells, *J. Biol. Chem. 257,* 2775 (1982).
133. W. C. Russel, B. Precious, S.R. Martin and P.M. Bayley, *EMBO J. 2,* 1647 (1983).
134. N. Ramesh and S.K. Brahmachari, *FEBS Lett. 164,* 33 (1983).
135. B. Malfoy, B. Hartmann and M. Leng, *Nucleic Acids Res. 9* 5659 (1981).
136. G.V. Fazakerley, *Nucleic Acids Res. 12,* 3643 (1984).
137. M. Behe, G. Felsenfeld, S. Chen Szu and E. Charney, *Biopolymers 24,* 289 (1985).
138. M. Robert-Nicoud, T.M. Jovin and D.M. Soumpasis, *in preparation.*
139. R.V. Gessner, G.J. Quigley, A.H.J. Wang, G.A. van der Marel, J.H. van Boom and A. Rich, *Biochemistry 24,* 237 (1985) and these proceedings.
140. T.M. Jovin, *Book of Abstracts Fourth Conversation in Biomolecular Stereodynamics SUNYA,* R.H. Sarma Ed., 39 (1985).
141. D. M. Soumpasis, *Proc. Natl. Acad. Sci. USA 81,* 5116 (1984).
142. J.B. Matthew and F.M. Richards, *Biopolymers 23,* 2743 (1984).
143. B.J. Klein and G.R. Pack, *Biopolymers 23,* 2801 (1984).
144. M.D. Franck-Kamenetskii, A.V. Lukashin and V.V. Anshelevich *Book of Abstracts Fourth Conversation in Biomolecular Stereodynamics SUNYA,* R.H. Sarma Ed., 36, (1985).
145. D.M. Soumpasis, *in preparation.*
146. D.M. Soumpasis, J. Wiechen, T.M. Jovin, *J. Mol. Biol.,* submitted (1985).
147. R. Klement, E. von Kitzing, T.M. Jovin and D.M. Soumpasis, *Book of Abstracts Fourth Conversation in Biomolecular Stereodynamics, SUNYA,* R.H. Sarma Ed. 38 (1985) and *in preparation.*
148. P.A. Kollman, P. Weiner, G. Quigley and A. Wang, *Biopolymers 21,* 1945 (1982).

Biomolecular Stereodynamics IV, Proceedings of the Fourth Conversation in the Discipline Biomolecular Stereodynamics, State University of New York, Albany, NY, June 04-09, 1985, Eds., Ramaswamy H. Sarma & Mukti H. Sarma, ISBN 0-940030-18-7, Adenine Press, ©Adenine Press 1986.

Environmental Influences on DNA Superhelicity. Ionic Strength and Temperature Effects on Superhelix Conformation in Solution

George W. Brady,[†] Michael Satkowski,[*] David Foos[*] and Craig J. Benham[§]

[†]Wadsworth Center for Laboratories and Research
New York State Department of Health
Albany, NY 12201
[*]Department of Physics, Rensselaer Polytechnic Institute
Troy, NY 12181
[§]Department of Mathematics
University of Kentucky, Lexington, KY 40506

Abstract

The techniques of small-angle x-ray scattering and analysis which have recently been developed by the authors are used to investigate the influence of ionic strength and temperature on the superhelical conformation of native COP608 plasmid DNA in solution. For salt concentrations below 0.1 M the superhelicity is partitioned between twisting and writhing in the ratio $\Delta Tw/Wr = 2$. Near the physiological salt concentration $[Na^+] = 0.2$ M a cooperative transition is observed in which the pitch angle of the toroidal superhelix is drastically decreased. This results in an almost complete relaxation of writhe. At salt concentrations in excess of the threshold for this transition the superhelical partitioning occurs in the ratio $\Delta Tw/Wr > 25$. Energetic considerations support the suggestion that this transition results from cooperative, superhelical B $\Rightarrow$ Z transconformation reactions at susceptible sites. With temperature, a bimodal relaxation of W_r is observed. Below 22°C the writhing number of this structure is near −8. Between 21.5°C and 30°C a relaxation of tertiary structure occurs to values near $W_r = -4.5$. Between 30°C and 55°C the magnitude of W_r decreases in a slow, approximately linear manner. Near T = 60°C a sudden, almost total relaxation of tertiary structure occurs. Because the linking numbers Lk of the DNA molecules remain fixed, these apparently cooperative transitions in W_r must reflect a repartitioning of the superhelicity ΔLk away from tertiary structure and towards some form of torsional deformation ΔTw. The observed relaxation of W_r with increasing temperature near T = 23°C is shown to be consistent with the onset and extension of superhelicity stabilized helix-coil transitions.

[†]To whom reprint requests should be addressed.

Introduction

A previous communication presented the results of an x-ray scattering study of the conformation of natively superhelical COP608 plasmid DNA in solution (1). Using a specifically constructed small-angle apparatus including a position-sensitive detector, x-ray scattering patterns were obtained which were free of slit smearing effects. These patterns were shown to be consistent with a toroidal-helical superhelix conformation but not with an interwound one. The close agreement seen between theoretical and experimental scattering profiles allowed accurate evaluation of the helical parameters of pitch angle and contour length per turn c of the first-order superhelix. This permitted exact computation of the writhing number W_r of the toroidal helix. The partitioning of molecular superhelicity, $\Delta LK = Lk = Lk_o$ (the difference between the actual molecular linking number and its value in the relaxed state (2) between torsional deformations $\Delta Tw = Tw - Tw_o$ and writhe, W_r, was estimated to be $\Delta Tw/W_r \approx 2$ for natively supercoiled COP608 plasmid DNA under experimental conditions. The intercalations of PrTS was observed to relax molecular writhing W_r. This relaxation involved primarily a decrease in the pitch angle of the superhelix, with only small changes seen in the number of toroidal turns.

These methods have now been applied to study the effects of NaCl concentration and temperature on superhelical conformations in solution. Ionic strength may influence superhelical structure in several ways. First, the unstressed duplex rotation angle ψ is known to vary with both monovalent and divalent cation concentration (2). This alters the relaxed total twist Tw_o, which affects the relaxed linking number Lk_o and thus also the superhelicity ΔLk of a closed circular molecule. Next, the effective mechanical properties of the molecular secondary structures—that is, the bending stiffness A and torsional stiffness C—also vary with ionic strength (3,4). Finally, increasing the ionic strength is known to favor the Z-form structure in susceptible sequences (5,6). Transitions to Z-form involve large local changes in twist. In a negatively supercoiled molecule, the B-Z transition acts to localize a portion of the superhelical deformation as twist, which relaxes the molecule elsewhere. Such transitions become favored at equilibrium when the cost in free energy of performing them is exceeded by the relaxation which they provide (7,8). Because increasing the ionic concentration reduces the free energy cost of B-Z transitions, this transconformation reaction may occur at less extreme negative superhelicities under conditions of higher salt concentrations.

Several previous investigations have been performed of the influence of ionic strength on superhelical DNA tertiary structure in solution (9-12). These early studies used hydrodynamic techniques or light scattering (13) to detect changes in the effective molecular radius of gyration, which were suggested to reflect variations in writhe. In contrast, the present techniques permit direct determinations of the superhelix parameters of pitch angle and contour length per turn, from which W_r may be evaluated.

The sedimentation velocity of native superhelical DNA was seen in these previous studies to increase with salt concentrations in the range $.05 \leq [Na^+] \leq 2M$, although the behavior exhibited was complex. At low ionic strengths ($[Na^+] < .01M$), locally denatured regions were hypothesized to cause the observed sharp drop in the value of the sedimentation coefficient r. In solutions having $[Na^+] \leq .05M$, both intrastrand and interstrand electrostatic repulsion appeared to affect the distribution of tertiary structures. At high ionic strengths ($[Na^+] \geq 1.0M$) Upholt et al. (10) found that increasing $[Na^+]$ induces a corresponding increase in the magnitude of the superhelix density. This is due to an increase in the average unstressed duplex rotation angle ψ, as suggested by Wang (14). More recently Anderson and Bauer (2) have presented a systematic study of the variations in the duplex rotation angle ψ with counterion concentration and species. This work confirms the conclusion that ψ increases with increasing salt.

Regarding secondary structure, Peck and Wang (15) and Wells and coworkers (16) have reported that B $\Rightarrow$ Z transitions may be induced in solution by negative superhelicity at physiological ionic strengths (viz. $[Na^+] = 2M$; (17). The B $\Rightarrow$ Z transition has been shown to occur at less extreme negative superhelix densities under conditions of higher salt concentrations, in accord with theoretical predictions (18,19). This transconformation reaction induces relaxation of the molecule elsewhere, hence may be reflected by a decrease of the magnitude of writhing.

Temperature variations are known to affect the secondary structure of linear or nicked circular molecules in two important ways. First, the average duplex rotation angle ψ of the relaxed B-form decreases with elevated temperature. This effect has been measured using both ethidium bromide titration and band counting methods (20). Secondly, increased temperature destabilizes the B-form conformation relative to locally denatured structures. This effect is sequence-specific, occuring at lower temperatures in A+T-rich regions (21). In a negatively supercoiled molecule these two influences together may cause important alterations of secondary and/or tertiary structure with temperature. Although the linking number Lk of such a molecule is fixed, its superhelicity ($\Delta Lk = Lk - Lk_o$) will depend upon temperature due to the variations in the relaxed angular twist ψ, which in turn affects Lk_o. Since ψ decreases with temperature, the superhelicity ΔLk of a negatively supercoiled molecule becomes less extreme at higher temperatures, other influences remaining fixed. Because superhelicity couples secondary and tertiary structure through the relationship

$$\Delta Lk = W_r + \Delta Tw$$

this smooth variation of ΔLk may be reflected in corresponding alterations of tertiary structure.

Several previous experimental investigations have been made of the influence of temperature on superhelical DNA structures. Numerous lines of evidence indicate that negatively supercoiled molecules undergo a temperature-dependent premelting reaction at temperatures below that required to initiate denaturation in nicked or linear molecules (22-26). In particular, local strand separations in A+T-rich regions may be stabilized at this lower temperature by local conformational transitions to alternative secondary structures having smaller twist rates than the B-form (27).

Theoretical calculations suggest that these local premelting transitions will be increasingly favored at equilibrium under conditions of more extreme negative superhelicity, and may occur at temperature significantly lower than those needed to initiate denaturation in a relaxed molecule. As the temperature is increased beyond this threshold, the expected number of transformed base pairs at equilibrium increases smoothly.

Materials and Methods

The x-ray scattering apparatus used has been described previously (1). Briefly, the power source is a Rigaku 12 kW rotating node operated at 45 kV and 190 ma. The primary beam is collimated by two pinholes 0.75 mm in diameter spaced 100 cm apart in an evacuated chamber. It then passes through the sample, contained in a flat sample holder with mice windows. The sample path length is 1 mm. The diffraction pattern is recorded by a position sensitive detector placed at the end of an evacuated path 150 cm from the sample. With such small pinholes and large sample-to-detector slit corrections are not necessary.

This apparatus permits the acquisition of diffraction patterns free of all corrections, except for a minor one associated with the finite beam width. This causes an overlap of two or three channels of the detector and is not significant.

The DNA used in these experiments is the negatively supercoiled COP608 plasmid (4,258 bp), a 179 bp deletion of the PT181 plasmid of Staphylococcus Aureus (34). It was purified by the method of Zasloff (35). The solutions were brought up to the experimental ionic strengths by dialysing overnight against solutions of the appropriate NaCl concentration in a Bethesda Research Laboratories microdialysis system. The concentration of DNA used in the experiments was 11.6 mg/ml. The experimental temperature was 21°C.

The sample holder was made from machined aluminum and fitted with thin (1 mil) mica windows. Two Peltier couples (Melcor Industries, Trenton, NJ) were thermally bonded to the sides of the holder. Depending on the polarity of the applied voltage the couples will either heat or cool the sample. An iron-constantan thermocouple (Omega Industries) embedded in the sample holder measures the temperature of the sample. Associated circuitry controls the temperature to ± 1°C, and the programmable range is from 0°C to 120°C.

Results

Figure 1 shows a plot of the observed scattering from solutions of natively superhelical COP608 DNA at different NaCl concentrations. These scattering patterns have been corrected for background as described previously (1). The counting time was four hours. The scatter in the data points could have been smoothed out by using longer counting times. However, this risks nicking the DNA, which would relax writhing and degrade the quality of the resulting data. Gel electrophoresis patterns were obtained before and after each scattering experiment. Densitometer tracings of the resulting gels showed no significant changes, so nicking effects were minimized for the x-irradiation times used (1).

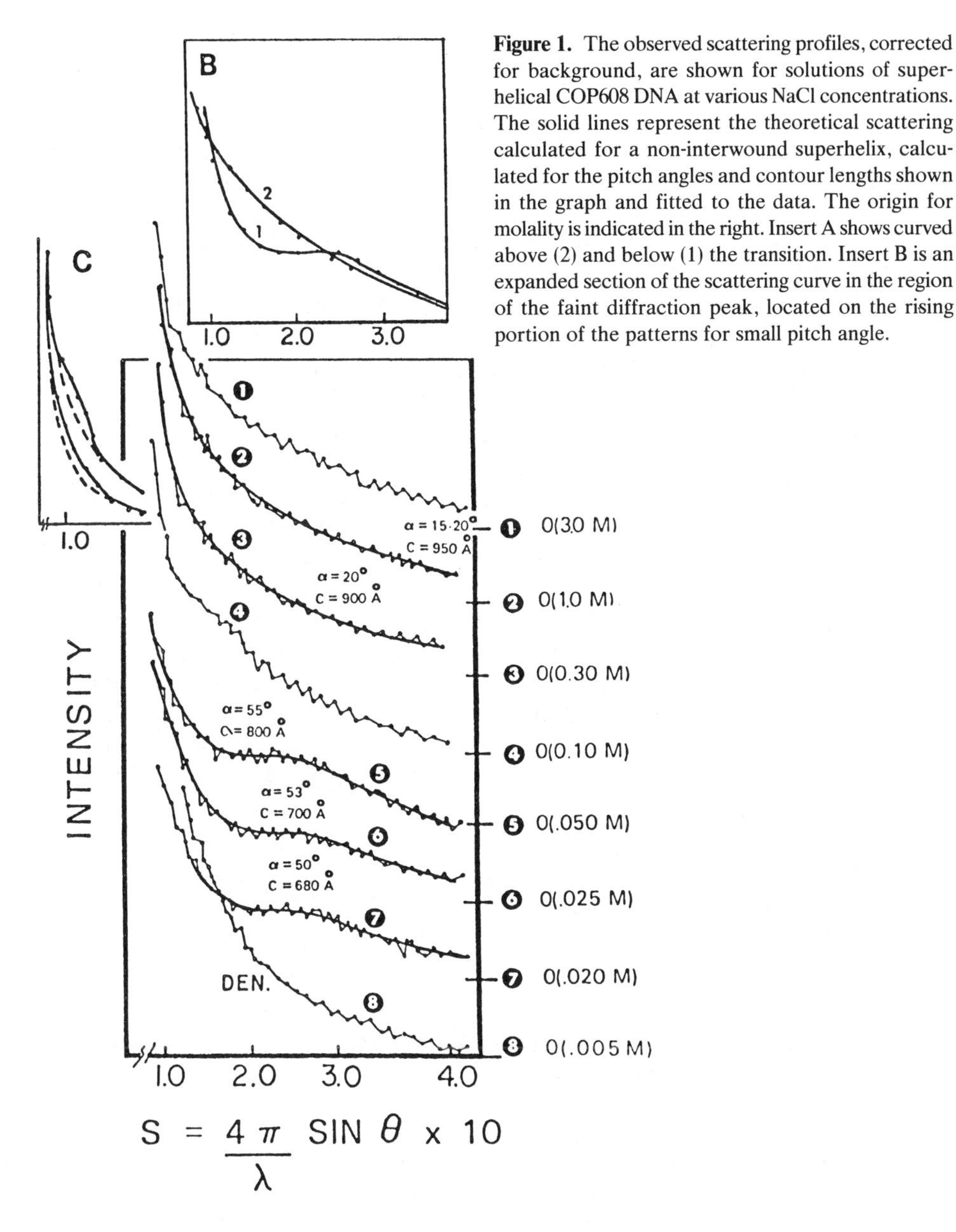

Figure 1. The observed scattering profiles, corrected for background, are shown for solutions of superhelical COP608 DNA at various NaCl concentrations. The solid lines represent the theoretical scattering calculated for a non-interwound superhelix, calculated for the pitch angles and contour lengths shown in the graph and fitted to the data. The origin for molality is indicated in the right. Insert A shows curved above (2) and below (1) the transition. Insert B is an expanded section of the scattering curve in the region of the faint diffraction peak, located on the rising portion of the patterns for small pitch angle.

The curves of Figure 1 show the patterns obtained from a system of identical, independently scattering superhelical COP608 DNA molecules. They are interpreted using the equation of Debye (31,32).

$$I(s) = \frac{\rho^2}{2} \int_0^L dl_1 \int_0^L \frac{\sin sr\,(l_1,l_2)}{sr(l_1,l_2)}\, dl_2. \tag{1}$$

Here I is the scattering intensity from a sample of randomly oriented filaments, while l_1 and l_2 are two copies of the contour length parameter. Also, ρ is the effective electron density of the DNA and $s = 4\pi/\lambda \sin\theta$, where λ is the wavelength of the x-rays used (CuKα, λ = 1.54 Å) and θ is the scattering half-angle. The variable $r(l_1,l_2)$ is the distance in space between any two points at positions λ_1 and λ_2 along the DNA filament and L is the total molecular length. The scattering from helical supercoils may be computed by reducing Debye's equation to a single integral which can be integrated numerically (33). These calculated curves are fitted to the experimentally measured diffraction patterns (corrected for background) as previously described (1). The shape of the calculated curve which fits a given experimental curve determines the pitch angle a of the first-order superhelix. The angular scale at which the scattering pattern occurs is inversely related to the contour length of the helix. It follows that the pitch angle a and the contour length per turn c may both be deduced from the experimentally measured diffraction patterns (1,33).

The values of a, c and the total molecular length L together determine a unique toroidal helix (1), whose writhing number may be computed.

The superhelix parameters a and c as well as the total number n (= L/c) of toroidal turns as deduced from the observed scattering patterns are given in Table I below for various values of salt concentration [NaCl]. At low ionic strengths the superhelical pitch angle a is seen to increase slightly with [Na^+]. At salt concentrations near 0.2 M an apparently cooperative transition in tertiary structure occurs which is reflected by a drastic decrease in a. The transition is strikingly evident in a plot of the writhing number vs. Log_{10} [Na] given in Figure 2.

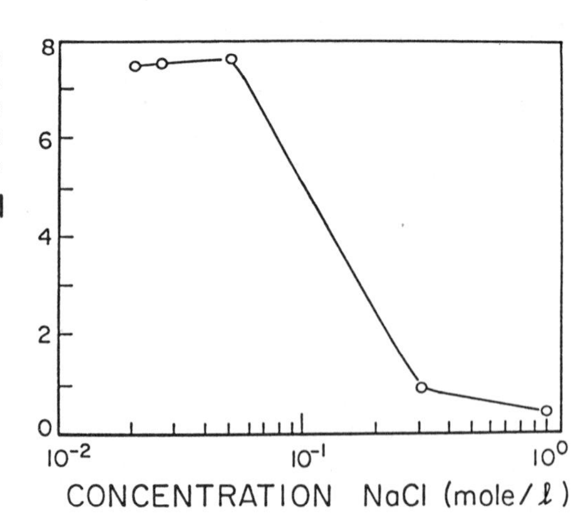

Figure 2. The computed writhing number W_r is plotted as a function of salt concentration. A cooperative relaxation of W_r is observed to occur in the range .05 M < [Na^+] < .2 M. This relaxation involves primarily a decrease in the pitch angle of the superhelix.

The writhing number of a toroidal helix having N turns and ratio of radii $\tau = r/R$ is given by the following expression (1):

$$W_r(N,\tau) =$$

$$\frac{-N}{2\pi} \int_0^{2\pi} \frac{\tau \cos N\theta \, |\sqrt{(1 + \tau \cos N\theta)^2 + {}^2N^2} + (1 + \tau \cos N\theta)| + \tau^2 N^2}{(1 + \tau \cos N\theta) \, |\sqrt{(1 + \tau \cos N\theta)^2 + \tau^2 N^2} + (1 + \tau \cos N\theta)| + \tau^2 N^2} \, d\theta. \quad (2)$$

Here the variable of integration is θ, the angle subtended by an arc of the circular central axis. The values of W_r shown in Figure 2 have been computed using this integral from the measured parameters in Table I. At low ionic strengths ΔLk is partioned between writhing and torsional deformations in approximately the ratio 1:2, as has been reported previously. However, at high ionic strength this distribution is dramatically shifted. The writhing is largely relaxed, so the superhelicity occurs almost exclusively as torsional deformation. This cooperative transition in tertiary structure results from the decrease in pitch angle observed to occur near salt concentrations 0.2 M. Roughly seven turns of writhe are repartitioned to twist in this manner.

Table I
Helix Properties

[NaCl]M	α (deg)	c (Å)	n (turns)	W_r
.02	50° ± 2°	680 ± 10	21	−7.49
.025	53° ± 2°	700 ± 10	20	−7.51
.05	55° ± 3°	800 ± 10	18	−7.66
.3	20° ± 7°	900 ± 15	16	−0.95
1.0	15° ± 7°	950 ± 15	15	−0.51

This repartitioning of superhelicity can be brought about through local conformational transitions to alternative secondary structures. Theoretical considerations suggest that such transitions may be stabilized by negative DNA superhelicity (29,30). The only alternative conformation which is favored by increasing salt concentration is the Z-form structure (5,6). If the observed relaxation of writhe occurs exclusively through superhelical B-Z transitions, about 34 bp would have to revert to Z-form to absorb 6.5 turns of writhe. (About half a turn of this relaxation is thought to occur at the junction regions between the B-form and the Z-form sites.) A computer analysis of the COP608 sequence reveals at least 25 sites of length six base pairs or more whose alternating purine-pyridimine base sequence suggests susceptibility to this transition. Assuming that only perfect alternating sequences can undergo this transconformation reaction, then transition at the four longest sites would suffice to produce the observed relaxation. The free energy required for this transition has been found to be 5 kcal/mole-bp for a B-Z junction and 0.44 kcal/mole-bp for transition at $[Na^+] = .2$ M (17). The transformation of 34 base pairs in four sites to

Z-form requires a free energy of approximately 65 kcal/mole. As the salt concentration increases, this free energy requirement decreases. At 0.2 M salt the observed relaxation of curvature contributes a favorable free energy, whose magnitude may be estimated using the measured value of the molecular bending stiffness at this salt concentration, that is 2.15×10^{-19} erg-cm (36). This favorable free energy is also about 65 kcal/mole. Its value remains approximately constant, independent of salt concentration in this range. As the salt concentration is increased, a point will be reached where the decreasing energetic costs of performing this conformational transition are exceeded by the constant energetic relief they provide through relaxation of bending. Beyond this threshold, the locally transformed structure becomes favored at equilibrium (30). These energetic considerations support the suggestion that the observed relaxation of writhe results from cooperative B-Z transitions near $[Na^+] = .2$ M.

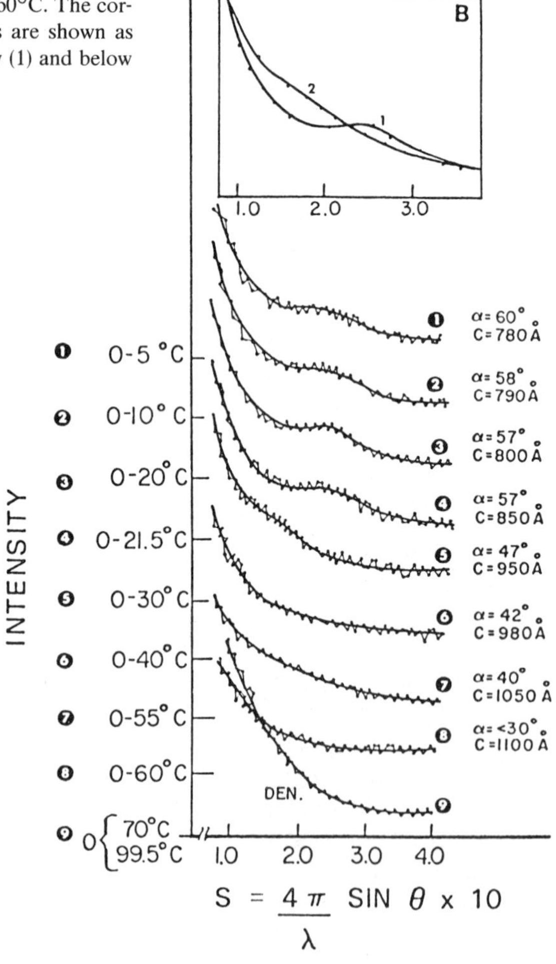

Figure 3. The observed x-ray scattering, corrected for background, is plotted as a function of the angular parameter s for temperatures between 5°C and 60°C. The corresponding theoretical scattering profiles are shown as solid lines. The insert shows curves below (1) and below (2) the transition at 21.5°C.

Figure 3 shows the observed scattering patterns at various temperatures, corrected for background. Up to a temperature of 21.5°C the curves show a pronounced maximum at s ≈ 0.24. Above 25°C the maximum fades abruptly while at the same time shifting rapidly to smaller angle. The abruptness of the change (at 21.5°C) is reminiscent of the transition observed in the ionic strength curves of Figure 1. Its cause cannot be ascribed to thermally induced structural irregularities. These are normally accounted for by using a Gaussian or box distribution of distances *r* in the Debye equation. The alteration of plus and minus values in the sin *rs* function leads to significant cancellation of portions of the added terms in the distribution (37) and thus the data can accommodate up to ± 10% variation in variation of the distance parameter. This class should include all reasonable thermally induced fluctuations. Further, such temperature effects, normally handled by a Debye-Waller treatment, would predict a continuous broading and gradual disappearance of the maximum with temperature. Thus it would be difficult to reconcile any other explanation but a transitional one for the behavior of the curves at 21.5.

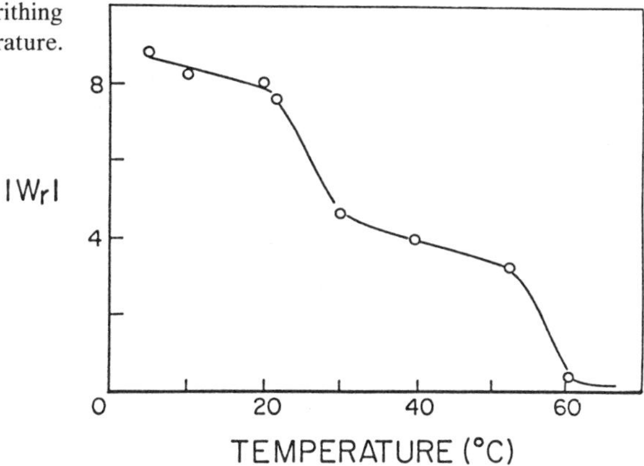

Figure 4. The magnitude $|W_r|$ of the writhing number is shown as a function of temperature.

In Figure 3 the smooth curves represent the theoretical scattering from the best fitting helix. The corresponding contour length and pitch angle are listed on the figure. From these parameters the writhing number W_r of the toroidal helix may be computed. Figure 4 plots the magnitude of the writhing, $|W_r|$, as a function of temperature. This graph displays a clearly bimodal pattern of relaxation of tertiary structure. In the low temperature regime the writhing number is almost constant. The slight relaxation of W_r which is observed between 5°C and 20°C can be attributed to the decrease in Lk_o (and hence in ΔLk) with increasing temperature, as described by equation (2) above. In the temperature interval between 20°C and 30°C the magnitude of the writhing number is seen to decrease to approximately half its low-temperature value in an apparently cooperative manner. Between 30°C and 55°C a slight additional, approximately linear relaxation of writhe is observed. At the high temperature of 60°C an almost total relaxation of tertiary structure is seen, which probably results from the onset of large-scale denaturation as the helix-coil transition temperature is approached. These observations are consistent with the interpretation that local, superhelically stabilized denaturations occur as the temperature increases, which act in part to relax tertiary turns. We analyze equilibrium denaturation in this system using the superhelical transition theory which has been developed previously (27-29).

The linking number of the native COP608 plasmid has been estimated to be Lk = 387 (1) corresponding to a superhelix density of $\sigma = -.055$ at 27°C. The relaxed linking number Lk_o at various values of temperature is given by equation (2) above. From these quantities the superhelicity $\Delta Lk = Lk = Lk_o$ may be found. Subtraction of W_r from ΔLk yields the total torsional deformation ΔTw which is the parameter of importance in the transition theory. In this analysis all local denaturation is assumed to occur at the A+T-richest region of the molecules, which for the COP608 plasmid is a sequence of 130 bp, 80% of which are AT's. The free energy per base pair b needed to denature such a region is

$$b = \Delta H \left| 1 - \frac{T}{T_m} \right|, \tag{3}$$

where $T_m \approx 348°K$. The cooperatively free energy required to initiate transition is $a \approx 8$ kcal/mole. The effective torsional stiffness of the denatured conformation is approximately $C_c = 3.6 \times 10^{-13}$ erg/bp/(rad)^2 (38). The torsional stiffness of the B-form has recently been measured (39) to be $C_h = 7.1 \times 10^{-12}$ erg/bp/(rad)^2.

The free energy of the state containing n sequential coil base pairs, with the B-form regions torsionally deformed an amount τ_h and the coil regions similarly deformed an amount τ_c, is

$$F_n = \frac{nC_c\tau^2_c}{2} + \frac{(N-n)C_h\tau^2_h}{2} + a + bn, \tag{2}$$

while that of the competing, entirely B-form state is:

$$F_o = \frac{NC_h\tau_h^2}{2} = 2\pi^2q^2C_h/N. \tag{3}$$

Here $q = \Delta Tw$ is the total torsional deformation of the molecule. The equilibrium distribution of a population of identical molecules among these two states is given by

$$p_o/p_n = \exp[(F_n - F_o)/kT]. \tag{4}$$

This theory may be applied to the analysis of the temperature dependence of local denaturation in the COP608 plasmid. The Torsional deformation used in this analysis is $q = \Delta Tw = \Delta Lk - W_r$. For temperature values intermediate between those at which scattering experiments were performed, the value of q was estimated by interpolation. The results of these computations are shown in Figure 5 below, where the probability of transition and the expected number n of denatured base

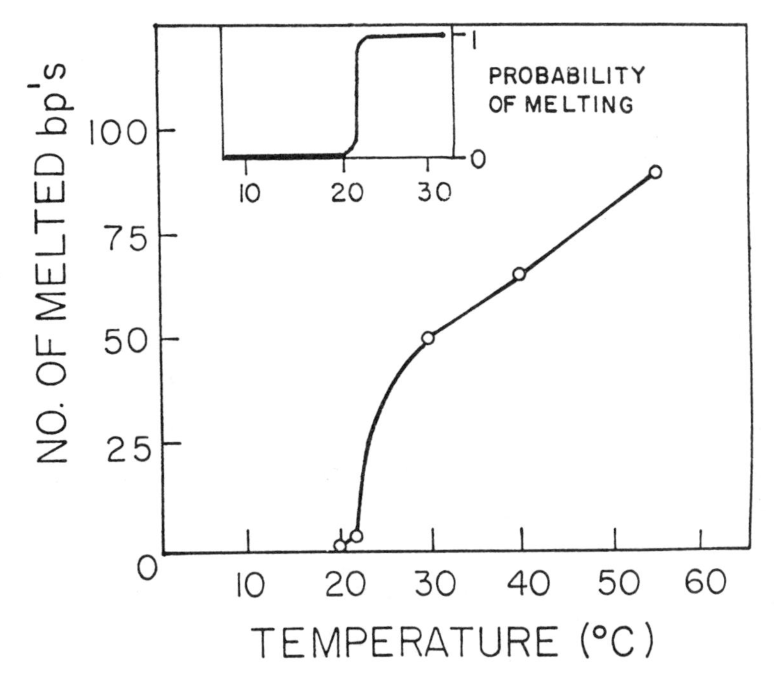

Figure 5. These graphs display the equilibrium expected number n of denatured base pairs and the probability p of transition as a function of temperature. These curves have been computed from superhelical transition theory as described in the text.

pairs are plotted as functions of temperature. These results accord closely with the changes in tertiary structure detected by the scattering experiments. The onset of transition occurs near T = 23°C, where an abrupt drop in the magnitude of W_r is seen. Beyond this threshold the number of denatured base pairs increases smoothly until near the transition temperature, where an abrupt increase is seen (40).

Acknowledgment

The research reported herein was supported in part by the National Science Foundation Grants PCM-8041337 to George W. Brady and PCM-8403523 to Craig J. Benham.

References and Footnotes

1. Brady, G.W., Fein, D.B., Lambertson, H., Grassian, V., Foos, D. and Benham, C.J., *Proc. Natl. Acad. Sci. USA 80,* 741-744 (1983).
2. Anderson, P. and Bauer, W., *Biochemistry 17,* 594-601 (1978).
3. Hagerman, P.J., *Biopolymers 20,* 1503-1535 (1981).
4. Millar, D.P., Robbins, R.J. and Zewail, A.H., *J. Chem. Phys. 76,* 2080-2094 (1982).
5. Pohl, F. and Jovin, T., *J. Mol. Biol. 67,* 375-396 (1972).
6. Wang, A.H., Quigley, G.J., Kolpak, F.J., Crawford, J.L., van Boom, J.H., van der Marel, G. and Rich, A., *Nature 282,* 680-686 (1979).

7. Benham, C.J., *Nature 286,* 637-638 (1980).
8. Benham, C.J., *Cold Spring Harbor Symposium on Quantitative Biology 47,* 219-227 (1982).
9. Kiger, J.A., Young, E.T. and Sinsheimer, R.L., *J. Mol. Biol. 33,* 395-413 (1968).
10. Upholt, W.B., Gray, H.B. and Vinograd, J., *J. Mol. Biol. 61,* 21-38 (1971).
11. Triebel, H. and Reinert, K.E., *Biopolymers 10,* 827-837 (1971).
12. Boettger, M. and Kuhn, W., *Biochim. Biophys. Acta 254,* 407-411 (1971).
13. Campbell, A.M., *Biochem. J. 171,* 281-283 (1978).
14. Wang, J.C., *J. Mol. Biol. 43,* 25-39 (1969).
15. Peck, L.J. and Wang, J.C., *Proc. Natl. Acad. Sci. USA 80,* 6202-6210 (1983).
16. Klysik, J., Stirdivant, S.M., Singleton, C.K., Zacharias, W. and Wells, R.D., *J. Mol. Biol. 168,* 51-71 (1983).
17. Klysik, J., Stirdivant, S.M., Larson, J.E., Hart, P.A. and Wells, R.D. *Nature 290,* 672-677 (1981)
18. Singleton, C.K., Klysik, J., Stirdivant, S.M. and Wells, R.D., *Nature 299,* 312-316 (1982).
19. Zacharias, W., Larson, J.E., Klysik, J., Stirdivant, S.M. and Wells, R.D., *J. Biol. Chem. 257,* 2775-2782 (1982).
20. Bauer, W.R., *Ann. Rev. Biophys. Bioeng. 7,* 287-313 (1978).
21. Poland, D. and Scheraga, H.A., *Theory of Helix-Coil Transitions in Biopolymers,* Academic Press, NY (1970).
22. Lau, P.P. and Gray, H.B., *Nuc. Acids. Res. 6,* 331-357 (1979).
23. Kowalski, D. and Sanford, J.P., *J. Biol. Chem. 257,* 7820-7825 (1982).
24. Burke, R.L. and Bauer, W.R., *Nuc. Acids Res. 5,* 4819-4836 (1978).
25. Vinograd, J., Lebowitz, J., and Watson, R., *J. Mol. Biol. 33,* 173-197 (1968).
26. Palecek, E., *Prog. Nuc. Acids Res. and Mol. Biol. 18,* 151-213 (1976).
27. Benham, C.J., *J. Mol. Biol. 150,* 43-68 (1981).
28. Benham, C.J., *Proc. Natl. Acad. Sci. USA 76,* 3870-3874 (1979).
29. Benham, C.J., *Biopolymers 19,* 2143-2164 (1980).
30. Benham, C.J., *Biopolymers 21,* 679-696 (1982).
31. Debye, P., *Ann. Physik 46,* 809-824 (1915).
32. Schmidt, P. in *Small Angle x-ray Scattering,* H. Brumberger, editor (Gordon Breach, NY), 17-31 (1965).
33. Benham, C.J., Brady, G.W. and Fein, D.B., *Biophys. J. 29,* 351-366 (1980).
34. Novick, R.P. and Bouanchaud, D., *Ann. N.Y Acad. Sci. 182,* 279-294 (1971).
35. Zasloff, M., Ginder, G.D. and Felsenfeld, G., *Nuc. Acids Res. 5,* 1139-1152 (1978).
36. Hagerman, P.D., *Biopolymers 20,* 1503-1535 (1981).
37. Brady, C.W., Cohen-Addad, C. and Lyden, E.F.X. , *J. Chem. Phys. 51,* 4309-4319 (1969).
38. Crothers, D. and Spatz, H., *Biopolymers 10,* 1949-1972 (1971).
39. Shore, D. and Baldwin, R.L., *J. Mol. Biol. 170,* 983-1008 (1983).
40. A full account of this work will appear in the Journal of Molecular Biology.

Biomolecular Stereodynamics IV, Proceedings of the Fourth Conversation in the Discipline Biomolecular Stereodynamics, State University of New York, Albany, NY, June 04-09, 1985, Eds., Ramaswamy H. Sarma & Mukti H. Sarma, ISBN 0-940030-18-7, Adenine Press, ©Adenine Press 1986.

This manuscript is dedicated to the memroy of Paul Flory

The Effects of Base Sequence and Morphology Upon the Conformation and Properties of Double Helical DNA

Wilma K. Olson, A. R. Srinivasan, Maria A. Cueto, Ramon Torres, Rachid C. Maroun, Janet Cicariello and Jeffrey L. Nauss
Department of Chemistry
Rutgers, The State University
New Brunswick, New Jersey 08903

Abstract

By focusing on base morphology rather than the chain backbone, it is becoming possible to unravel the conformational complexities of the nucleotide repeating unit and also to understand the effects of primary chemical sequence upon overall nucleic acid structure. The numerous torsions in the sugar-phosphate backbone offer multiple ways of linking adjacent bases in a particular arrangement. The solutions are providing insight into the flexibility of the chain backbone and the likely conformational pathways between different helical forms. The energetically preferred arrangements of specific base sequences are yielding distinctive morphological patterns. These preferences, in turn, have pronounced effects on the macroscopic properties of long chains in both regular helical and flexible states. The variations in base morphology resulting from the higher order folding of the B-DNA double helix further suggest the important influence of primary sequence upon nucleic acid properties and interactions.

Introduction

The elaborate architecture of its monomer repeating units makes the nucleic acid a difficult system to comprehend at the detailed structural level. The sugar-phosphate chain backbone includes five acyclic torsions as well as a flexible furanose ring, itself determined (assuming fixed bond lengths) by five independent angular variables (1). The glycosyl torsion χ linking the sugar to the heterocyclic base completes the set of angular parameters. Even a very crude description of local flexibility generates an untold number of three-dimensional spatial arrangements. A simple dimer permitted to adopt a single rotational isomeric state in each of its sterically accessible torsional ranges, for example, generates more than 2500 distinct conformational states (2). If modeled more realistically at 10°-increments over the allowed angular ranges, the number of three-dimensional arrangements exceeds 10^{10}, and, if extended to the polymeric level, the figure approaches 10^{10N} where N is the number of chain repeating units.

This conformational complexity is better understood in terms of the local morphology of the nucleic acid bases. Neighboring base side groups may be arranged in only a few different ways, such as with planes parallel or perpendicular, with residues closely stacked or opened vertically to permit the intercalation of a foreign ligand, with rings overlapped or slipped horizontally to a "breathing" state, etc. This simplified view has long been used to interpret various solution properties of the nucleic acids (3-10) and more recently has been adapted to detailed conformational analyses of crystalline DNA (11-18). The relative orientation of adjacent base pairs can be described in terms of so-called rolling, tilting, and twisting motions. The roll and tilt are rotations about orthogonal axes in the mean base pair plane, while the twist is a rotation about the normal to the plane.

A base morphological approach should also facilitate understanding the effects of primary chemical sequence upon overall properties of the nucleic acid chain. Until now, detailed treatments of polynucleotide size and shape (19-25) have focused on the role of the sugar-phosphate backbone. While some theoretical schemes (8,10,26-29) have included the bases, they have not offered a means to account for the precise structural features of individual chain units. Moreover, no attempts have been made to estimate the likelihood of the chain backbone accommodating the presumed fluctuations in local structure.

This work is a survey of recent attempts from this laboratory to understand the role of base morphology and sequence on the local structure and macroscopic properties of double helical DNA. A scheme has been devised to identify the sugar-phosphate conformations most likely to link a given dimeric base pair arrangement. The various solutions are closely interrelated and are useful in understanding the intrinsic flexibility of the chain backbone and the likely pathways between different helical forms. A series of potential energy calculations have been initiated to study the possible effects of base sequence on local structure, particularly the tendency of certain residues to distort the perfect parallel arrangement of stacked bases in helical structures. These structural perturbations have been incorporated in regular helical chains, demonstrating the important influence of primary chemical sequence. Local base morphology has also been introduced in statistical mechanical computations of chain extension and flexibility, and the effect of various sequences on cyclization tendencies has been examined. Finally, a differential geometric approach which describes the local morphological changes associated with a selected chain trajectory has been undertaken. Conformational features of a relaxed closed circular and a doubly interwound form of B-DNA are described in detail.

Backbone Linkage

Mathematically, the relative disposition of adjacent bases is determined by six independent variables—three translational and three orientational changes (30-33). The additional parameters in the nucleic acid backbone are therefore able to provide multiple ways of achieving a particular geometric arrangement. For each predetermined position of sequential bases in a single-stranded dinucleotide or for each arrangement of base pairs in a dimer duplex, there should be several ways to attain chain closure.

A simple way to identify the conformations which link a given base-base arrangement is to treat the glycosyl and sugar torsions (both the ring puckering and the exocyclic C5′-C4′ angle) as independent variables and to use the resulting O3′···O5′ distances as closure determinants. The latter distance must conform to the known geometry of phosphate chemical bonding. In the calculations described here this parameter is restricted to values between 2.47 and 2.56Å corresponding to a fixed P-O bond length of 1.60Å and a flexible O-P-O valence angle of 103±3° (the range of values found by us in a survey of 33 high resolution phosphate structures from the Cambridge Crystallographic Database (34)). If found in the appropriate range, the O3′···O5′ atom pair can frequently be bridged by an intervening phosphorus atom with correct C-O-P valence angles (119±3° as determined by the above survey).

Constrained model building techniques like this have been used for many years to generate the structures of nucleic acids (31-33,35-39). The computations are usually coupled to some sort of potential energy function that identifies the most likely sugar-phosphate backbones to link a given base pair morphology. The resulting structures are necessarily dependent upon the choice of potential and the predictions are often quite different from one study to the next (2). In general only the optimum solution is reported and alternate low energy linkages are ignored. The solution is usually determined by an energy minimization procedure and is closely tied to the choice of preliminary variables and the method of optimization.

The B-DNA backbone geometries reported here, in contrast, derive from an exhaustive search of all possible combinations of sugar and glycosyl conformations. The sugar puckerings (P_1 and P_2) span 20 states on a pseudorotational pathway of 38° amplitude, while the exocyclic C5′-C4′ (ψ) and glycosyl torsions (χ_1 and χ_2) cover 36 states separated by 10°-increments between 0° and 350°. Of the more than 18 million cases examined, nearly 3,600 permit chain closure with acceptable valence angle geometries. Of these, 18 describe solutions within 5 kcal/mole of the lowest single-stranded energy state and 124 within 10 kcal/mole. Energies are computed on the basis of a semiempirical potential function (40,41) that accounts for nonbonded (Lennard-Jones, van der Waals, electrostatic, and charge-induced polarizability), torsional, and gauche effects. The influence of solvent and counterion condensation is treated indirectly through modification of atomic parameters.

The 18 low energy backbones of B-DNA are included in the conformation rings in Figure 1. The independent parameters are plotted in part (a) and the dependent parameters in part (b). Most of the low energy solutions are closely related to the X-ray fiber diffraction model (noted by asterisks) from which the base positions were taken (42). Almost equally likely on the basis of energetic interactions and frequency of occurrence, however, are so-called Watson-Crick backbone solutions with the exocyclic C5′-C4′ torsion switched from the *gauche*$^+$ (60±30°) to the *trans* (180±30°) range. As is evident from the figure, this change of conformation is

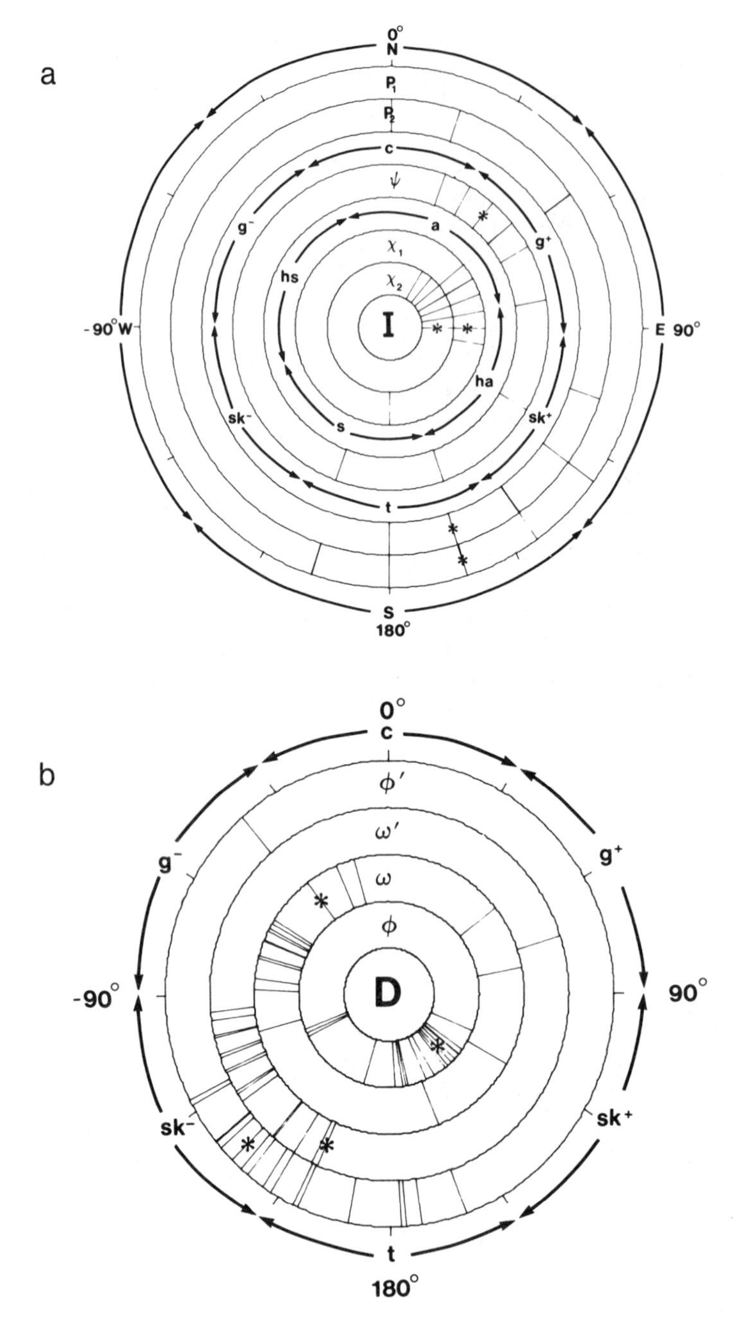

Figure 1. Conformation rings describing 18 low energy backbones of B-DNA. Rotational ranges are designated along the periphery of the respective groups of rings. Torsions associated with the standard B-DNA reference are noted by asterisks in the appropriate rings. See text for angle conventions and further description of terms. (a) The independent parameters I describing sugar puckering (P_1,P_2), exo-cyclic C5′-C4′ sugar torsion (ψ), and glycosyl orientation (χ_1,χ_2). (b) The dependent torsions D (ϕ', ω', ω, and ϕ) describing the C3′-O3′-P-O5′-C5′ backbone sequence.

accompanied by large (150-180°) angular variations in the dependent torsion around the P-O5′ bond. This set of changes is the same correlated angular pathway reported in earlier analyses (43,44) of the local flexibility and long-range stiffness of nucleic acid helices. Smaller changes (of 20-40°) in the other dependent torsions, while not so noticeable in the figure, are required to preserve the rigid orientation of base pairs required here. If only the C5′-C4′ and P-O5′ torsions are allowed to vary, the bases are translated with respect to one another while preserving their parallel angular alignment (43).

A third low energy B-DNA dimeric arrangement with alternating *syn/anti* glycosyl torsions is also evident from Figure 1. Interestingly, the chain backbone is simultaneously altered to a conformation with mixed S/N (C2′-endo/C3′-endo) sugar puckering and the C3′-O3′-P-O5′-C5′-C4′ exocyclic torsions in a $g^-g^+g^+tt$ arrangement. This backbone sequence is identical to that of the CpG segments of Z-DNA (45). The *syn/anti* glycosyl sequence, however, is just the opposite of that found in the left-handed structure. This unusual backbone geometry is compared with the two standard B-DNA backbones in Figure 2. Unlike the standard dimers,

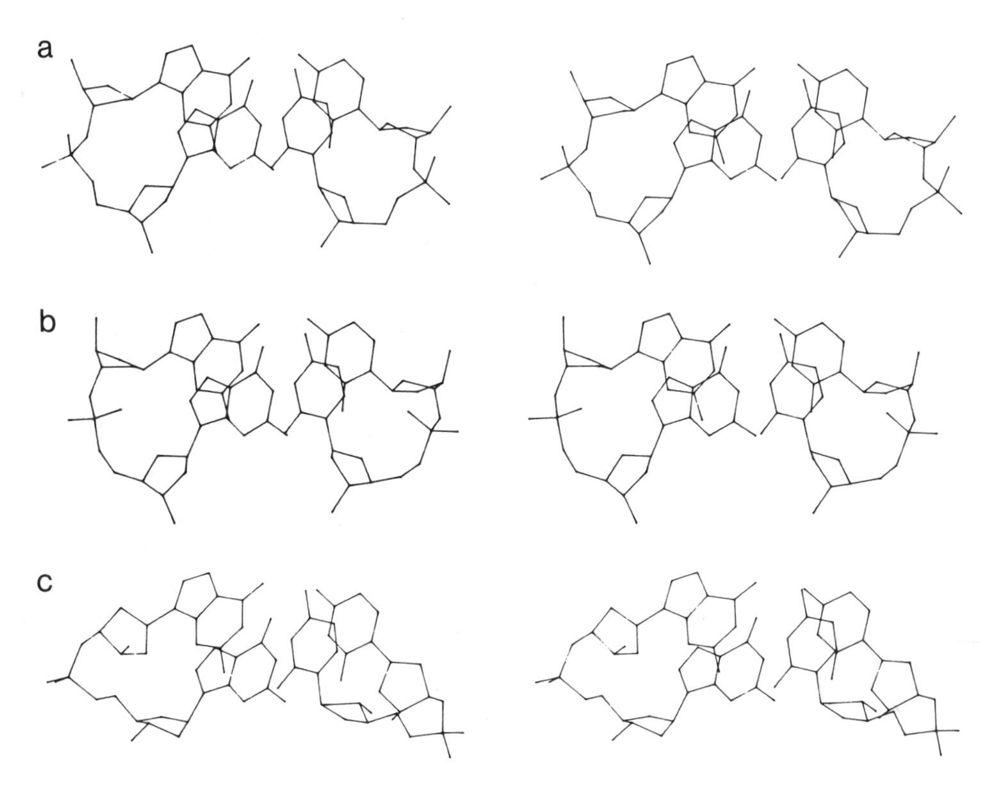

Figure 2. Stereo views of three different low energy B-DNA backbone geometries with identical base pair morphologies. (a) Standard B-DNA duplex (0.2 kcal/mole) with P_1P_2 = S,S; $\chi_1\chi_2 = a,a$; $\phi'\omega'\omega\phi\psi = sk^-\ sk^-\ g^-\ sk^+\ g^+$. (b) Watson-Crick type conformation (3.1 kcal/mole) with P_1P_2 = S,S; $\chi_1\chi_2 = a,a$; $\phi'\omega'\omega\phi\psi = t\ sk^-\ t\ sk^-\ t$. (c) *Syn/anti* conformation (3.8 kcal/mole) with P_1P_2 = S,N; $\chi_1\chi_2 = s,a$; $\phi'\omega'\omega\phi\psi = g^-\ g^+\ g^+t\ t$.

the *syn/anti* sequence is not a feasible building block of a B-DNA double helix. The *syn* glycosyl conformation is not easily accommodated in the 5′-position of the preceding dimeric unit as required for polymer formation. The lowest energy structure with the necessary *syn* value of χ and C2′-endo (S) puckering is more than 4,000 kcal/mole above the lowest energy state of this survey.

The *syn/anti* conformer is separated from the standard B-DNA backbones by several correlated torsional changes which are somewhat higher in energy. The 3′-glycosyl torsion is found to change gradually from the *syn* to the *anti* range in backbone solutions of energy between 5 and 10 kcal/mole. Moreover, as χ is decreased in magnitude, the ω' torsion about the succeeding O3′-P bond is found to increase from the g^+ to the t range. The resulting g^-tg^+tt acyclic torsional sequence (involving the $\phi'\omega'\omega\phi\psi$ angles) is readily converted by correlated variations in the C3′-O3′ and P-O5′ rotations to the *ttttt* conformational sequence which is only a minor perturbation of one of the standard backbone arrangements (43,46). The pathway is sufficiently complicated and the transition energies are high enough to preclude rapid interconversion between the standard B-DNA geometries and the *syn/anti* conformer. Under certain conditions the barriers may possibly be lowered. The potential utility of this pathway in initiating B- to Z-DNA conformational changes at the 3′-end of the double helix is under current investigation.

Local Sequence Effects

There is increasing evidence that the pattern of local base stacking is a sequence dependent property of the nucleic acid chain. Local variations of base orientation and displacement are apparently dictated by the distribution of purines and pyrimidines in the A- and B-DNA crystallographic sequences (11,12,15,47-49). While the rules governing the observed patterns are not yet well understood, steric arguments (13-18) have already proven quite useful in accounting for some structural features.

In solution short chains containing stretches of the $A_N \cdot T_N$ duplex are known to be curved rather than rodlike (50-55). As evident from Table I and from arguments in the following section, this long range curvature may also be linked to specific local base stacking patterns. The base stacked conformations resulting from a series of steepest descent energy minimizations of pTpTpTp and pApApAp are described in terms of the geometric interactions of adjacent bases. Included in the table are the bending angle Λ described by the normals of neighboring base planes, the average distance $\langle D \rangle$ between pairs of corresponding atoms in adjacent chain units, and the average vertical distance $\langle Z \rangle$ between atoms of one base and the plane of its neighbor. If the bases are perfectly parallel and overlapping, $\langle D \rangle$ and $\langle Z \rangle$ are identical to one another. In general, however, $\langle D \rangle$ is greater than $\langle Z \rangle$ and their difference is a rough measure of base-base overlap (41).

The various conformations in Table I are categorized in terms of rotational domains rather than specific torsion angles since there are several examples of each base-stacked geometry. The ranges of Λ, $\langle D \rangle$, and $\langle Z \rangle$ and the numbers (#) of

Table I

Energetically Preferred Stacking Geometries of ApA and TpT Dimeric Sequences of pApApAp and pTpTpTp

	Chain Conformation								Base Stacking Geometry								
		(C3′ — O3′ —		P — O5′ — C5′ — C4′)					ApA					TpT			
Category	χ_1	P_1	ϕ'	ω'	ω	ϕ	ψ	P_2	χ_2	#	Λ,deg	⟨D⟩,Å	⟨Z⟩,Å	#	Λ,deg	⟨D⟩,Å	⟨Z⟩,Å
I.	*a*	N	*t*	g^-	g^-	*t*	g^+	N	*a*	4	3.1-8.3	3.4-3.6	3.1-3.2	6	10.4-12.2	3.5-3.6	3.0-3.1
II.	*a*	N	*t*	g^-	g^+	*t*	g^-	N	*a*	6	8.1-17.5	3.7-3.8	3.0-3.2	6	15.7-23.6	3.7-3.9	2.8-3.0
III.	*a*	N	*t*	g^-	*t*	*t*	*t*	N	*a*	3	7.9-17.4	3.8-4.0	3.0-3.2	4	16.3-20.4	3.8-4.0	2.9-3.0
IV.	h*a*/*a*	N	*t*	c/g^+	sk^-/t	g^+	*t*	N	*a*	7	2.2-21.4	3.3-3.9	2.9-3.3	6	17.0-24.4	3.5-3.8	2.9-3.1
V.	*a*	N	*t*	c/g^+	g^-	g^-	g^-	N	*a*	2	5.3-6.0	3.3-3.4	3.0-3.1	2	10.5-11.6	3.7-3.9	2.9-3.0
VI.	*a*	S	*t*	g^-	g^-	*t*	g^+	S	*a*	5	1.8-5.7	3.6-3.7	3.1	4	7.1-9.1	3.7	3.0-3.1
VII.	*a*	S	*t*	g^-	g^+	*t*	g^-	S	*a*	5	3.0-9.1	3.7-4.0	3.0-3.1	4	11.4-24.5	3.8	3.0-3.2
VIII.	*a*	S	*t*	g^-	*t*	*t*	*t*	S	*a*	5	2.4-5.4	3.6-3.8	3.0-3.1	4	8.4-10.8	3.8	3.1
IX.	*a*	S	*t*	*c*	*t*	g^+	*t*	S	*a*	4	2.9-10.6	3.6-3.8	3.1	4	9.4-11.8	3.6-3.7	3.0-3.1
X.	*s*	S	*t*	*c*	*t*	g^+	*t*	S	*s*	1	4.9	3.5	−3.1	4	7.9-22.1	3.5-3.7	−3.1-−3.5
XI.	*a*	S	*t*	t/sk^-	sk^+/t	g^-	*t*	S	*a*	1	2.4	3.6	3.1	2	7.7-14.0	3.6-4.1	2.8-3.0

conformational examples are reported for eleven different categories. A detailed listing of internal torsions of the 89 low energy arrangements used to generate this table is available upon request. Each structure used is within 5 kcal/mole of the lowest average dimeric energy conformation of either sequence. Since computations were carried out on trimers rather than dimers, the average dimeric energy is estimated to be half of the minimized trimeric value. Conformational ranges of the chain backbone angles are described in terms of *trans* (t = 180±30°), *gauche*$^{\pm}$ ($g^{\pm}$ = ±60±30°), *cis* (c = 0±30°), and *skew*$^{\pm}$ ($sk^{\pm}$ = ±120±30°) states. Those of the glycosyl bond are labeled *anti* (a = 30±45°), high *anti* (ha = 120±45°), *syn* (s = −150±45°), and high *syn* (hs = −60±45°) with respect to the planar *cis* arrangement of O1′-C1′-N1-C6 (pyrimidines) or O1′-C1′-N9-C8 (purines) as the 0° reference state. Sugar puckering P_i (i = 1,2) is labeled either N or S to reflect the rigid C3′-endo and C2′-endo puckering geometries included in the calculations. Only homogeneously puckered sequences are treated. The minimizations are carried out with respect to the backbone and glycosyl torsions as independent variables using a simple variation of the steepest descent method of Powell (56). No *a priori* restrictions are placed on base-base positioning.

As evident from the table, TpT segments are consistently more inclined than ApA dimers of the same conformational type. In every example the bending angle Λ between adjacent thymidine base planes is greater than the corresponding angle between neighboring adenine residues. The optimized overlap of adjacent residues, as measured by the difference between $\langle D \rangle$ and $\langle Z \rangle$, is also generally less ideal for thymidine than adenine pairs. This difference is almost always further from zero for TpT than the corresponding ApA conformer. The perturbation of TpT stacking geometries can be traced to repulsive interactions of the relatively large 5-methyl groups on the two bases. In contrast to the relatively constant (vertical) thickness of adenine, thymine is a structure with a noticable bump (due to the 5-methyl) at one end. Successive T's therefore cannot be aligned in a perfectly parallel stack. The van der Waals radius of the methyl group is taken to be 1.95Å in these energy calculations compared to values of 1.70, 1.55, and 1.50Å for the ring sp^2 carbons, nitrogens, and oxygens, respectively.

Interestingly, this proclivity toward bending in the single strand is found to persist in the nonbonded interactions of successive A•T base pairs of the A_N•T_N duplex. The van der Waals energy $E(\rho,\lambda)$ is reported in Figure 3 in terms of the roll angle ρ about the long axis parallel to the N1(pyrimidine)•••N9(purine) vector and the tilt angle λ about the dyad axis of an ideal Watson-Crick base pair. The distance between base pair centers is fixed at 3.4Å and the angular twist τ about the vertical axis at 36° as in the B-DNA duplex. The orientation of base i + 1 is then varied at 5°-increments of ρ and λ with respect to base i and the resulting energies are reported on the contour surface. While the optimum arrangement denoted by an asterisk at ρ,λ = (5°,0°) is only slightly skewed, states distorted as much as −10° or 15° in ρ and ±5° in λ are found within 1 kcal/mole of the energy minimum. The tendency toward bending, however, is somewhat more pronounced for positive rather than negative values of roll, the respective contributions to the total partition

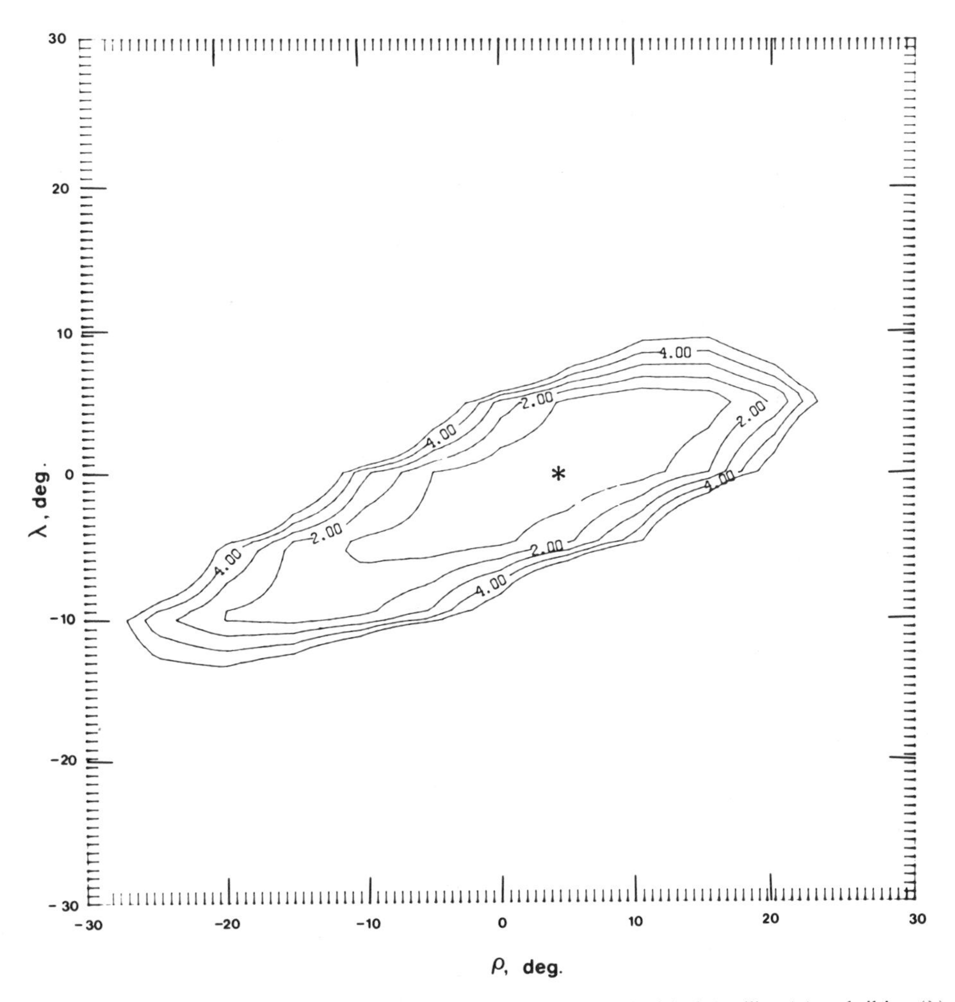

Figure 3. Contour diagram of the van der Waals energy associated with the rolling (ρ) and tilting (λ) motions of adjacent A•T base pairs in the B-DNA duplex. The angular twist τ is fixed at 36° and the distance between base pair centers at 3.4Å. Contours are drawn at 1 kcal/mole intervals from 1 to 5 kcal/mole relative to the lowest energy conformation located at $*$.

function being 0.70 and 0.30. The variations in roll are also strongly correlated with those in tilt. As ρ is increased in value, λ is correspondingly increased and vice versa. The allowed variations in roll are roughly three times those in tilt in qualitative accordance with earlier theoretical studies (57-59) and with observations on crystalline B-DNA (11,12). The roughly comparable ease of rolling or tilting the base pairs at low values ($\pm 5°$) of ρ and λ, however, is in sharp disagreement with the previous energy predictions. The conformational energies reported in Figure 3 are, of course, approximate and are apt to change to some degree when the interactions of the intervening backbone and surrounding solvent atoms are considered. The procedure used to introduce the chain backbone in the earlier studies, on the other hand, has been criticized (60) for using inadequate steps to search certain areas of conformation space and for therefore missing some low energy states.

Regular Polymeric Effects

The slight tendency for residues to adopt locally bent arrangements can lead to pronounced effects on chain structure at the polymeric level. If all residues of B-DNA are rolled by 10° and tilted by 5°, for example, the 10-fold duplex converts to a helix with 9.46 residues per turn (n) and with 96% of the standard vertical extension (e.g., local step height h = 3.25 vs h° = 3.40Å for B-DNA). The central axis of the duplex also shifts by a distance r = 1.56Å leaving a small central hole and an overall curvature $\kappa = r^2/(r^2 + (nh/2\pi)^2) = 0.09$ in the new helical arrangement.

The effects of local perturbations upon nucleic acid structure can be further magnified if the bent residues are interspersed with standard parallel units in the double helix. The overall chain bending is a function of both the composition and the sequence of modified units. As illustrated in Figure 4, the maximum angle between terminal base planes in a stretch of consecutive bent linkages is achieved at half helical turns (e.g., n/2, 3n/2, 5n/2, etc.) of the modified duplex. The shortest such stretch in the ρ,λ = (10°,5°)-bent B-DNA described here is five residues in length, the integral value closest to 4.73 (e.g., 9.46/2). The resulting angle Λ between the normals of terminal base planes in this sequence is 34.6°, compared to a value of 11.2° in an isolated dimeric bend of the same conformational type.

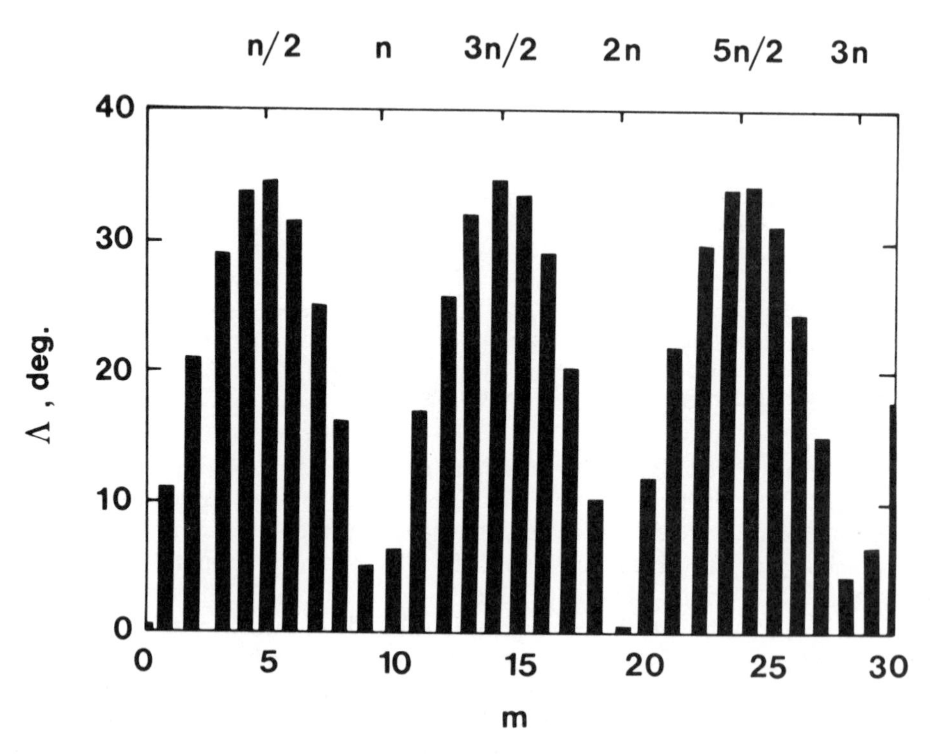

Figure 4. Bend angle Λ between terminal base pair planes in a perturbed B-DNA helix containing m nucleotide repeating residues. The local rolling and tilting, ρ,λ = (10°,5°), generates a modified duplex with n = 9.46 residues per helical turn.

The influence of this 5-unit bend (B'_5) on overall polynucleotide structure is described in Figure 5. The relative extension h/h^o and the curvature κ are plotted as a function of the number of standard B-DNA units m in the alternating copolymeric repeating sequence $-B_mB'_5-$. The data are reported for values of m between 0 and 100 residues. Both chain extension and curvature are sinusoidal functions of m with local maxima and minima occurring at the half- and integral helical turn values of B-DNA reported here. As evident from the figure, chain extension is found to decrease and curvature to increase at selected chain lengths (m=5, 15, 25, etc.). The resulting structures, however, are not appreciably altered from standard B-DNA, the greatest chain compression being only 15% (e.g., h/h^o=0.85) and the maximum curvature being 0.28 for the 5-unit bent sequences with an intervening 95-residue stretch of the B-DNA duplex.

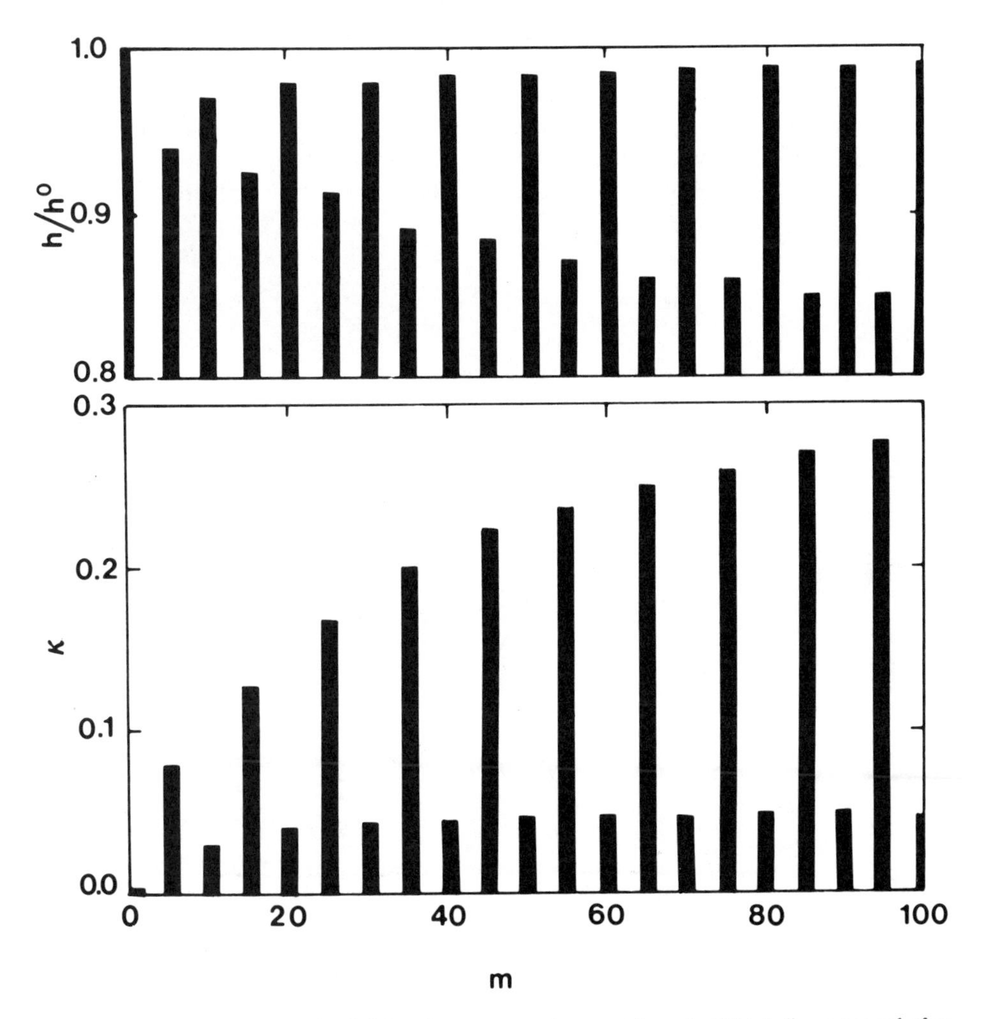

Figure 5. Relative chain extension h/h^o and curvature κ in a copolymeric DNA helix composed of m standard B-DNA repeating units ρ,λ,τ = (0°,0°,36°); h^o = 3.4Å interspersed with 5 bent units where ρ,λ,τ = (10°,5°,36°).

Much more dramatic structural effects result if the chain contains a so-called inverted repeat of bent nucleotide links. The same nonbonded interactions which force an AA•TT dimeric segment of B-DNA to roll and/or tilt in a given direction may cause the inverted TT•AA dimer to vary in the opposite rotational sense. Moreover, when placed in juxtaposition, these inverted bends dramatically alter the overall bending of terminal base planes compared to that in a stretch of identically perturbed units. The $B'_5B''_5$ sequence generated by 5 consecutive $\rho,\lambda = (10^\circ,5^\circ)$ bends and 5 more $\rho,\lambda = (-10^\circ,-5^\circ)$ links, for example, opens the angle between terminal base planes from 34.6° in B'_5 or B''_5 to 68.8° in the 10-mer repeat.

The relative extension and curvature of the $-B'_5B''_5-$ chain sequence are reported as a function of the degree of local bending in Figure 6. For simplicity, only variations in roll angle are included with positive values assigned to B′ bends and negative values to B″ bends. While the effect on overall structure is limited for small values of ρ ($\pm 5^\circ$), the results associated with greater values of ρ are quite dramatic. The 5-residue inverted repetition of a $\pm 17^\circ$ roll angle is found to decrease h/h° to 0.17 and increase κ to 0.96. Such a structure is related to the condensed superhelical models of DNA that wrap around the protein core of the chromatin nucleosome (61) where $h/h^\circ = 0.11$ and $\kappa = 0.99$ as well as to perfectly closed circular duplexes where $h/h^\circ = 0$ and $\kappa = 1$.

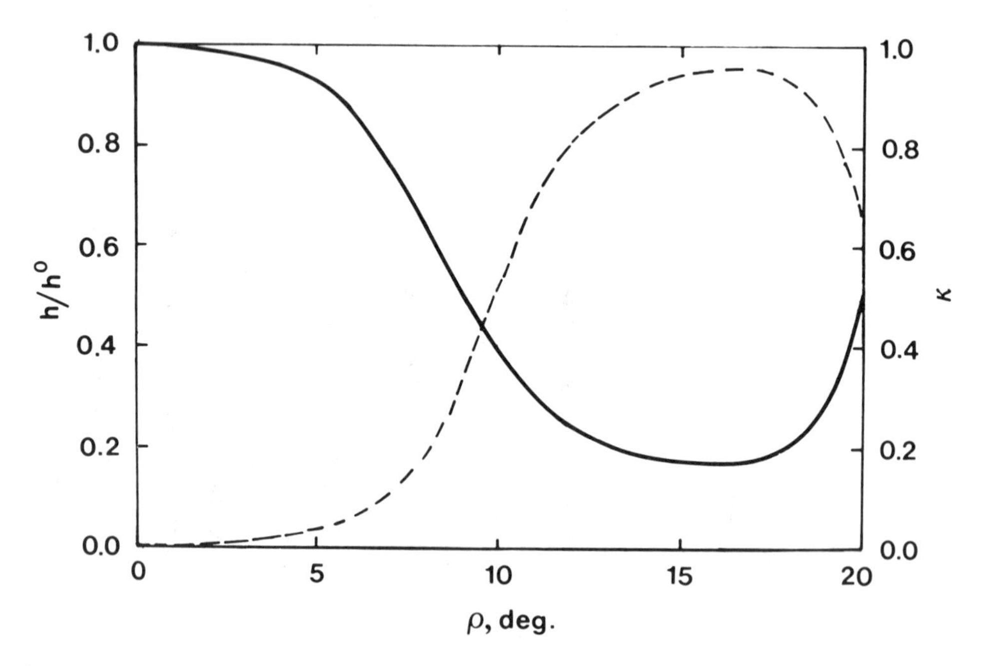

Figure 6. Relative chain extension h/h° (solid curve) and curvature κ (dashed curve) as a function of roll angle ρ in a copolymeric B-DNA duplex containing the inverted repeating unit $-B'_5B''_5-$. The five B′ bends entail a positive value of ρ and the five B″ bends a negative value.

Duplex Flexibility

The overall extension and flexibility of the nucleic acid double helix also depends upon the nature and degree of local chain bending. Chain dimensions can be related to local motions of the polynucleotide by a simple virtual bond scheme that considers the relative orientation and displacement of adjacent nucleotide units. The end-to-end vector **r** connecting chain termini, for example, is a sum of successive virtual bond vectors, one such vector $\mathbf{v}_i$ generally taken per nucleotide repeating unit. When the chain is modeled in terms of base pairs, the vectors can be drawn from a central point on one unit to a corresponding point on the next. In the case of regular B-DNA, these vectors are 3.4Å in length and collinear with the double helical axis (8,26). When the B-DNA is subjected to local rolling and tilting motions, successive virtual bond vectors are no longer collinear but are oriented at an angle Λ (where $\cos \Lambda = \cos \rho \cos \lambda$) with respect to one another.

In the calculations described below, virtual bond vectors are taken along the local helical axis that would be generated if the designated residues were part of a regular B-DNA. The separation between adjacent residues is fixed at 3.4Å, thereby discounting the possibility of any translation of bases. The x-and y-axes are located in the base pair plane from which the virtual bond originates, with the x-axis running parallel to the N1(pyrimidine)•••N9(purine) long axis and the y-axis coincident with the base pair dyad. The orientation of a given virtual bond vector $\mathbf{v}_{i+1}$ with respect to its predecessor $\mathbf{v}_i$ is then obtained by premultiplication with the transformation matrix $\mathbf{T}_{i,i+1}(\tau,\rho,\lambda)$ where

$$\mathbf{T}_{i,i+1}(\tau,\rho,\lambda) = \mathbf{Z}(\tau)\,\mathbf{X}(\rho)\,\mathbf{Y}(\lambda) \qquad (1)$$

and $\mathbf{Z}(\tau)$, $\mathbf{X}(\rho)$, and $\mathbf{Y}(\lambda)$ are matrices of right-handed rotation about the twist (z), roll (x), and tilt (y) axes of residue i+1. The relative orientation of more distant units is obtained from a serial product of appropriate transformation matrices with **r** then given by

$$\mathbf{r} = \sum_{i=1}^{N} \mathbf{v}_i = \mathbf{v}_1 + \mathbf{T}_{12}\,\mathbf{v}_2 + \mathbf{T}_{12}\,\mathbf{T}_{23}\mathbf{v}_3 + \cdots + \mathbf{T}_{12}\cdots\mathbf{T}_{N-1,N}\mathbf{v}_N \qquad (2)$$

The average value of **r** defined in the above manner with respect to a common internal coordinate frame in the first residue of the chain is the so-called persistence vector $\mathbf{a} = \langle \mathbf{r} \rangle$ (62). Since the virtual bond length is fixed in these calculations, **a** is a function of the average matrix products $\langle \mathbf{T}_{i,i+1} \cdots \mathbf{T}_{j-1,j} \rangle$. If the orientations of adjacent units are independent of one another, the average product can be replaced by the product of average matrices $\langle \mathbf{T}_{i,i+1} \rangle \cdots \langle \mathbf{T}_{j-1,j} \rangle$, each factor determined by the energetic interactions of the designated pair of residues. The relative extension of the A_N•T_N duplex subject to the rolling and tilting of the AA•TT dimer reported in Figure 3 is described by the solid curve in Figure 7. The ratio of the magnitude of the persistence vector (a) to the maximum end-to-end separation r_{max} = 3.4N Å is plotted as a function of chain length N (e.g., number of base pairs + 1). As expected for a system with very limited local motions, the average extension is very close to that of a perfectly rodlike polymer at short chain lengths. Only when the DNA is extremely long (>2^{10}-2^{11} residues) is the average chain displacement small compared to a rigid model.

The decrease in relative mean extension of the DNA duplex with increase in N is simultaneously accompanied by an increase in overall chain flexibility. The average fluctuations in structure are described in Figure 7 in terms of the sum of the three diagonal elements of the second-order displacement tensor $\langle \rho^{x2} \rangle$ (dashed curve). The vector ρ is simply the difference $\mathbf{r} - \mathbf{a}$ and ρ^{x2} the product of ρ with its transpose, $\rho\rho^T$ (62). With $\mathbf{r}$ given by (u,v,w,), the trace elements of $\langle \rho^{x2} \rangle$ are $\langle u^2 \rangle - \langle u \rangle^2$, $\langle v^2 \rangle - \langle v \rangle^2$, and $\langle w^2 \rangle - \langle w \rangle^2$. Since $\langle r^2 \rangle_o = \langle \mathbf{r} \cdot \mathbf{r} \rangle$ and $\mathbf{a} = \langle \mathbf{r} \rangle$, the parameter reported in Figure 7 ($\mathrm{tr}\langle \rho^{x2} \rangle / \langle r^2 \rangle_o$) can therefore be reduced to $1 - a^2/\langle r^2 \rangle_o$. At short chain lengths, where the DNA is subject to rigid rod behavior, the fluctuations from the average structure are quite small and the square magnitude of the persistence vector a^2 is roughly equivalent to the mean-square unperturbed end-to-end distance $\langle r^2 \rangle_o$. The relative fluctuations in structure are therefore essentially zero. At very long chain lengths, in contrast, where the chain obeys Gaussian statistics (22,23), the fluctuations are predominant with $\langle r^2 \rangle_o \gg a^2$. The decrease in a/r_{max} is accordingly mirrored by an increase in $\mathrm{tr}\langle \rho^{x2} \rangle / \langle r^2 \rangle_o$. At intermediate chain lengths, where the two ratios are roughly equivalent, the chain is most apt to form closed circular species.

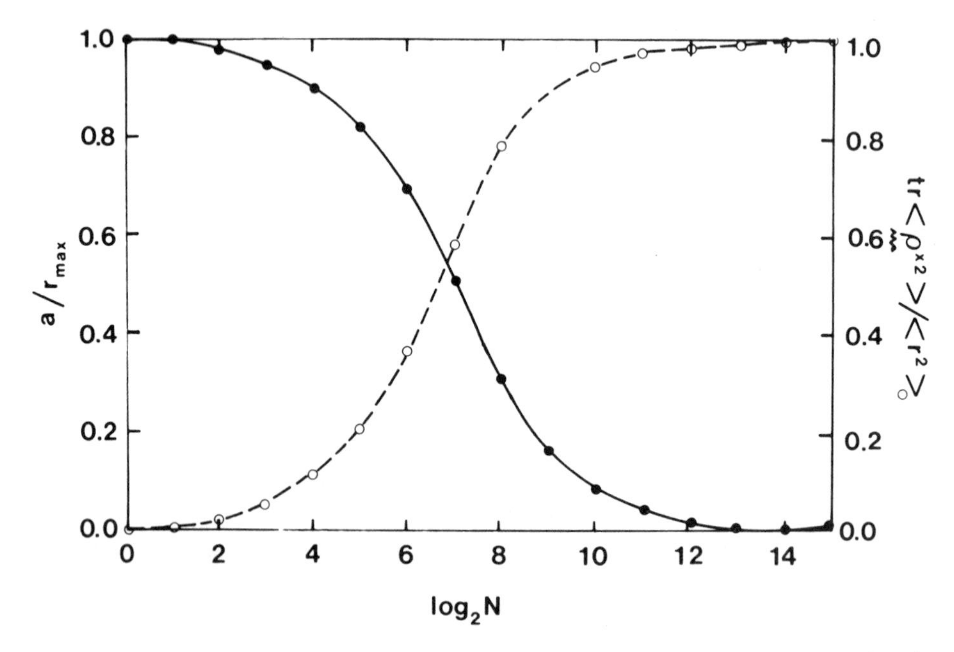

Figure 7. Relative mean extension (a/r_{max}, solid line) and relative chain flexibility ($\mathrm{tr}\ \langle \rho^{x2} \rangle / \langle r^2 \rangle_o$, dashed line) as a function of the number of repeating residues N in an $A_N \cdot T_N$ duplex subject to the local rolling and tilting motions detailed in Figure 3. Data points computed for integral powers of 2 and connected by a smooth curve.

A rough measure of the loop closure probability of the $A_N \cdot T_N$ duplex is presented in Figure 8. The likelihood of chain closure Q is estimated from the integral

$$Q = \int_{-\mathbf{a}-\Delta}^{-\mathbf{a}+\Delta} \left(\frac{1}{2\pi}\right)^{3/2} \left(\frac{1}{\langle \rho_1^2\rangle\langle \rho_2^2\rangle\langle \rho_3^2\rangle}\right)^{1/2} \exp\left[\frac{-1}{2}\left(\frac{\rho^1{}_2}{\langle \rho_1^2\rangle} + \frac{\rho^2{}_2}{\langle \rho_2^2\rangle} + \frac{\rho^3{}_2}{\langle \rho_3^2\rangle}\right)\right] d\rho \qquad (3)$$

where $-\mathbf{a}$ is the vector locating the chain origin with respect to the persistence vector and Δ is the allowed three-dimensional displacement about this point (typically $\pm 10\%$ of $\mathbf{a}$). The term appearing inside the integral is a three-dimensional ellipsoidal Gaussian distribution function. The vector ρ is reexpressed in terms of the principal components, $\langle \rho_i^2\rangle$ where i = 1-3, of the diagonalized form of $\langle \rho^{x2}\rangle$. The chain length is varied, as above, by powers of 2 from 2^0 to 2^{15} (i.e., 1-32,768) residues. According to this estimate, $A_N \cdot T_N$ duplexes containing approximately $2^7 = 128$ base pairs are most likely to form closed circular structures. Such species are much smaller than the circles found in earlier studies of DNA closure (63,64) but are in reasonable agreement with the 120-160 residue circles observed in greatest amount in recent cyclization studies of $A_N \cdot T_N$ rich duplexes (65). Interestingly, the predicted maximum in Q is found at the chain length where relative chain extension and the relative fluctuations in overall structure are roughly equivalent in magnitude. A crude estimate of cyclization tendencies may therefore be provided by this so-called "equivalence" length. A length of $2^{6.8} = 111$ base pairs is obtained by linear extrapolation of the extension/flexibility data in Figure 7.

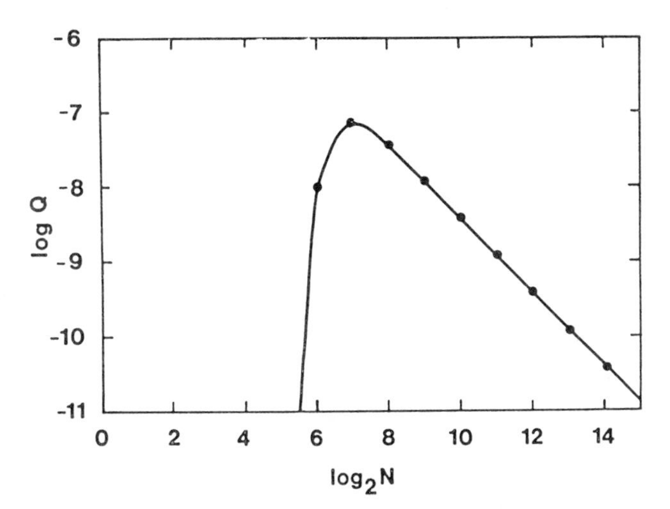

Figure 8. Chain length dependence of cyclization in the $A_N \cdot T_N$ duplex. Data points computed for integral powers of 2 and connected by a smooth curve.

A more precise description of cyclization tendencies requires the computation of additional moments of ρ (21,66) as well as the determination of angular correlation factors (23,25,67). The rough guidelines described above, however, serve as useful tools in examining the effect of changes in local nucleotide mobility and sequence on the overall flexibility and cyclization rates of the DNA duplex. Some examples of these appear in Table II where the limiting magnitude of the persistence vector a_∞ and the extrapolated "equivalence" length N_e are reported for selected potential energy schemes and primary sequence combinations. The conformational schemes under investigation include the inverted van der Waals energy maps associated with the AA•TT and TT•AA dimeric sequences (labeled M and M′, respectively) and various elliptical Gaussian functions describing different degrees of rolling and tilting.

Table II
Duplex Dimensions and Flexibility as a Function of Selected Conformational Energy Schemes

van der Waals Energies*			
Repeating Sequence	Z	a_∞,Å	$\log_2 N_e$
-M-	5.6	286	6.8
-M′-	5.6	286	6.8
-MM′-	5.6	286	6.8
$-M_5M'_5$	5.6	98	6.3

Gaussian Energies‡				
$\langle\rho^2\rangle^{-1}$,rad^{-2}	$\langle\lambda^2\rangle^{-1}$,rad^{-2}	Z	a_∞,Å	$\log_2 N_e$
100	100	4.9	578	7.8
75	100	5.6	496	7.6
50	100	6.9	387	7.3
75	75	6.5	434	7.4
50	75	8.0	348	7.1
50	50	9.7	291	6.9

$^*Z = \sum_\rho \sum_\lambda \exp(-E_{vdw}(\rho,\lambda)/RT)$

$^\ddagger Z = \sum_\rho \sum_\lambda \exp(-\rho^2/2\langle\rho^2\rangle)\exp(-\lambda^2/2\langle\lambda^2\rangle)$

The computed values of a_∞ and N_e in the table are identical for the $A_N{\bullet}T_N$ and $T_N{\bullet}A_N$ duplexes as well as for a conformational copolymer composed of alternating AA•TT and TT•AA type bends. The limiting dimensions in all three cases are less than currently accepted values of the persistence length of DNA in dilute aqueous salt solution but are in close agreement with values observed at higher salt concentrations (68,69). The copolymer is treated here in terms of the alternating sequence of averaged transformation matrices $\langle \mathbf{T}_{AA\bullet TT}\rangle$ $\langle \mathbf{T}_{TT\bullet AA}\rangle$, the former associated with the AA•TT energy map and the latter with the TT•AA surface. As anticipated from the above rigid models of an alternating sequence of five bent linkages, a flexible duplex constructed from a similar 5-mer pattern of energy surfaces is markedly condensed. Average chain extension for this duplex is roughly a third of that found for the corresponding homopolymers and the alternating copolymer and loop closure is favored at chain lengths as small as 80 residues. The local flexibility of individual repeating units as measured by the residue partition function Z, however, is constant throughout the structure and identical to that of the homopolymers and the alternating copolymer.

The effects of local chain stiffness upon chain properties are illustrated with the Gaussian surfaces in the table. The probability of a given fluctuation in ρ or λ is described in terms of the average standard deviations $\langle\rho^2\rangle$ and $\langle\lambda^2\rangle$ of these angles. The probabilities are assumed to be independent of one another and are described with respect to the $\rho_o\lambda_o = (0°,0°)$ reference state. Computations are carried out at 5°-angular increments in ρ and λ to permit direct comparison with the van der Waals surfaces. As evident from the data, macroscopic flexibility is sensitive to the type as well as to the degree of local chain bending. The Gaussian model in Table II with the identical partition function as the AA•TT energy map is considerably more extended than the A_N•T_N duplex and also less likely to form small circular structures. There is essentially no local bias in chain bending in the model to favor condensed forms. The effective 5 kcal/mole energy barrier on the Gaussian surface is found at $\rho = (10\langle\rho^2\rangle/RT)^{1/2}$ and $\lambda = (10\langle\lambda^2\rangle/RT)^{1/2}$ which translates with this example at 298°K to $\rho = \pm 24°$, $\lambda = \pm 27°$. The map is almost circularly symmetric and is considerably broader than the AA•TT energy map with the same Z. The equality in Z is apparently a consequence of the larger 1 kcal/mole energy regions on the van der Waals surface compared to the Gaussian model.

As the Gaussian surfaces are modified, the chain dimensions and flexibility follow a more predictable pattern. As local stiffness diminishes, both a_∞ and N_e increase in value. There is an inverse correlation of these values with Z in the table. Interestingly, the model which best matches the persistence length and the equivalence chain length of the A_N•T_N duplex is highly flexible at the local level with a partition function of 9.7 and an effective 5 kcal/mole boundary of ±33° in ρ and λ at 298°K.

Cyclic DNA

An alternative approach to understanding the effects of chain sequence on macromolecular flexibility involves deforming the DNA duplex along a specified trajectory and extracting the resulting conformational patterns. One can take advantage of the moving coordinate frame used in the differential geometric description of a space curve (70). The trajectory of the closed circular duplex illustrated in Figure 9 follows the movement of the tangent to the curve. The base pairs lie in the normal planes determined by the principal normal and binormal at the given points. The spacing between base pairs depends upon the number of repeating residues and the overall radius of the circle. In this example the DNA contains 80 residues evenly spaced at 3.4Å arc lengths around the circular "helix" axis. The bending angle Λ between successive base planes is therefore fixed at 4.5° (= 360°/80).

The distribution of Λ into rolling and tilting motions depends upon the chosen variation of the local twist density T, a topological parameter which describes the

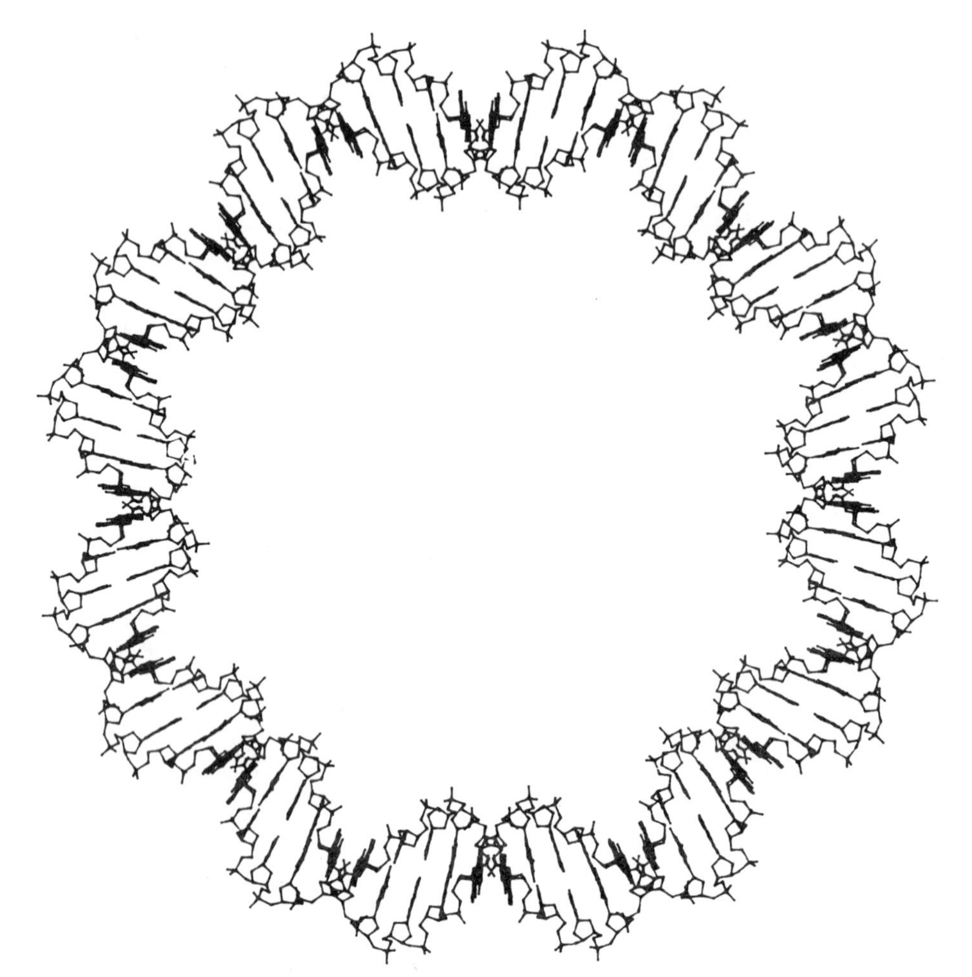

Figure 9. Computer-generated representation of the bases and backbone in a relaxed closed circular B-DNA double helix containing 80 base-paired residues.

relative oscillation of a reference vector in adjacent normal planes with respect to the local helical axis.

$$T_i = (1/2\pi)(\cos \Omega_i \sin \Omega_{i+1} - \sin \Omega_i \cos \Omega_{i+1} \cos \Lambda) \tag{4}$$

The parameter Ω in this expression is the cylindrical angle used to measure the oscillation in successive normal planes. With the DNA presumed to be a tenfold helix and the structure free of higher order supercoiling, the total twist is fixed by chain closure constraints at 8 helical turns (70,71). While Ω is not necessarily required to increase uniformly, the parameter is varied here at fixed increments of 36° from 0° to 2844°. The expression $(1/2\pi) \sum_i \Omega_i$ equals 8 as required for regular repetition of helical structure around the circle.

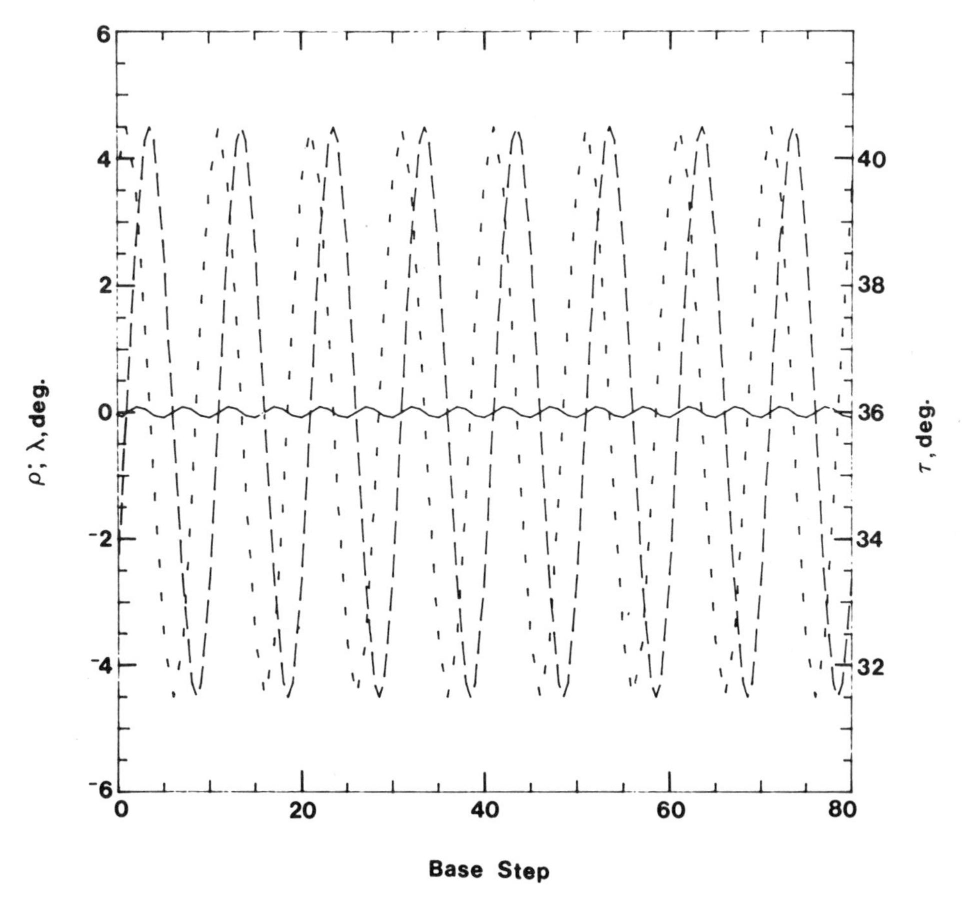

Figure 10. Observed variations in roll ρ (— —), tilt λ (-------), and twist τ (——) between adjacent base pairs as a function of relative position (base pair step) in the closed circular duplex illustrated in Figure 9.

The variations in ρ and λ that result from the above procedure are detailed in Figure 10. The angles are found to fluctuate in a sinusoidal fashion with an amplitude of 4.5° and a phase angle difference between curves of 2.5 residues. While the predicted variations in tilt are equivalent to those in roll, the perturbations are small enough to be of comparable energy. A smooth bending model of this sort is not easily discounted by arguments (16,57-59) that the DNA base pairs are more likely to roll than to tilt. The mean base-base van der Waals stacking energy of the poly dG•poly dC structure illustrated in Figure 9 is 1.0 kcal/mole higher than that of a perfectly linear duplex with 3.4Å base pair spacing. The energy is reduced to 0.5 kcal/mole in a 140-residue circle and 0.1 kcal/mole in a 200-residue circle where the amplitudes of ρ and λ are respectively 2.6° and 1.8°.

It is also important to emphasize that the variations in τ from one residue to the next are not constant in the circular structure despite the choice of fixed Ω. The variations are, nevertheless, quite small ranging from a low of 35.9° to a high of

36.1° over a 5-residue repeating sequence of the 80-residue circle and summing to the same total as the Ω_i. The τ_i can be fixed only by varying the Ω_i in the generation of chain structure. The twist that relates successive base pair coordinate frames is also slightly different from the topological twist density $2\pi T_i$ owing to the finite arc lengths between successive residues along the circle.

The backbone of the circular structure is the lowest energy solution which allows successive base pairs to assume the positions dictated by the differential geometric approach. Each dimer is not necessarily in its lowest energy form but is in the most favored state that allows it to link with the optimum backbone solutions for its neighbors. Specifically, the glycosyl torsions and sugar puckering at each end of the dimer are forced to adopt the same values of χ and P found at the opposite ends of the dimer fragments which abut it.

The advantages of a differential geometric approach to chain cyclization become apparent in the treatment of more complex trajectories such as the interwound duplex illustrated in Figure 11. It is impossible to differentiate a supercoiled structure like this with nonzero writhe (72) from a relaxed circular form of the same chain length on the basis of the separation and orientation of chain ends. A comparable statistical mechanical analysis must be limited to smaller subfragments of the supercoil which are then linked in succession.

The model in Figure 11 was constructed from two 80-residue left-handed superhelical trajectories and two intervening hairpin loops containing 20 residues apiece. Adjacent base planes are separated by arc lengths of 3.5Å and oriented at an angle $\Lambda = 2.2°$ in the 160 superhelical chain units. The dimensions of the hairpin loops are chosen to minimize fluctuations in base pair separation and orientation, the arc lengths ranging from 3.5 to 4.0Å and Λ varying between 7.9 and 13.4°. The parameter Ω describing the oscillation of a reference vector in successive normal planes is again fixed at 36°. Because the structure has a writhe of approximately 2, the total twist of the structure is reduced from the 20 turns of a relaxed circular duplex of the same chain length to roughly 18. The average unwinding per residue is therefore about $2 \times -360°/200$ residues $= -3.6°$/residue.

The resulting variations of ρ, λ, and τ along the interwound pathway are plotted as a function of chain sequence in Figure 12. The changes in ρ and λ are simple sinusoidal functions with an appreciably larger amplitude in the hairpin ends (10-12°) than in the superhelical core (2.2°) of the structure. More interesting, however, are the variations in τ which exhibit minor fluctuations ($\pm 0.02°$) about a mean value of 32.1° in the superhelical core but which follow a decided pattern of $\pm 13°$ underwinding, overwinding, and underwinding at the hairpin ends. Moreover, the overwinding and underwinding occur over 5 base pair stretches with the average twist at 36°. The large changes in τ can be eliminated if the structure generating angle Ω is assigned variable increments along the hairpin turns. The fluctuations in ρ and λ, however, are then more irregular. Interestingly, the structure with large variations in τ is favored compared to that with fixed τ by the van der Waals interactions between adjacent base pairs.

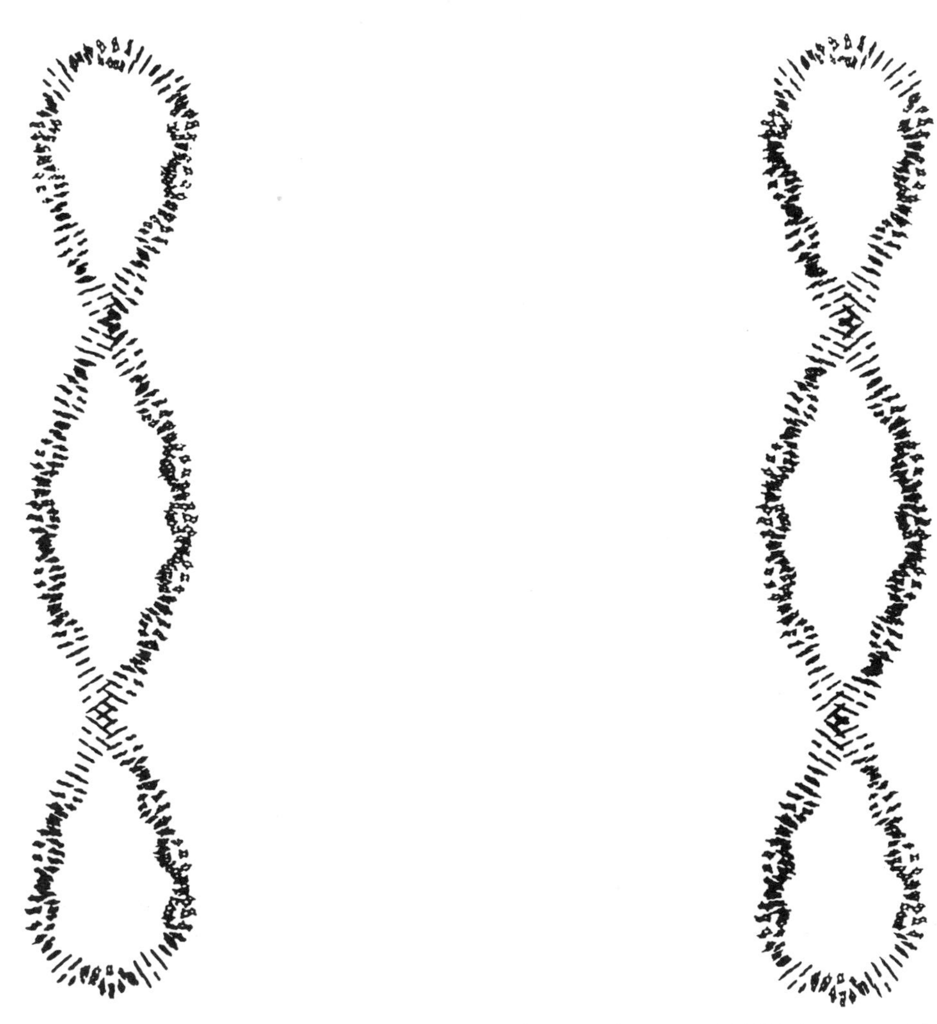

Figure 11. Stereo representation at the base pair level of a doubly interwound B-DNA duplex containing 200 residues.

Discussion

The above series of calculations illustrate how the primary sequence of nucleic acid bases affects the conformation of the DNA helix and how these local perturbations of base stacking translate into large scale effects at the macromolecular level. One of the most striking examples involves the thymine bases which possess a bulky methyl group at the C5 position. Steric interactions involving this group prevent neighboring residues from assuming a perfectly parallel base stacked arrangement in single stranded structures. Adjacent thymines are consistently more inclined and further separated than the corresponding adenines in trimeric fragments subjected to energy minimization from the same starting conformation. Interestingly, this bending is less pronounced in the standard B-DNA duplex where adjacent

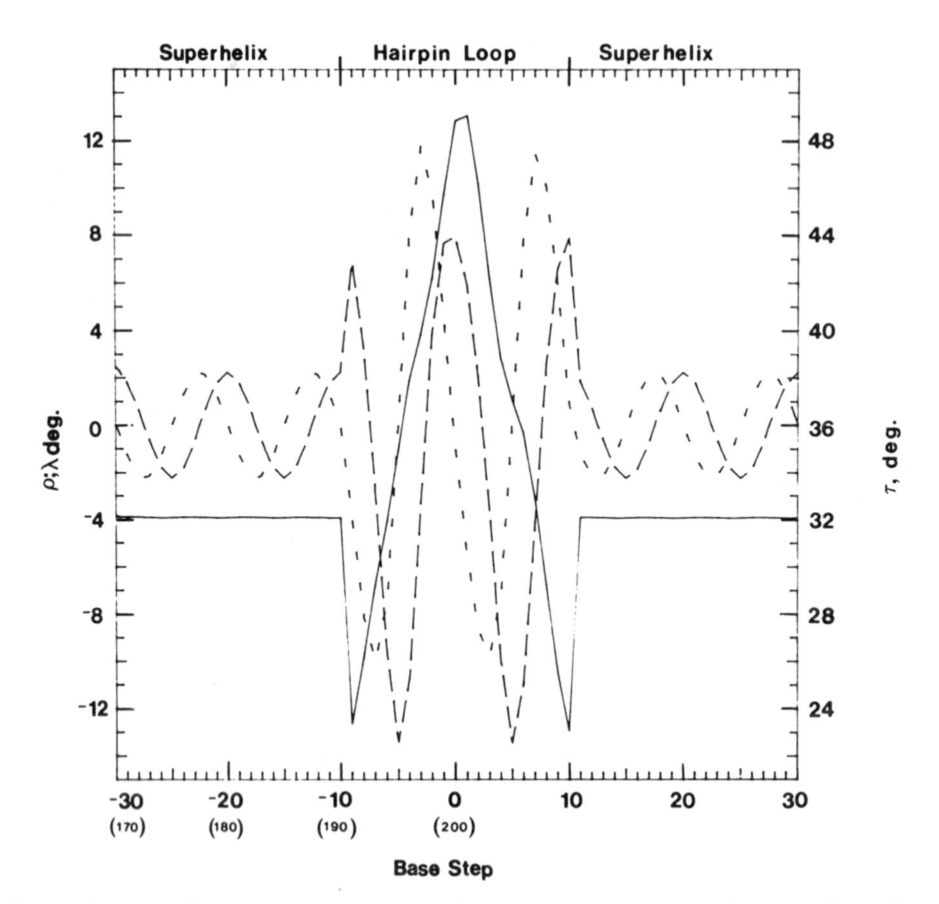

Figure 12. Observed variations in roll ρ (— —), tilt λ (-------), and twist τ (——) between adjacent base pairs as a function of relative position (base pair step) in the interwound duplex illustrated in Figure 11. Data reported for a single hairpin turn and 40 residues of the superhelical fragments which intersect it.

thymines are offset by the 36° twist angle constraint. The single-stranded preferences reappear as the duplex unwinds without intercalative opening to 0° and the base planes overlap totally.

The nucleic acid base sequence effects, however, are much less pronounced than those associated with the different amino acid side groups of a protein. The four common bases are similar chemical species, all of which possess a strong tendency to stack like coins in long vertical arrays. Alternative unstacked arrangements are generally so much higher in energy that it is unnecessary to consider their occurrence within the double helical backbone. Moreover, the unstacking of a single residue is accompanied by enormous viscous drag and inertial effects (73). In order to avoid such large scale bending, nearby residues are forced to adopt compensating unstacked arrangements (43). The nucleic acid folding problem is accordingly far simpler than the protein folding question. Nucleic acid folding is more analogous to the higher order folding of isolated α-helices or β-sheets. Furthermore, because the local nucleotide units are so restricted in their motions, nucleic acid folding is spread over a much larger molecular scale than in the proteins.

Recent studies (17,74) have emphasized how minor base-sequence dependent structural perturbations can alter the helical properties of the B-DNA duplex. The changes involve every residue of the chain with the double helix as a whole converted to a new helical form with altered pitch, radius, and residues per turn. This work has additionally considered the effects of regularly perturbed residues within the standard tenfold structure. The largest macromolecular changes occur when a given dimeric perturbation persists over a half helical turn of the altered duplex. Moreover, if the half turns alternate with an analogous sequence of bends of the opposite sign, the DNA will curve into a compact superhelical arrangement. Such a conformational pattern might characterize an inverted primary repeating sequence such as $A_5T_5{\cdot}T_5A_5$. Sequences that exhibits similar tendencies to roll or tilt in opposite directions could also adopt such a chain conformation. The likelihood of ten consecutive units bending in the desired fashion, however, is quite small in this case. If each unit could adopt only two bent states of opposite sign with equal ease, the probability of the inverted repeating decamer arrangement is only $1/2^{10} \sim 10^{-3}$.

Statistical mechanical approaches further demonstrate that even a slight energetic preference for local bending in a particular direction condenses the flexible DNA duplex. Chains composed of energetically similar repeating residues behave analogously to regularly bent helices. Those with centrosymmetric energy surfaces are more extended than those with asymmetrically positioned surfaces of identical topology just as standard B-DNA is more extended than a helix with regularly rolled and/or tilted units. The regular repetition of inverted energy surfaces, five at a time, also depresses average chain dimensions compared to the homopolymeric chain and increases the likelihood of cyclization at very short chain lengths. These two parameters mirror the decrease in step height (h/h°) and the increase in curvature (κ) associated with superhelix formation in a regular structure.

The DNA condensation problem has also been examined from the perspective of differential geometry. The B-DNA duplex has been deformed along a closed circular trajectory and the local variations of roll, tilt, and twist deduced from the model. The bending of the chain is accomplished in this instance by a correlated sequence of positive and negative rolling and tilting motions. The required perturbations are quite small and likely to be of low energy, although previous studies (57-59) have emphasized the energetic preference of rolling over tilting motions at low levels of bending. A more complex closed circular interwound trajectory has also been examined. The unwinding of the duplex is accomplished primarily in the superhelical core of the structure where every residue has a twist angle of roughly 32° and the roll and tilt angles fluctuate in the sinusoidal manner observed in the relaxed circle. The 180°-turns at the ends of the interwound species are characterized by sizable variations in all three angular parameters. The variations in twist ($\pm 13^{\circ}$) reported here are larger than those observed in X-ray crystallographic examples (11,12,15).

These fluctuations, however, can be reduced in magnitude in a larger interwound structure containing more residues. The superhelical density of the 200-residue structure illustrated in Figure 12 is much larger than that found in naturally occurring systems (72). The same pattern of conformational changes, however, would be repeated in a longer chain of the same geometric proportions. Twist is also thought to be a base sequence dependent property (11,12). The necessary variations of roll, tilt, and twist required to form a hairpin bend could very likely be translated from a given chemical sequence of bases. Specific regulatory sites on DNA might be expected to possess such unique tertiary conformations.

Finally, a base morphological approach is not only useful for understanding the long range structure of DNA but also for comprehending the conformational complexities of the local sugar-phosphate backbone. By searching for the multiple low energy combinations which link a given base pair arrangement it is becoming possible to understand the mechanism of interconversion between closely related states and to model complex conformational transitions. Various routes between different conformers are now being tested at the backbone as well as at the base pair level.

Acknowledgment

Sponsorship of this research by the U. S. Public Health Service under grants GM 20861 and GM 34809 and the Charles and Johanna Busch Memorial Fund is gratefully acknowledged. The authors are also indebted to Dr. Suse Broyde for assistance in the early stages of the energy minimization work and to Dr. Irwin Tobias for initial suggestion of and continued interest in the differential geometric modeling studies. Computer resources were generously supplied by the Center for Computer and Information Services of Rutgers University.

References and Footnotes

1. W. K. Olson, *J. Am. Chem. Soc. 104,* 278-284 (1982).
2. W. K. Olson, in *Topics in Nucleic Acid Structure Part II,* Ed. S. Neidle, Macmillan Press, London, pp. 1-79 (1982).
3. L. S. Lerman, *J. Mol. Biol. 3,* 18-30 (1961).
4. M. P. Printz and P. H. von Hippel, *Proc. Natl. Acad. Sci. USA 53,* 363-370 (1965).
5. M. Leng and G. Felsenfeld, *J. Mol. Biol. 15,* 455-466 (1966).
6. J. Brahms, A. M. Michelson, and K. E. van Holde, *J. Mol. Biol. 15,* 467-488 (1966).
7. D. Poland, J. N. Vournakis, and H. A. Scheraga, *Biopolymers 4,* 223-235 (1966).
8. J. A. Schellman, *Biopolymers 13,* 217-226 (1974).
9. H. Teitelbaum and S. W. Englander, *J. Mol. Biol. 92,* 55-78 and 79-92 (1975).
10. M. D. Barkley and B. H. Zimm, *J. Chem. Phys. 70,* 2991-3007 (1979).
11. R. E. Dickerson and H. R. Drew, *J. Mol. Biol. 149,* 761-786 (1981).
12. A. V. Fratini, M. L. Kopka, H. R. Drew, and R. E. Dickerson, *J. Biol. Chem. 257,* 14686-14707 (1982).
13. C. R. Calladine, *J. Mol. Biol. 161,* 343-352 (1982).
14. R. E. Dickerson, M. L. Kopka, and H. R. Drew, in *Conformation in Biology,* Eds. R. Srinivasan and R. H. Sarma, Adenine Press, New York, pp. 227-257 (1982).

15. Z. Shakked, D. Rabinovich, O. Kennard, W. B. T. Cruse, S. A. Salisbury, and M. A. Viswamitra, *J. Mol. Biol. 166,* 183-201 (1983).
16. R. E. Dickerson, *J. Mol. Biol. 166,* 419-441 (1983).
17. C. R. Calladine and H. R. Drew, *J. Mol. Biol. 178,* 773-782 (1984).
18. C.-S. Tung and S. C. Harvey, *Nucleic Acids Res. 12,* 3343-3356 (1984).
19. W. K. Olson and P. J. Flory, *Biopolymers 11,* 1-66 (1972).
20. W. K. Olson, *Macromolecules 8,* 272-275 (1975).
21. R. Yevich and W. K. Olson, *Biopolymers 18,* 113-145 (1979).
22. W. K. Olson, *Biopolymers 18,* 1213-1233 (1979).
23. W. K. Olson, in *Stereodynamics of Molecular Systems,* Ed., R. H. Sarma, Pergamon Press, New York, pp. 297-314 (1979).
24. W. K. Olson, *Macromolecules 13,* 721-728 (1980).
25. N. L. Marky and W. K. Olson, *Biopolymers 21,* 2329-2344 (1982).
26. J. A. Schellman, *Biophys. Chem. 11,* 321-328 (1980).
27. P. J. Hagerman and B. H. Zimm, *Biopolymers 20,* 1481-1502 (1981).
28. S. D. Levene and D. M. Crothers, *J. Biomol. Str. Dynam. 1,* 429-435 (1983).
29. J. Shimada and H. Yamakawa, *Macromolecules 17,* 689-698 (1984).
30. N. Gō and H. A. Scheraga, *Macromolecules 6,* 273-281 (1973).
31. V. G. Tumanyan and N. G. Esipova, *Biopolymers 14,* 2231-2246 (1975).
32. V. B. Zhurkin, Yu. P. Lysov, and V. I. Ivanov, *Biopolymers 17,* 377-412 (1978).
33. K. J. Miller, *Biopolymers 18,* 959-980 (1979).
34. F. H. Allen, S. Bellard, M. D. Brice, B. A. Cartwright, A. Doubleday, H. Higgs, T. Hummelink, B. G. Hummelink-Peters, O. Kennard, W. D. S. Motherwell, J. R. Rodgers, and D. G. Watson, *Acta Cryst. B35,* 2331-2339 (1979).
35. S. Arnott and A. J. Wonacott, *Polymer 7,* 157-166 (1966).
36. S. Arnott, S. D. Dover, and A. J. Wonacott, *Acta Cryst. B25,* 2192-2206 (1969).
37. J. L. Sussman, S. R. Holdbrook, G. M. Church, and S.-H. Kim, *Acta Cryst. A33,* 800-804 (1977).
38. A. Jack and M. Levitt, *Acta Cryst. A34,* 931-935 (1978).
39. E. R. Taylor and K. J. Miller, *Biopolymers 23,* 2853-2878 (1984).
40. A. R. Srinivasan and W. K. Olson, *Fed Proceed. 39,* 2199 (1980).
41. E. R. Taylor and W. K. Olson, *Biopolymers 22,* 2667-2702 (1983).
42. S. Arnott, unpublished data.
43. W. K. Olson, in *Biomolecular Stereodynamics I,* Ed., R. H. Sarma, Adenine Press, New York, pp. 327-343 (1981).
44. W. K. Olson, in *Intramolecular Dynamics,* Eds., J. Jortner and B. Pullman, D. Reidel Publishing Co., Dordrecht, pp. 525-536 (1982).
45. A. H.-J. Wang, G. J. Quigley, F. J. Kolpak, G. van der Marel, J. H. van Boom, and A. Rich, *Science 211,* 171-176 (1981).
46. W. K. Olson, A. R. Srinivasan, N. L. Marky, and V. N. Balaji, *Cold Spring Harbor Symp. Quant. Biol. 47,* 229-241 (1983).
47. M. A. Viswamitra, O. Kennard, P. G. Jones, G. M. Sheldrick, S. Salisbury, L. Falvello, and Z. Shakked, *Nature 273,* 687-688 (1978).
48. A. H.-J. Wang, S. Fujii, J. H. van Boom, and A. Rich, *Proc. Natl. Acad. Sci. USA 79,* 3968-3972 (1982).
49. A. H.-J. Wang, S. Fujii, J. H. van Boom, G. A. van der Marel, S. A. A. van Boeckel, and A. Rich, *Nature 299,* 601-604 (1982).
50. C.-H. Lee and E. Charney, *J. Mol. Biol. 161,* 289-303 (1982).
51. J. C. Marini, S. D. Levene, D. M. Crothers, and P. T. Englund, *Proc. Natl. Acad. Sci. USA 79,* 7664-7668 (1982).
52. J. C. Marini, P. N. Effron, T. C. Goodman, C. K. Singleton, R. D. Wells, R. M. Wartell, and P. T. Englund, *J. Biol, Chem. 259,* 8974-8979 (1984).
53. E. N. Trifonov, in *Nucleic Acids: The Vectors of Life,* Eds., B. Pullman and J. Jortner, D. Reidel Publishing Co., Dordrecht, pp. 373-385 (1983).
54. H.-M. Wu and D. M. Crothers, *Nature 308,* 509-513 (1984).
55. P. G. Hagerman, *Proc. Natl. Acad. Sci. USA 81,* 4632-4636 (1984).
56. M. J. D. Powell, *Computer J. 7,* 155-162 (1964).

57. V. B. Zhurkin, Yu. P. Lysov, and V. I. Ivanov, *Nucleic Acids Res. 6,* 1081-1096 (1979).
58. V. B. Zhurkin, Yu. P. Lysov, V. L. Florentiev, and V. I. Ivanov, *Nucleic Acids Res. 10,* 1811-1830 (1982).
59. N. B. Ulyanov and V. B. Zhurkin, *J. Biomol. Str. Dynam. 2,* 361-385 (1984).
60. Yu. A. Neyfack and V. G. Tumanyan, *Biopolymers 23,* 2419-2440 (1984).
61. T. J. Richmond, J. T. Finch, B. Rushton, D. Rhodes, and A. Klug, *Nature 311,* 532-537 (1984).
62. P. J. Flory, *Proc. Natl. Acad. Sci. USA 70,* 1819-1823 (1973).
63. D. Shore, J. Langowski, and R. L. Baldwin, *Proc. Natl. Acad. Sci. USA 78,* 4833-4837 (1981).
64. D. Shore and R. L. Baldwin, *J. Mol. Biol. 170,* 957-981 (1983).
65. E. Trifonov, in *Biomolecular Stereodynamics IV,* Eds., R. H. Sarma and M. H. Sarma, Adenine Press, New York, pp. 35-44 (1986).
66. P. J. Flory and D. Y. Yoon, *J. Chem. Phys. 61,* 5358-5365 (1974).
67. P. J. Flory, U. W. Suter, and M. Mutter, *J. Am. Chem. Soc. 98,* 5733-5739 (1976).
68. Z. Kam, N. Borochov, and H. Eisenberg, *Biopolymers 20,* 2671-2690 (1981).
69. N. Borochov and H. Eisenberg, *Biopolymers 23,* 1757-1769 (1984).
70. W. B. Fuller, *Proc. Natl. Acad. Sci. USA 68,* 815-819 (1971).
71. F. H. C. Crick, *Proc. Natl. Acad. Sci. USA 73,* 2639-2643 (1976).
72. W. R. Bauer, *Ann. Rev. Biophys. Bioeng. 7,* 287-313 (1978).
73. M. R. Pear, S. H. Northrup, and J. A. McCammon, *J. Chem. Phys. 73,* 4703-4704 (1980).
74. A. Prunell, I. Goulet, Y. Jacob, and F. Goutorbe, *Eur. J. Biochem. 138,* 253-257 (1984).

Biomolecular Stereodynamics IV, Proceedings of the Fourth Conversation in the Discipline Biomolecular Stereodynamics, State University of New York, Albany, NY, June 04-09, 1985, Eds., Ramaswamy H. Sarma & Mukti H. Sarma, ISBN 0-940030-18-7, Adenine Press, ©Adenine Press 1986.

On loopfolding in nucleic acid hairpin-type structures[29]

C.A.G. Haasnoot, C.W. Hilbers,
Biophysical Chemistry, University of Nijmegen,
Toernooiveld, 6525 ED Nijmegen, The Netherlands

G.A. van der Marel, J.H. van Boom,
Gorlaeus Laboratories, State University
P.O.B. 9502, 2300 RA Leiden, The Netherlands

U.C. Singh, N. Pattabiraman and P.A. Kollman,
Department of Pharmaceutical Chemistry, University of California
San Francisco, California 94143

Abstract

In a series of studies, combining NMR, optical melting and T-jump experiments, it was found that DNA hairpins display a maximum stability when the loop part of the molecule comprises four or five nucleotide residues. This is in contrast with the current notion based on RNA hairpin studies, from which it had been established that a maximum hairpin stability is obtained for six or seven residues in the loop. Here we present a structural model to rationalize these observations. This model is based on the notion that to a major extent base stacking interactions determine the stability of nucleic acid conformations. The model predicts that loop folding in RNA is characterized by an extension of the base stacking at the 5′-side of the double helix by five or six bases; the remaining gap can then easily be closed by two nucleotides. Conversely, loop folding in DNA is characterized by extending base stacking at the 3′-side of the double helical stem by two or three residues; again bridging of the remaining gap can then be achieved by one or two nucleotides.

As an example of loop folding in RNA the anticodon loop of yeast $tRNA^{Phe}$ is discussed. For the DNA hairpin formed by $d(ATCCTAT^4TAGGAT)$ it is shown that the loop structure obtained from molecular mechanics calculations obeys the above worded loop folding principles.

Introduction

The first picture that comes to mind when discussing nucleic acid conformations is that of the illustrious double helix (1). Of course, it is irrefutable that the Watson-Crick duplex and its descendants had (and still have) a profound influence on the biomolecular way of thinking and therefore fully deserve this imperative status. Of late, single-crystal X-ray diffraction studies have further widened our view and not only disclosed the existence of aberrant helices like the lefthanded Z-DNA (2), but also showed that the statutory B-DNA helix is not a regular and very uniform structure: in fact, the local structure of the B-DNA helix appeared to be dependent upon the base sequence (3-5). Notwithstanding this, from a conformational point of view the overall rod-like shape of the double helix is rather monotonous, especially when compared to proteins in which the polypeptide chain can fold into a wide variety of conformations. However, the conformational space of the polynucleotide chain is certainly not confined to the canonical double helix. In fact, one of the important conformational features of nucleic acids is their intrinsic flexibility which may result in significant modifications of the chain extension.

This characteristic is most prominent in the RNA genus of nucleic acid molecules. Due to their single stranded nature, RNA molecules tend to fold back on themselves to form intramolecular duplex regions thereby originating structural elements like hairpins, internal loops and bulges. In turn, the latter structural elements may interact with each other and thus lead to a very complicated three-dimensional overall structure for the molecule at issue.

A classical example of such an intricate architecture is set (6) by the tRNA molecule: its secondary (cloverleaf) structure is determined by four intramolecular double helical regions (which involves the formation of three hairpins), but it is the pairwise stacking of these double helical stems together with the interactions between the DHU- and TψC-loops that give rise to the L-shaped overall structure of the molecule. For other classes of RNA molecules (ribosomal RNA, plant virus RNA, etc.) the postulated secondary structures abounds with hairpins, internal loops, etc., but detailed information about the interaction between these structural elements (i.e. tertiary structure) is lacking.

For the DNA genus of nucleic acid molecules the situation is somewhat more complicated and now two situations have to be discriminated. In the case of single stranded DNA a conformational behaviour as described above for RNA molecules is to be expected. Indeed, hairpin structures are known to occur in some single stranded phage DNA's and appear to play an important role in gene expression.

In the case of double stranded DNA molecules the dominance of the common double helix is of course overwhelming and therefore the abundance of structural elements like hairpins, bulges, etc., is expected to be very low in these molecules. Still, there is growing evidence that so-called cruciform structures (i.e. hairpins) occur in supercoiled DNA sequences with short inverted repeats (7-11). As of yet the biological relevance of these short inverted repeats is unsettled but it is noted that these palindromic base sequences have been found within many biological important regions of DNA molecules such as promotor/operator regions, replication origins, transcription termination sites, etc.

Thus, the above discerned structural elements (i.e. hairpins, bulges, internal loops) may be considered important fundamental building blocks in nucleic acid structure and a thorough understanding of their physical and structural properties is called for. In order to raise such detailed structural and thermodynamical information regarding DNA hairpin loops, we started a systematic investigation of partly self-complementary oligodeoxynucleotides (12-17). The present paper resumes our findings so far and relates these to existing knowledge on RNA hairpins.

Loops stabilize the duplex stem in DNA hairpins

Our studies concentrated on a series of synthetic homologous DNA fragments which can be delineated by the generic base sequence d(ATCCTAT$_n$TAGGAT), n=0−7 (cf. Fig. 1). For n=0 the fragment is fully self-complementary, for n> 0 there is a non-complementary part of n thymidines in the middle of the sequence which intervenes the (self) complementary parts of the fragments. Of course, the complementary ends of the latter fragments can still interact and may form duplexes either intramolecularly or intermolecularly. In the case of intramolecular duplex formation this will lead to a (monomeric) hairpin structure; for the intermolecular case a dimeric species with an internal loop will result; see Fig. 1.

(a) d(A-T-C-C-T-A-T$_n$-T-A-G-G-A-T)

(b)

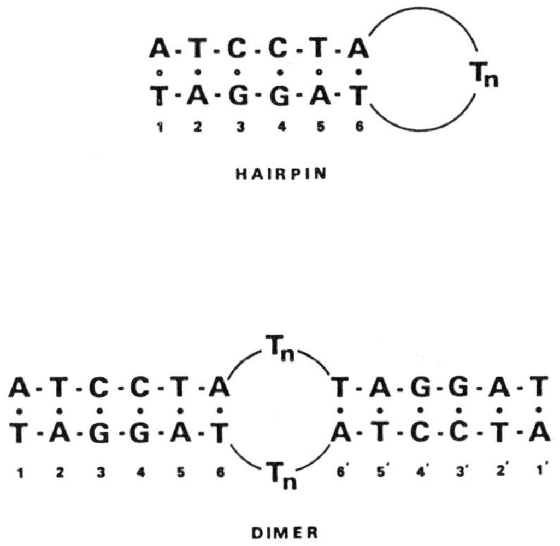

A-T-C-C-T-A
T-A-G-G-A-T
T$_n$
1 2 3 4 5 6

HAIRPIN

A-T-C-C-T-A T$_n$ T-A-G-G-A-T
T-A-G-G-A-T T$_n$ A-T-C-C-T-A
1 2 3 4 5 6 6' 5' 4' 3' 2' 1'

DIMER

Figure 1. (a) Generic base sequence of the series of oligodeoxynucleotides studied. (b) Secondary structures that can be adopted by the homologous molecules given under (a). The prefix d (for deoxy) is omitted for the sake of clarity; base pair numbering used throughout the text is indicated. Note the pairwise equivalency of the base pairs in the dimeric form.

Legends For Color Folios

Figure 4. Far Left Panel Top. Perspective view of a (regular) B-DNA double helix (basepairs colored blue) of which one of the strands is elongated at its 3′-end with three nucleotides stacked in a single helical B-type way (bases colored yellow). The terminal phosphate of this single stranded region is connected to the 5′-terminal phosphate of the opposite strand by means of a dashed line. This line corresponds to the shortest phosphate-phosphate distance in Fig. 2.

Figure 5. Far Left Panel Bottom. Perspective view of a regular A-RNA double helix (basepairs colored blue) of which the 5′-terminal strand is elongated with five nucleotides stacked in an A-type single helical way (bases colored yellow). The dashed line indicates the remaining gap of ca. 13 Angstrom that can easily be bridged by two nucleotides (see text).

Figure 7. Middle Panel Top. Perspective view of the anticodon loop of yeast $tRNA^{Phe}$. This loop presents an example of loop folding in A-RNA. The strand with the 5′-end at the bottom of the anticodon stem is elongated by five nucleotides of which the bases (indicated in yellow) continue to stack as in an A-type helix. The remaining gap is bridged by two nucleotides (bases colored red). Note the similarity between the base stacking pattern in this anticodon loop and that in Fig. 5.

Figure 8. Middle Panel Bottom. Perspective view of the energy-minimized structure of the hairpin formed by the hexadecanucleotide d(ATCCTAT$_4$TAGGAT). The base pairs in the stem are colored blue. It is seen that the 3′-end of the ATCCTA-part of the hairpin stem (in front of the picture, facing the reader) is elongated with three nucleotides (of which the thymine bases are colored yellow) which continue the stacking pattern of the double helix. The remaining thymidine residue of the loop (colored red) fits nicely in the hole formed by the above mentioned three stacked thymidines.

Figure 9. Far Right Panel Top. Packing of the loop in d(ATCCTAT$_4$TAGGAT). Dots indicate the Van der Waals surfaces of the bases in the loop and in the top A•T(6) base pair of the stem.

Figure 10. Far Right Panel Bottom. Perspective view of the solvent (H_2O) accessible surface of the hexadecamer hairpin shown in Fig. 8. The sugar-phosphate backbone is indicated in yellow, the bases are colored blue and the exchangeable amino- and iminoprotons are indicated in green and red, respectively. Note that only two iminoprotons in the loop region are but partially accessible for water molecules.

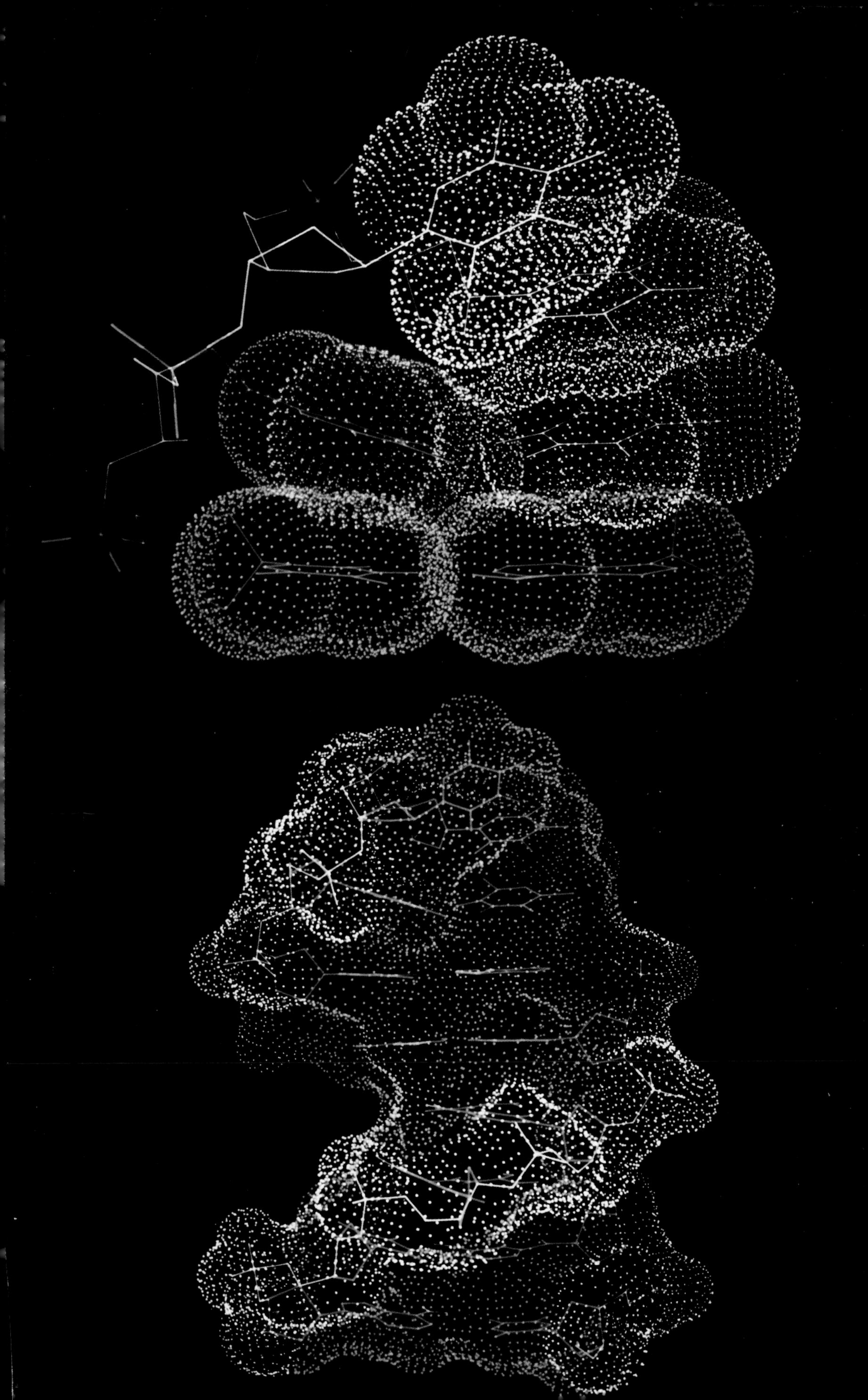

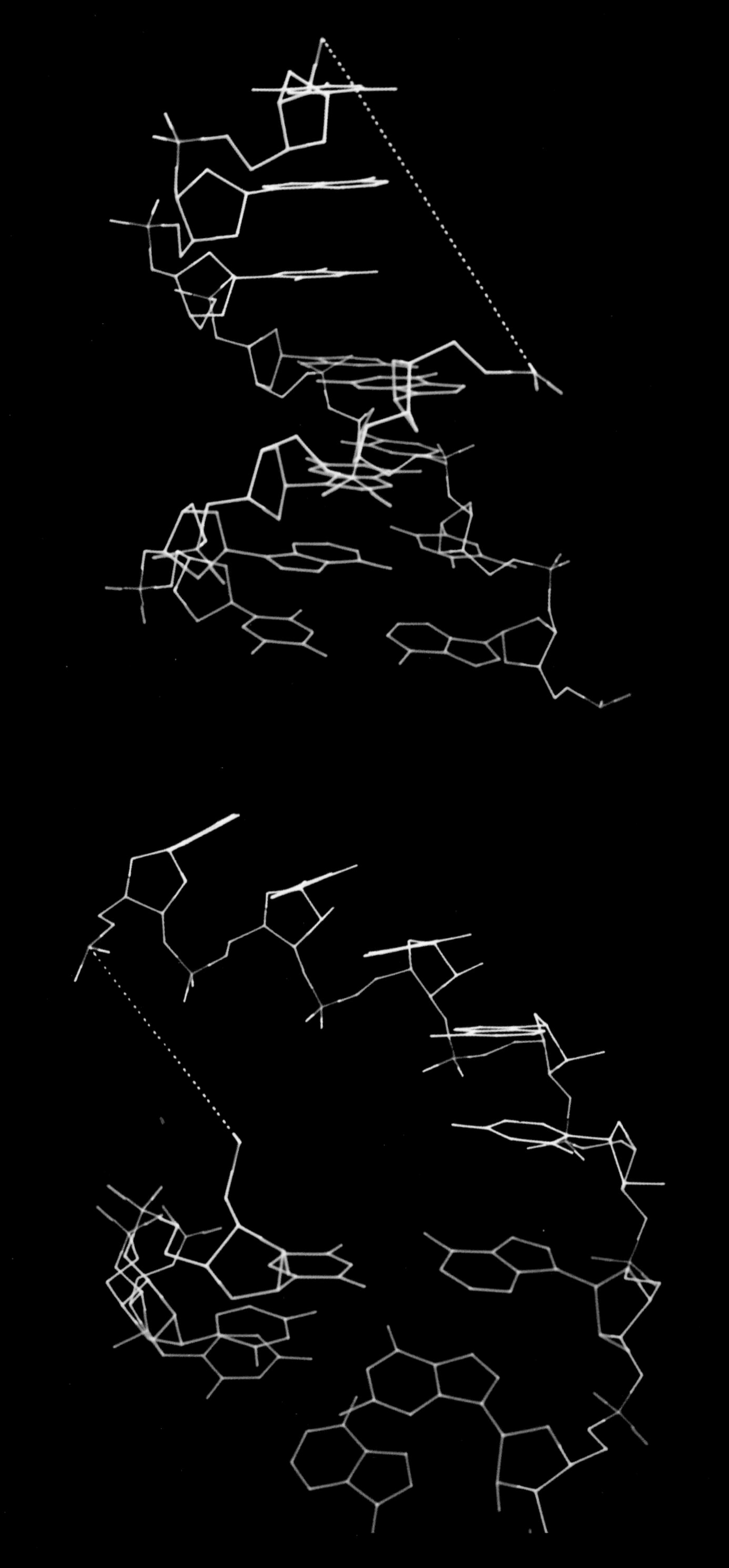

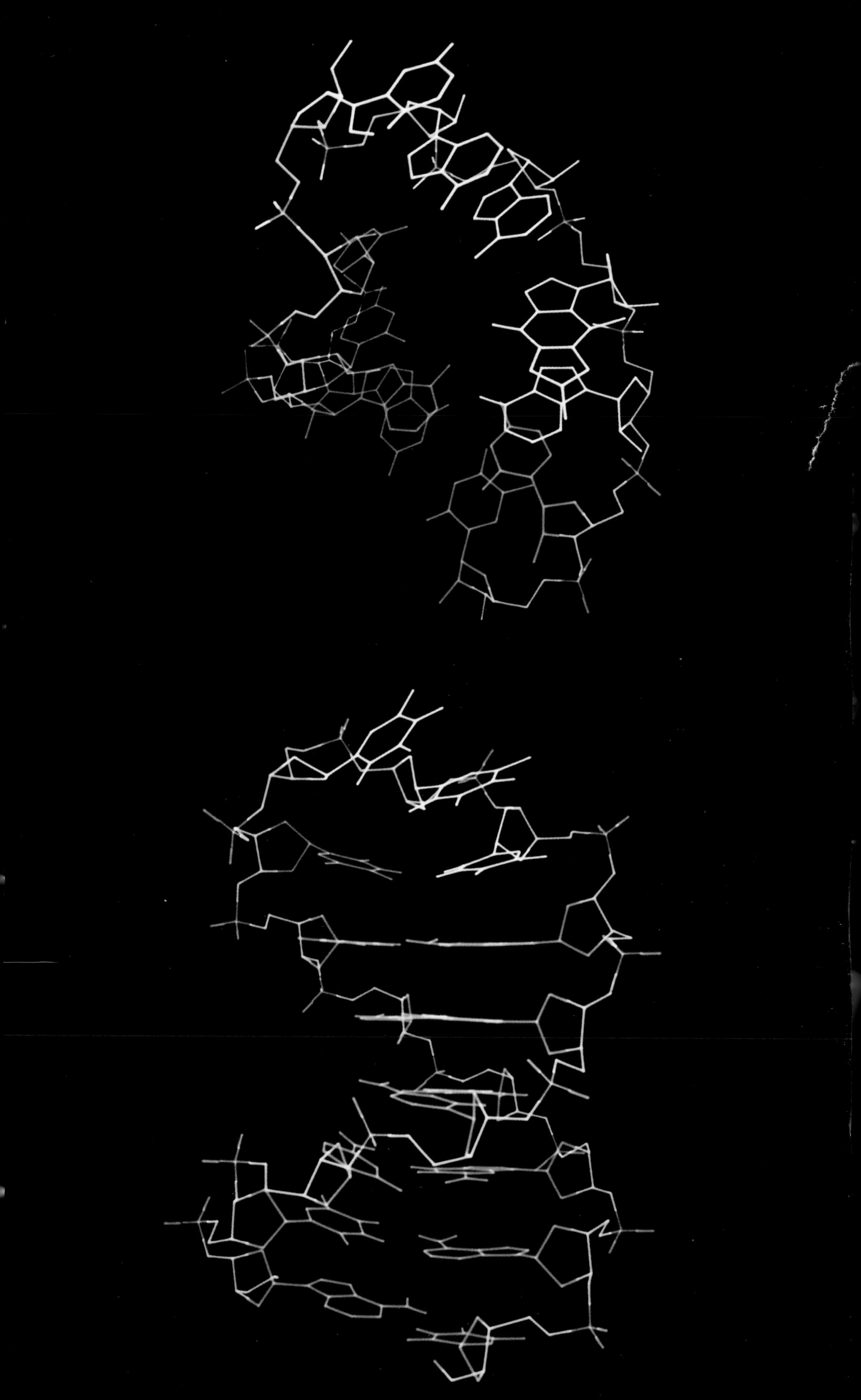

During the "Third Conversation in the Discipline—Biomolecular Stereodynamics" organized at SUNY Albany in 1983, we presented (15) solid evidence that for $n \geq 4$ the aforementioned DNA fragments adopt exclusively a hairpin-like conformation, both at high (millimolar) and low (micromolar) oligomer concentrations. Analysis of the remaining (n=0−3) fragments in the series revealed (16,17) that under suitable conditions also these smaller (n=0−3) oligonucleotides may take up a hairpin structure (albeit often in equilibrium with the dimeric internal loop structure). Especially for the case of the fully self-complementary fragment (n=0) the latter finding was somewhat surprising, but not unprecedented: studies of the fully self-complementary poly d(TA) (Ref. 18) and d(CGCGAATTCGCG) (Ref. 19) showed that the latter molecules are also able to form hairpin structures in solution. Be this as it may, our DNA fragments d(ATCCTAT$_n$TAGGAT) n=0−7, thus offered the opportunity to study a series of hairpins with a wide variety of loop sizes.

The thermodynamic parameters characterizing the thermally induced hairpin-to-coil transition were determined for all fragments (n=0−7) by UV-melting- and T-jump techniques; the results are summarized in Table I. In addition, we also determined the transition enthalpy of the duplex formed by the complementary hexanucleotides (d(ATCCTA) and d(TAGGAT). By measuring the T_m of the duplex-to-coil transition as a function of the hexanucleotide concentrations, the enthalpy for d(ATCCTA)•d(TAGGAT) melting was found to be $\Delta H^\circ = -33$ kcal/mole. The latter value agrees nicely with the prediction that can be made for ΔH° on basis of available base stacking enthalpies in double helical DNA molecules (20; predicted $\Delta H^\circ = -35.3$ kcal/mole).

Since the base sequence of the duplex d(ATCCTA)•d(TAGGAT) corresponds to the base sequence of the double helical stem in our hairpins, this hexanucleotide dimer may serve as a yardstick in evaluating the influence of the various loops on the stability of our hairpins. Thus, the data in Table I show that for hairpins with $n \geq 5$ the enthalpy of hairpin formation is approximately constant ($\Delta H^\circ \simeq -43$ kcal/mole) and about 10 kcal/mole more favourable than would be expected on basis of the hairpin stem duplex (i.e. the reference dimer d(ATCCTA)•d(TAGGAT) alone. In other words, the loops consisting of 5 or more thymidine residues contribute ca. −10 kcal/mole to the overall $\Delta H^\circ_{hairpin}$. The origin of this negative enthalpy contribution of the loops cannot be determined from the data at hand, but it seems reasonable to hold stabilization of base pair A•T(6) and stacking interactions within the loops or between the loop and the top of the hairpin stem responsible for this phenomenon (cf. also ref. 15 and 17).

As can be inferred from $\Delta H^\circ_{hairpin}$ determined for d(ATCCTAT$_4$TAGGAT), cf. Table I, a loop consisting of four thymidine residues contributes only ca. −6 kcal/mole to the enthalpy content of the hairpin. This -in comparison with loops of 5 or more residues: reduced-contribution to the overall $\Delta H^\circ_{hairpin}$ may point to some steric crowding of the thymidine residues in the (tighter) loop of this hexadecamer.

In the DNA fragment with three intervening T's (n=3) the decreasing trend in overall hairpin enthalpy content is continued; now the $\Delta H^\circ_{hairpin}$ is only −3 kcal/

Table I

Melting temperatures, T_m, melting enthalpies, ΔH°_{exp}, and calcualted melting enthalpies, ΔH°_{calc} (see text) of the hairpins d(ATCCTAT$_n$TAGGAT)

n	HAIRPIN	ΔH°_{calc} (kcal/mole)	ΔHP°_{exp} (kcal/mole)	T_m (°C)
0	ATCC^{T}A • • • • TAGG$_{A}$T	−25	−22 to −25	52
1	ATCC^{T}A$_{T}$ • • • • TAGG$_{A}$T	−29	−29 to −32	54
2	ATCC^{TA}T • • • • TAGG$_{AT}$T	−29		
	or		−29 to −34	52
	ATCCT^{A}T • • • • • TAGGA$_{T}$T	−32		
3	ATCCT^{A}T T • • • • • TAGGA$_{T}$T	−36	−36 ± 1	52
4	ATCCTA^{T}T • • • • • • TAGGAT$_{T}$T	−39	−39 ± 1	53
5	ATCCTA^{T}T T • • • • • • TAGGAT$_{T}$T	−43	−43 ± 1	51
6	ATCCTA$^{T^{T}}$T • • • • • • TAGGAT$_{T_{T}}$T	−43	−42 ± 1	48
7	ATCCTA$^{T^{T}}$T T • • • • • • TAGGAT$_{T_{T}}$T	−43	−44 ± 1	43
	reference dimer			
	ATCCTA • • • • • • TAGGAT	−35.3	−33	

mole more favourable than the reference duplex stem. At first sight one is tempted to explain this finding in terms of a hairpin loop consisting of three thymidines with fairly heavy steric crowding. However, in the NMR spectra recorded for this pentadecamer fragment the Watson-Crick imino proton of base pair A•T(6) is lacking (16,17). This observation leads to an alternative explanation, namely that base pair A•T(6) is incorporated into the loop of the hairpin (thereby elongating the loop to five residues and concomitantly reducing the duplex stem of the hairpin to five base pairs). It is noted that this explanation is also in keeping with the enthalpic data: the stabilization brought about by loops of five residues amounts to ca. -10 kcal/mole (vide supra) whereas the breaking of an A•T base pair will cost (20) ca. 7 kcal/mole. The net result of such a trade off is an apparent overall change in ΔH° of the hairpin (in comparison with the reference dimer) by -3 kcal/mole which is in agreement with the observed value.

For the smallest DNA fragments ($n=0-2$) the recorded $\Delta H^{\circ}_{hairpin}$ are less favourable than ΔH° of the reference duplex stem (cf. Table 1). This is not unexpected since loops consisting of two or less thymidine residues are considered to be sterically impossible and therefore it follows that these hairpins have to sacrifice one or two base pairs in order to form a hairpin loop. Indeed, the enthalpic data ($\Delta H^{\circ}_{hairpin}$) determined for these smallest ($n=0-2$) hairpin fragments can be rationalized (16) on basis of the following equation:

$$\Delta H^{\circ}_{hairpin} = \Delta H^{\circ}_{stem} + \Delta H^{\circ}_{loop} + \Delta H^{\circ}_{broken}\ A{\cdot}T_{base\ pairs}$$

in which $\Delta H^{\circ}_{stem} = -33$ kcal/mole (i.e. the enthalpy content of the "unconstrained" reference dimer formed by the hexamers); $\Delta H^{\circ}_{loop} = -10$ kcal/mole (for loops consisting of five or more nucleotides) or -6 kcal/mole (for loops formed by four nucleotides); $\Delta H^{\circ}_{broken}\ A{\cdot}T_{base\ pairs} = +7$ kcal/mole for each A•T base pair that is incorporated into the loop. Melting enthalpies calculated with this equation are listed in Table I. Note that the equation is not only applicable to the smallest ($n=0-2$) fragments, but holds for the entire series of DNA hairpins studied ($n=0-7$).

The overall stability of the hairpins formed by the DNA fragments ($n=0-7$) as monitored by the melting temperature of the hairpin-to-coil transitions (T_m, cf. Table I) does not parallel the changes described above for $\Delta H^{\circ}_{hairpin}$: the T_m's of the shorter fragments ($n=0-4$) are more or less constant whereas T_m decreases for the longer DNA fragments ($n=5-7$). Since the enthalpy contents of the latter fragments ($n=5-7$) are identical (within experimental error), the observed decrease in overall stability of the larger hairpins is solely due to changes in the entropy content (ΔS°) of these hairpins. This is not unreasonable since -the duplex stem being equal in all these fragments- the nucleation entropy of hairpin formation will decrease (i.e become more unfavorable) as the number of nucleotides (= distance) between the corresponding C and G bases that form (15) the nucleation site increases. By the same arguments the smallest hairpins ($n=0-3$) will have a more favorable nucleation entropy which (at least in part) compensates the observed decrease in magnitude of $\Delta H^{\circ}_{hairpin}$ in these fragments so that the overall stability in terms of observed T_m's remains the same throughout the series $n=0-4$.

In all, the data presented in this section show that loops in DNA hairpin fragments contribute favorably to the overall enthalpy content of the hairpin structure. The results at hand indicate that a loop formed by four or five deoxynucleotides may even (over)compensate the breakage of an A•T base pair. Moreover, the stability of DNA hairpins is at its maximum for loop lengths of four to five residues and declines for larger loops (n=5−7).

The structure-stability relationship in DNA and RNA hairpins

Hairpin loop stabilities in RNA fragments were determined (21) over a decade ago and are nowadays standard textbook material (22). When the present results for DNA hairpins are compared with these RNA "standards", a striking variance is noted for the "optimum" loop length: in DNA hairpins the stability is at maximum when the loop comprises four to five deoxynucleotide residues, whereas in RNA hairpins loops with six or seven ribonucleotide residues appear to be the most stable. The question now forces itself upon one whether the differences in thermodynamic and conformational behaviour observed for DNA versus RNA hairpin fragments can be explained in terms of the differences in conformational space available and/or employed by the two classes of nucleic acids.

There is a consensus that in solution double helical RNA adopts an A-type duplex whereas DNA normally (i.e. under the solution conditions employed in our studies) prefers a B-type double helical form (23,24). Presuming that the duplex stem in hairpin fragments adheres to this precept (and preliminary NMR analysis of several hairpin fragments are in accordance with this presumption), the question can be rephrased: Does the difference between DNA and RNA hairpin fragments relate to the fact that in DNA hairpins the loop must bridge the gap between the ends of a B-type double helical stem whereas in RNA the loop must interconnect the ends of an A-type duplex stem? Of course, the first answer that comes to mind is that the distance between the 3′-end phosphorus atom on the one strand and the 5′-end phosphorus atom of the other strand (which defines the gap the loop has to span) might be larger for A-type helices than for B-type helices so that more nucleotide residues are required in an A-RNA loop compared to a B-DNA loop to close this gap. However, distance calculations performed for the standard fibre-diffraction structures of B-DNA and A-RNA (25) show that in both types of helices the interstrand phosphate-phosphate distance in a given pair is very similar (approximately 18 Angstrom). In other words it is not this interstrand distance in itself that is the underlying cause for the observed difference in "optimum" loop length for DNA versus RNA hairpins.

At this point an important point has to be brought into the discussion, namely the favorable contribution by the loop to the enthalpy content of the DNA hairpin as compared to the enthalpy content of the reference dimer, which we ascribe to stacking interactions within the loop and/or stacking interactions between loop bases and the top of the hairpin stem (vide supra). Obviously, any structural model to be proposed for the loop in nucleic acid hairpins should account for such stacking interactions. In a first approach we therefore considered the different influences that A- and B-type helices may have upon loop-folding when it is assumed that the base stacking pattern of at least one strand of the double helix stem is propagated into the loop. This analysis was carried out by calculating the interstrand distances within a double helix between a particular phosphorus atom in the one strand (P′) and the phosphorus atoms residing in the opposite strand; the results of these calculations are diagrammed in Figs. 2 and 3 for B-DNA and A-RNA respectively. Figure 2 shows that in the canonical B-DNA helix the shortest interstrand phosphorus-phosphorus distance (~12 Angstrom) occurs over the minor groove two to four nucleotides away from the "sampling" phosphorus atom P′. This implies that elongating the 3′-end of a B-type duplex stem with two to four nucleotides (stacked in a single helical B-type fashion on top of the double helix, see e.g. Fig. 4)

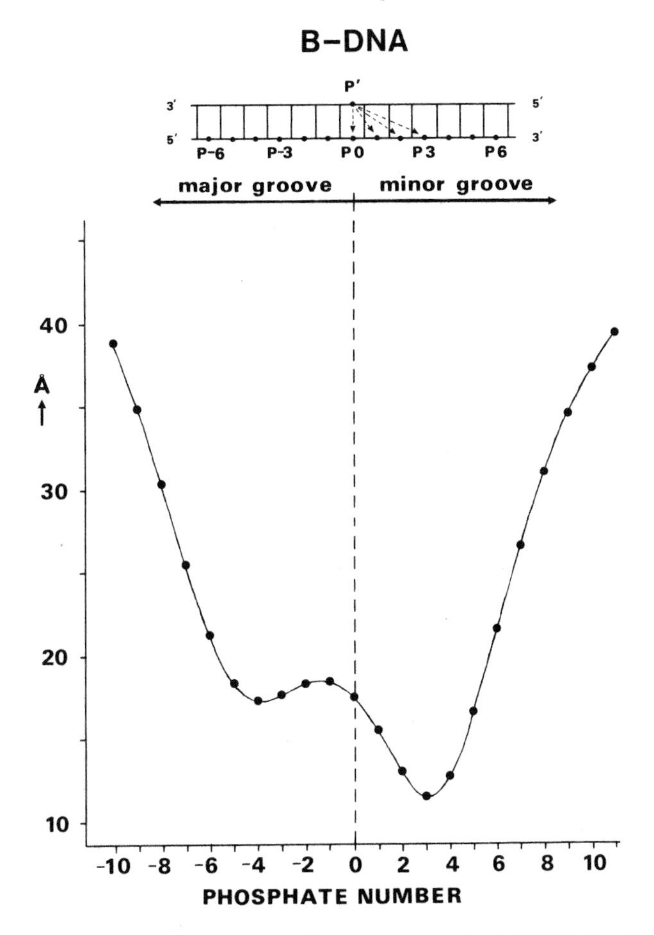

Figure 2. Interstrand phosphorus-phosphorus distances in a regular B-DNA helix. The curve represents the distances between a sampling phosphate (P′) in the one strand of the double helix and the phosphates (P_x) in the opposite strand. The numbering, x, of the phosphorus atoms in the opposite strand is indicated in the schematic two-dimensional representation above the curve. The distances are calculated on basis of the coordinates published by Arnott et al. (25).

reduces the distance between the 3′-end phosphorus atom of this strand and the 5′-end phosphorus atom located at the top of the duplex stem to approximately 12 Angstrom, a distance that can be spanned by one or two nucleotides. In this way short loops are created that bridge the ends of a B-type hairpin stem while still making optimal use of single stranded base-base interactions (note that Figure 2 also shows that if one continues the stacking on top of the 5′-end of the duplex a much larger gap has to be spanned by the "bridging" nucleotides).

In contrast, Figure 3 shows that in an A-RNA double helix the "sampling" phosphorus atom P′ is close (10-13 Angstrom) to a few phosphorus atoms across the major groove and located five to eight base pairs away in a linear structure notation. Therefore, the "natural" way to form a loop in RNA sequences is by elongating the 5′-end of the A-type double helical stem by, say, five nucleotides arranged in an

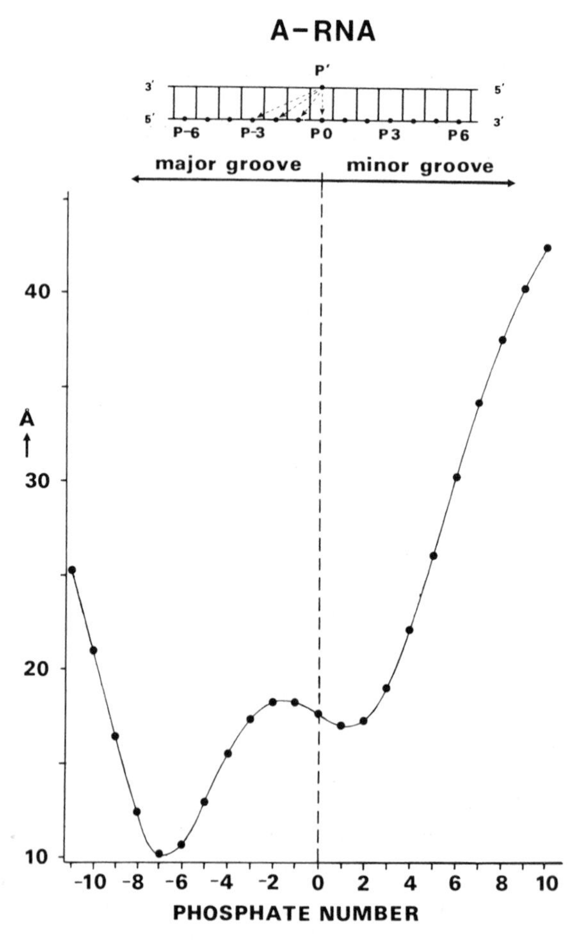

Figure 3. Interstrand phosphate-phosphate distances in a regular A-RNA double helix; cf. also legend of Figure 2.

A-type single helical stack (cf. Fig. 5). The gap that now remains between the 5′-end phosphorus atom of this stretch of nucleotides and the 3′-end phosphorus atom of the double helical stem is only ~11 Angstrom, a gap that is easily bridged by one or two nucleotides. The six- or seven-membered loop constructed in this way is stabilized by at least five single strand base-base stacking interactions which, of course, in turn will contribute to the overall stability of the RNA hairpin fragment thus formed.

Note that when smaller loops are built according to the above worded concept, they will be less favorable since in smaller loops there will be less single strand base-base interactions stabilizing the loop while the interstrand phosphate-phosphate distance to be spanned by the "bridging" nucleotides becomes larger. Moreover, such smaller loops will be tighter and therefore steric hindrance between e.g. the bulky base-moieties will increase. On the other hand, the expected increase in the probability of loop initiation with decreasing chain (= loop) length (i.e. nucleation entropy) may in part counterbalance these negative effects.

Of course, the same mechanisms are operative in RNA hairpins with more than seven nucleotides forming the loop: such longer loops will not suffer much from steric hindrance and may even have more single strand base-base interactions, but these positive effects are expected to be rather small and easily surpassed by the negative influence exerted by the concomitant change in nucleation energy. Taken together, we suggest that the above rationalizations account for the optimum stability observed for RNA hairpins with six- or seven-membered loops. Following a similar line of reasoning as given for the RNA hairpins it can be argued that a four- or five-membered loop will lead to the most stable hairpin fragments in DNA fragments (which is in perfect agreement with the experimental data, cf. the preceding section). We therefore conclude that the stability of nucleic acid hairpins is correlated to the conformation of the hairpin stem. Analysis of the canonical A-RNA and B-DNA structures makes it conceivable that the base stacking arrangement within a double helical stem is transmitted into the loop and thus governs the conformation and stability of the hairpin loop.

Conformational features of hairpin loops

The preceding section states the constraint that the double helical stem in nucleic acid hairpins imposes upon the loop conformation. This is schematically epitomized in Figure 6 for the "optimum" loops in RNA and DNA hairpins: in RNA the stacking pattern of the stem is propagated from the 5′ end of the stem into the loop for five nucleotides and the remaining gap (Fig. 6a) can then be bridged by one or two nucleotides. In contrast, the B-DNA hairpin continues its stacking pattern on the 3′ end of the stem for three nucleotides, followed by a change in direction of the nucleotide chain in order to close the remaining gap (Fig. 6c) with one or two nucleotides.

But not only schematically, also in full atom models these loop folding principles can be visualized. For the seven-membered RNA hairpin loop this is shown in Fig. 7

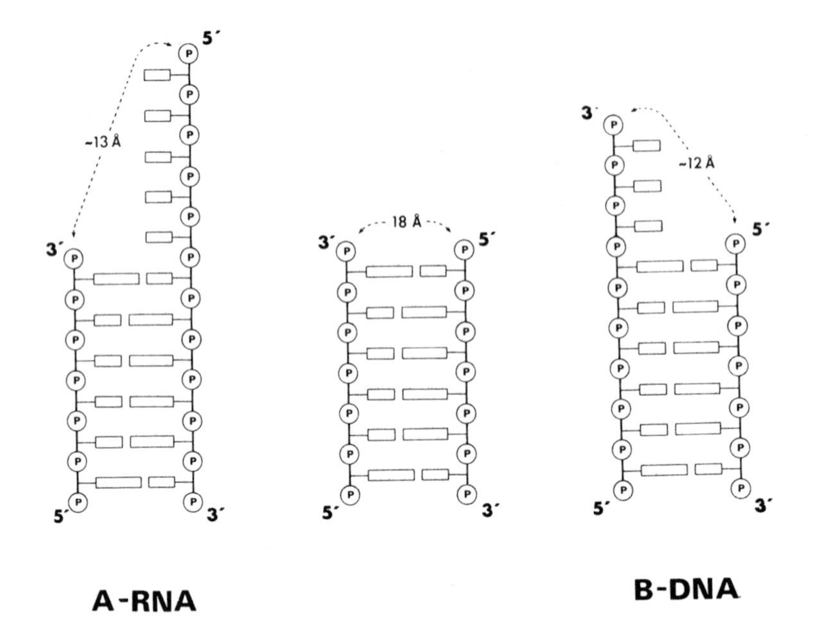

Figure 6. Schematic representation of the specific features important when considering hairpin loop formation in A-RNA and B-DNA. In A-type as well as B-type double helices the interstrand phosphate-phosphate distance within the same base pair amounts to approximately 18 Angstrom (middle). When in A-RNA one of the strands of the double helix is elongated at its 5′-end with five nucleotides arranged as in the regular A-RNA helix, then the remaining gap between the 3′- and 5′-terminal phosphates reduces to ca. 13 Angstrom (left). Analogously, when in B-DNA double helices one of the strands is elongated at its 3′-end with three nucleotides arranged as in a regular B-DNA helix then the remaining gap between the 3′- and 5′-terminal phosphates of the opposite strands reduces to ~12 A (right).

where the anticodon loop of $tRNA^{Phe}$ from yeast is depicted. In this tRNA molecule the nucleotides of the last base pair in the A-type double helical anticodon stem ($A_{31}\bullet\psi_{39}$, colored blue in Fig. 7) are interconnected via the ribonucleotide fragment r($-C_mUG_mAAYA$). The crystallographic data available for this molecule show that the sequence r($-G_mAAYA-$), colored yellow in Fig. 7, stacks in an A-type single helical fashion on top of ψ_{39} (5′-side of the anticodon stem), while the remaining two nucleotides of the fragment (C_{32} and U_{33}, colored red in Fig. 7) close the remaining gap between A_{31} and G_{34}.

Up till now no crystallographic structure determination that might serve as an example for loop structures in DNA hairpins has been reported. Therefore, we chose an alternative approach and generated -on basis of the loop folding principle described above- a crude model for the DNA hexadecamer d($ATCCTAT_4TAGGAT$) using computer graphics methods (program MIDAS; cf. Ref. 26). This crude model was refined by means of molecular mechanics/dynamics techniques (program AMBER; cf. Ref. 27); the energy-minimized structure of the hexadecamer hairpin is shown in Fig. 8. As expected, the double helical stem of the hairpin is of the B-type and the base stacking pattern is propagated from the 3′ end of the stem into the loop (yellow bases in Fig. 8). Note that, the single helical stack in this part of the loop deviates somewhat from the canonical B-DNA pattern in order to accommodate the short closure of the loop (only one nucleotide; colored red in Fig. 8). This is shown in more detail in Fig. 9 where a close-up of the loop (displaying also the Van der Waals radii of the bases) is given: it is seen that the bases are tilted with respect to what would have been expected on basis of a standard single strand B-type helix (cf. Fig. 4). One might speculate that it is this distortion (and its inherent steric crowding) that accounts for the somewhat reduced stabilizing effect of four-membered loops compared to five-membered loops (vide supra).

Figure 10 shows the solvent accessible surface (r_w = 1.4 Angstrom) calculated for this DNA hairpin model. Remarkably, only two out of the four imino protons of the loop thymidines are to some extent accessible to water (namely, T_8 and T_9; red dots in Figure 10), the other two imino protons (i.e. those belonging to T_7 and T_{10}) are completely buried within the loop. As such, this finding fully endorses the explanation given earlier for the observed retardation in the proton-exchange of the loop imino protons as was detected by NMR: the loop structure protects the imino protons in the loop from exchange with the solvent (15).

It should be mentioned that here a more general morphological difference between RNA and DNA hairpins is touched upon. In DNA hairpins the loop forms a "closed" structure in which the bases of the loop are more or less turned inwards, i.e. towards the center of the loop. In contrast, the seven-membered RNA hairpin loop adopts a more "open" structure in which a number of loop bases (especially the anticodon bases) are exposed and readily accessible from the "outside" (i.e. solvent). In view of the function of the anticodon loop, the latter characteristic is essential since it puts the anticodon triplet in a position that ensures facile base pairing with its cognate codon on the mRNA. Although for the time being one can only speculate about the function of hairpins in DNA, *mutatis mutandis* our results may indicate that such a possible function of DNA hairpins will not primarily involve the loop bases in tertiary (e.g. Watson-Crick type) interactions.

Discussion

In this paper we have developed a simple framework for logical thinking about the architecture of hairpin loops: the "natural" way to fold a loop in A-RNA hairpins is to propagate the stacking of the nucleotides at the 5′-end of the duplex stem, whereas in B-DNA hairpins the stacking should be continued at the 3′-side of the stem. Of course, within this framework, the bases will certainly try to optimize stacking, evade highly localized strain situations, etc. Therefore, it can be predicted that the *details* of the hairpin loop folding will depend on e.g. base sequence in the loop and perhaps even on the base sequence in the stem. In this respect it is good to note that the DNA fragments considered in this paper have loops consisting of only T-residues or alternating AT-sequences. Our MM-calculations show that the loop consisting of four T-residues is very tight. Therefore, in loops with (too) many purines, one or more of these purines might not always fit entirely into such a tight configuration. In such a case steric crowding might dictate deviations in the sense that e.g. the loop is enlarged by incorporating the top base pair of the stem into the loop.

For the sake of completeness we further note that the folding principles only hold for unperturbed hairpin loops, i.e. loops in which the loop bases are not engaged in interactions (such as base pairing, etc.) with bases from outside the loop.

With these provisos in mind, we conclude that the folding of unperturbed hairpin loops is dictated by the stacking pattern of the bases in the double helical stem of the hairpin. It is these stacking patterns that allows for shorter "optimum" loops in DNA when compared to RNA hairpins; the sugar-phosphate backbone appears just to follow suit. As such, the loop folding principles formulated in this paper link up nicely with (and are in a sense an extension of) the "base-centred explanation of the B- to A-transition in DNA" given by Calladine and Drew (28). There is no doubt that abstractions like these lay a foundation for developing a better understanding of nucleic acid conformational behavior in solution.

Acknowledgements

This work was supported by the Netherlands Foundation for Chemical Research (SON) with financial aid from the Netherlands Organization for the Advancement of Pure Research (ZWO). P. Bash and Dr. R. Langridge (Computer Graphics Laboratory, San Francisco) are gratefully acknowledged for expert help.

References and Footnotes

1. Watson, J.D. and Crick, F.H.C., *Nature 171,* 737 (1953).
2. Wang, A.H.-J., Quigley, G.J., Kolpak, F.J., Crawford, J.C., van Boom, J.H., van der Marel, G.A. and Rich, A., *Nature 282,* 680 (1979).
3. Dickerson, R.E. and Drew, H.R., *J. Mol. Biol. 149,* 761 (1981).
4. Calladine, C.R., *J. Mol. Biol. 161,* 343 (1982).
5. Dickerson, R.E., *J. Mol. Biol. 166,* 419 (1983).
6. Rich, A. & RajBhandary, U.L., *Ann. Rev. Biochem. 45,* 805 (1976).
7. Woodworth—Gutai, M. and Lebowitz, J., *J. Virol. 18,* 195 (1976).
8. Panayotatos, N. & Wells, R.D., *Nature 289,* 466 (1981).
9. Lilley, D.M.J., *Proc. Natl. Acad. Sci. USA 77,* 6468 (1980).
10. Lilley, D.M.J. *Nucl. Acids Res. 9,* 1271 (1981).
11. Singleton, C.K., *J. Biol. Chem. 258,* 7661 (1983).
12. Haasnoot, C.A.G., den Hartog, J.H.J., de Rooij, J.F.M., van Boom, J.H. and Altona, C., *Nature 281,* 235 (1979).
13. Haasnoot, C.A.G., den Hartog, J.H.J., de Rooij, J.F.M., van Boom, J.H. and Altona, C., *Nucl. Acids Res. 8,* 169 (1980).
14. Van Boom, J.H., van der Marel, G.A., Westerink, H., van Boeckel, C.A.A., Mellema, J.-R., Altona, C., Hilbers, C.W., Haasnoot, C.A.G., de Bruin, S.H. and Berendsen, R.G., *Cold Spring Harbor Symp. Quant. Biol., Vol XLVII,* 403 (1983).
15. Haasnoot, C.A.G., de Bruin, S.H., Berendsen, R.G., Janssen, H.G.J.M., Binnendijk, T.J.J., Hilbers, C.W., van der Marel, G.A. and van Boom, J.H., *J. Biomol. Struct. Dyns. 1,* 115 (1983).

16. Haasnoot, C.A.G., De Bruin, S.H., Hilbers, C.W., Van der Marel, G.A. & Van Boom, J.H. in *"Proceedings of the International Symposium on Biomolecular Structure and Interactions"* (Bangalore, India)—in the press (1985).
17. Hilbers, C.W., Haasnoot, C.A.G., De Bruin, S.H., Joordens, J.J.M., Van der Marel, G.A. & Van Boom, J.H. *Biochimie 67,* 685 (1985).
18. Scheffler, I.E., Elson, E.L. and Baldwin, R.L., *J. Mol. Biol. 36,* 291 (1968); *ibid 48,* 145 (1970).
19. Marky, L.A., Blumenfeld, K.S., Kozlowski, S. and Breslauer, K.J., *Biopolymers 22,* 1247 (1983).
20. Marky, L.A. and Breslauer, K.J., *Biopolymers 21,* 2185 (1982).
21. Tinoco, J., Borer, P.N., Dengler, B., Levine, M.D., Uhlenbeck, O.C., Crothers, D.M. & Gralla, J., *Nature New Biology 246,* 40 (1973).
22. Cantor, C.R. and Schimmel, P.R. in *Biophysical Chemistry. III. The behavior of biological macromolecules.* (Freeman, San Francisco), Chapter 23 (1980).
23. Pardi, A., Martin, F.H. and Tinoco, I., *Biochemistry 20,* 3986 (1981).
24. Haasnoot, C.A.G., Westerink, H.P., van der Marel, G.A. and van Boom, J.H., *J. Biomol. Struct. Dyns. 2,* 345 (1984).
25. Arnott, S., Campbell Smith, P.J. and Chandrasekaran, R. in *CRC Handbook of Biochemistry and Molecular Biology; Nucleic Acids Vol. II* (ed. Fasnar, G.D.), p. 411 (1976).
26. Ferrin, T.E. and Langridge, R., *Computer Graphics 13,* 320 (1980).
27. Weiner, S.J., Kollman, P.A., Case, D.A., Chandra Singh, U., Ghio, C., Alagona, G., Profeta, S. and Weiner, P., *J. Am. Chem. Soc. 106,* 765 (1984).
28. Calladine, C.R. and Drew, H.R., *J. Mol. Biol 178,* 773 (1984).
29. Reprinted from the *Journal* of *Biomolecular Structure of Dynamics 3,* 843-857 (1986),

Biomolecular Stereodynamics IV, Proceedings of the Fourth Conversation in the Discipline Biomolecular Stereodynamics, State University of New York, Albany, NY, June 04-09, 1985, Eds., Ramaswamy H. Sarma & Mukti H. Sarma, ISBN 0-940030-18-7, Adenine Press, ©Adenine Press 1986.

Structural Studies of DNA Fragments:[63] The G•T Wobble Base Pair in A, B and Z DNA; The G•A Base Pair in B-DNA

Olga Kennard
University Chemical Laboratory
Lensfield Road
Cambridge CB2 1EW, U.K.

Abstract

The crystal structures of five double helical DNA fragments containing non-Watson-Crick complementary base pairs are reviewed. They comprise four fragments containing G•T base pairs: two deoxyoctamers d(GGGGCTCC) and d(GGGGTCCC) which crystallise as A type helices; a deoxydodecamer d(CGCGAATTTGCG) which crystallises in the B-DNA conformation; and the deoxyhexamer d(TGCGCG), which crystallises as a Z-DNA helix. In all four duplexes the G and T bases form wobble base pairs, with bases in the major tautomer forms and hydrogen bonds linking N1 of G with O2 of T and O6 of G with N3 of T. The X-ray analyses establish that the G•T wobble base pair can be accommodated in the A, B or Z double helix with minimal distortion of the global conformation. There are, however, changes in base stacking in the neighbourhood of the mismatched bases. The fifth structure, d(CGCGAATTAGCG), contains the purine purine mismatch G•A where G is in the *anti* and A in the *syn* conformation. The results represent the first direct structure determinations of base pair mismatches in DNA fragments and are discussed in relation to the fidelity of replication and mismatch recognition.

Introduction

Biological processes are frequently controlled by a combination of finely tuned checks and balances, which, however, admit to an element of chance and randomness. Some of these processes, such as the programming of senescence, we can hardly guess at, while others, like the faithful transmission of genetic information and the modification of the genetic code leading to mutations have, to a degree, already been unravelled and charted.

The molecular basis of mutations is the modification of the regular DNA sequence, principally the introduction of non-Watson Crick complementary bases (mispairs), in the double helix. Mismatches can arise during replication, genetic recombination (1,2) or, in case of a G•T mismatch, deamination of 5 methyl cytosine (3). They have been observed to occur with varying frequency. The formation and detection

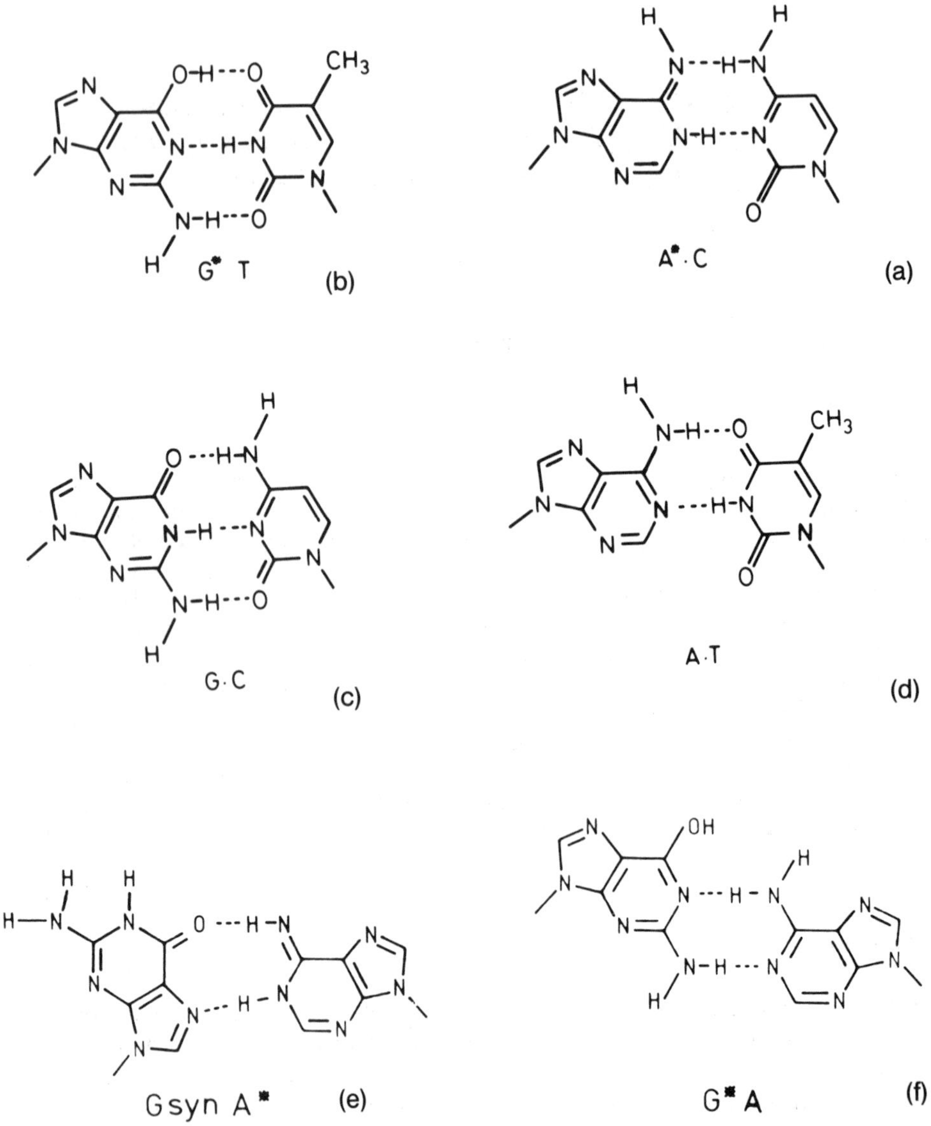

Figure 1. Schematic diagrams of various base pairing schemes. (a,b) Tautomer purine-pyrimidine pairs, (c,d) Watson-Crick base pairs, (e,f) Purine-purine tautomer pair. Bases in their rare tautomer forms are marked*.

of mismatched bases appears to be influenced by a variety of factors which include not only the nature of the polymerase or the repair enzyme but also the nature of the local DNA environment such as base sequence (4). An understanding of the way non-complementary bases form hydrogen bonded pairs and the influence of incorporating such mismatched base pairs on the local and global conformation of the DNA helix is thus of fundamental importance. This is a question which, until recently, could only be approached theoretically.

Two principal schemes have been suggested for hydrogen bonding between mismatched bases. Watson and Crick (5), in considering spontaneous mutation, proposed that bases may occasionally be incorporated in a rare tautomeric form implying the formation of A•C and G•T base pair mismatches illustrated in Figs. 1a, b. Such tautomeric base pairs are stereochemically very similar to standard Watson-Crick base pairs (Figs. 1c, d) and are likely to cause minimal changes in the conformation of the DNA double helix. Topal and Fresco (6) using chemical considerations and model building, have extended the tautomer hypothesis to account for purine-purine as well as purine-pyrimidine mismatches. They postulated that purine-purine base pairs would be too large to fit into the double helix unless one of the bases adopts a *syn* orientation. The A(*anti/imino*)•G(*syn*) base pair is illustrated in Fig. 1e. Topal and Fresco were able to correlate the estimated frequencies of minor tautomers (and *syn/anti* base pairs) with the observed frequencies (10^{-8} to 10^{-12}) of substitution mutations. It is in fact possible to postulate another tautomer base pair with both bases in the *anti* orientation (Fig. 1f).

An alternative hydrogen bonding scheme between non-complementary bases involved in codon-anticodon interactions was proposed by Crick (7) to account for the degeneracy of the genetic code. In this scheme the base pairs are displaced relative to their positions in the standard Watson-Crick base pairs, allowing the utilisation of different donor-acceptor groups for hydrogen bonding. The resultant base pairs for G•T and G•A, which are of most direct interest to the present communication, are illustrated in Figs 2a, b. Such base pairs were investigated theoretically by two groups. Mismatches in both A and B DNA were examined by Chuprina and Poltev (8). Poltev and Bruskov calculated the contribution of base interactions to mismatch stabilisation (9). They concluded that the G•U wobble pair is the most stable of all mismatches and causes minimal distortion of the double helix. They were not able to distinguish, in terms of stability, between the guanine-adenine base pairs when both bases are in the *anti* conformation (Fig. 2c) or when guanine is in the *anti* and adenine in the *syn* form (Fig. 2b). Calculations by Rein et al. (10), using different parameters, suggest that in B-DNA the G•T tautomer base pairs may be marginally more favourable than the wobble pair and the G•A(*syn*) pair is likely to be much more stable than the corresponding tautomer pair.

Major tautomer forms of G•U and A•G base pairs were observed in the crystal structures of various tRNA's (11-15). Their interactions with adjacent base pairs have been analysed (16). They have also been deduced from a solution study of poly d(G-T) (17). Fersht and co-workers, on the basis of kinetic experiments with DNA polymerase I, have presented evidence for the misinsertion of bases in their major tautomer form during replication (18).

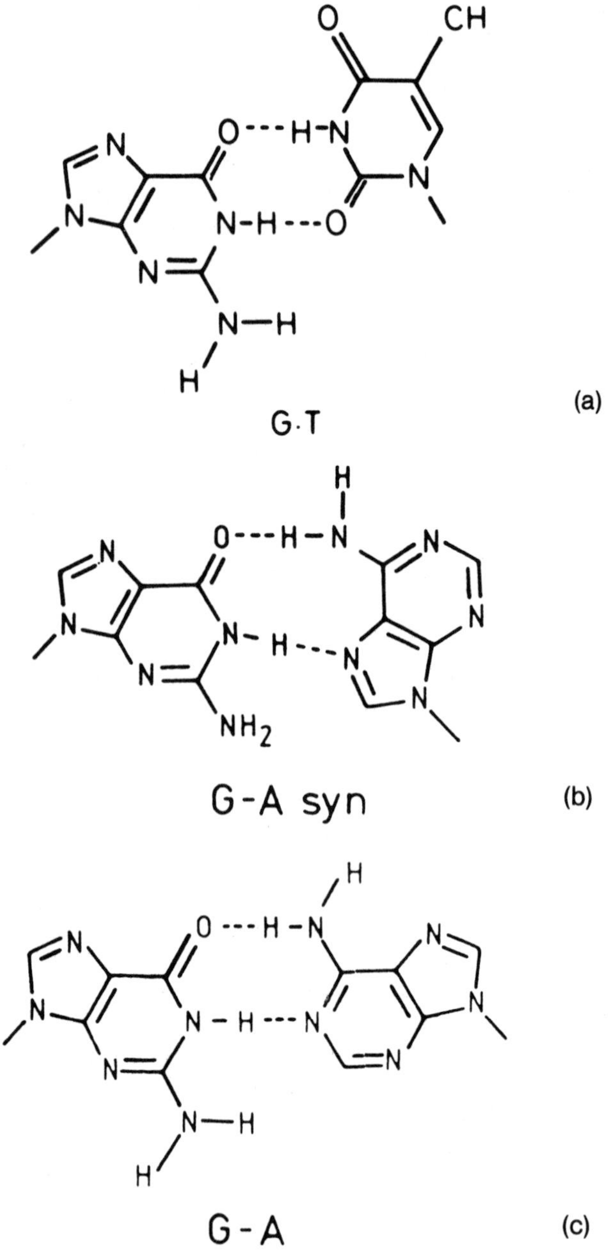

Figure 2. Schematic diagram of G•T and G•A base pairs with all bases in the major tautomer forms.

High resolution structural investigations of DNA fragments have only become possible in the past few years with the development of synthetic techniques for the preparation of defined DNA sequences in the quantities and purity required for NMR spectroscopy and for X-ray diffraction analysis of single crystals (19,20).

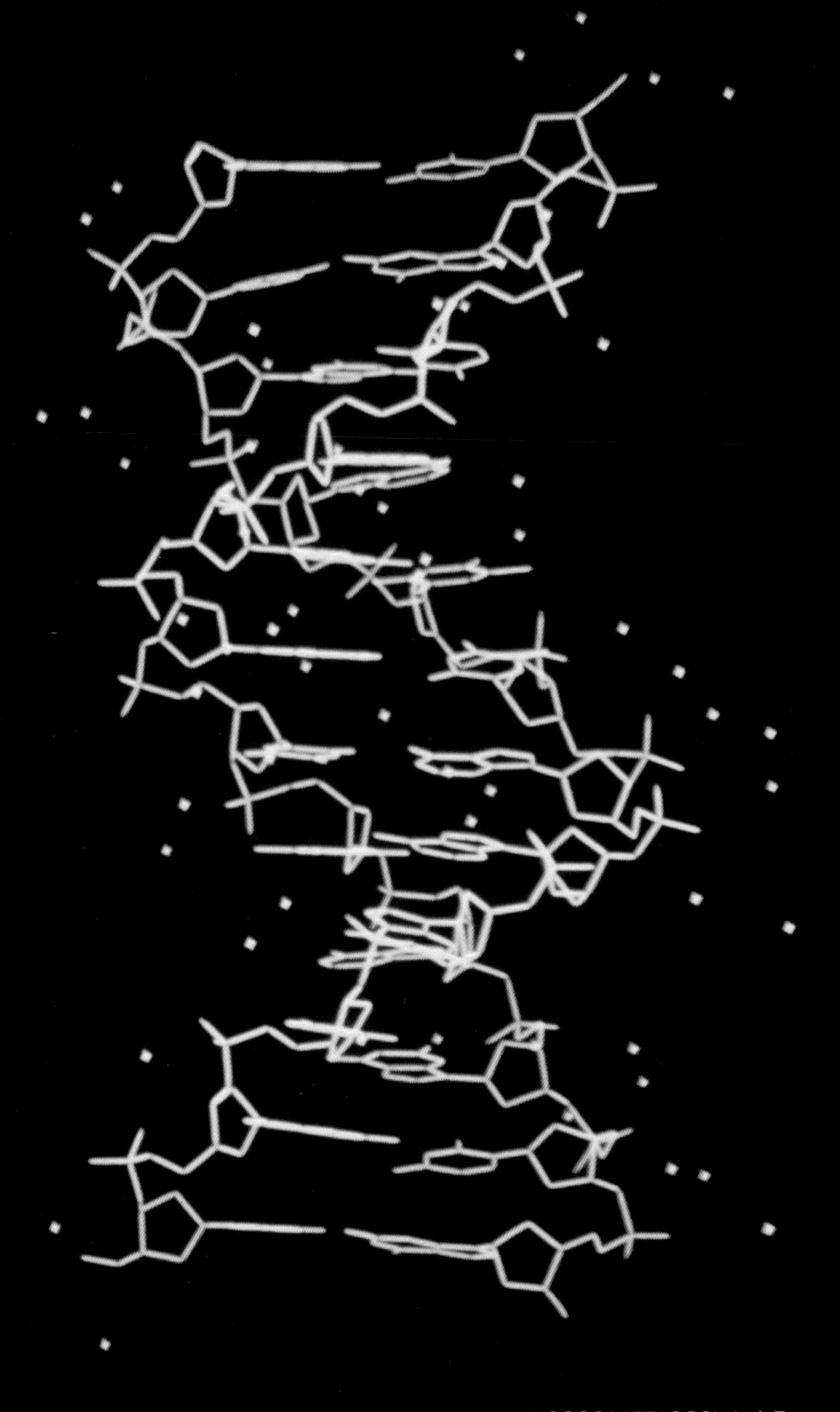

Figure 5. The G•A syn base pair in the B-DNA dodecamer d(CGCGAATTAGCG). *Left Top* Display of the G•A(*syn*) mismatch. Bases G(04) and A(21) are shown, together with the electron density in the plane of the base pairs (calculated from a $2F_o-F_c$ map). *Left Bottom* A sideways view of the structure and electron density in the neighbourhood of the mismatched bases. *Top Right* A view of the double helix and the associated solvent positions.

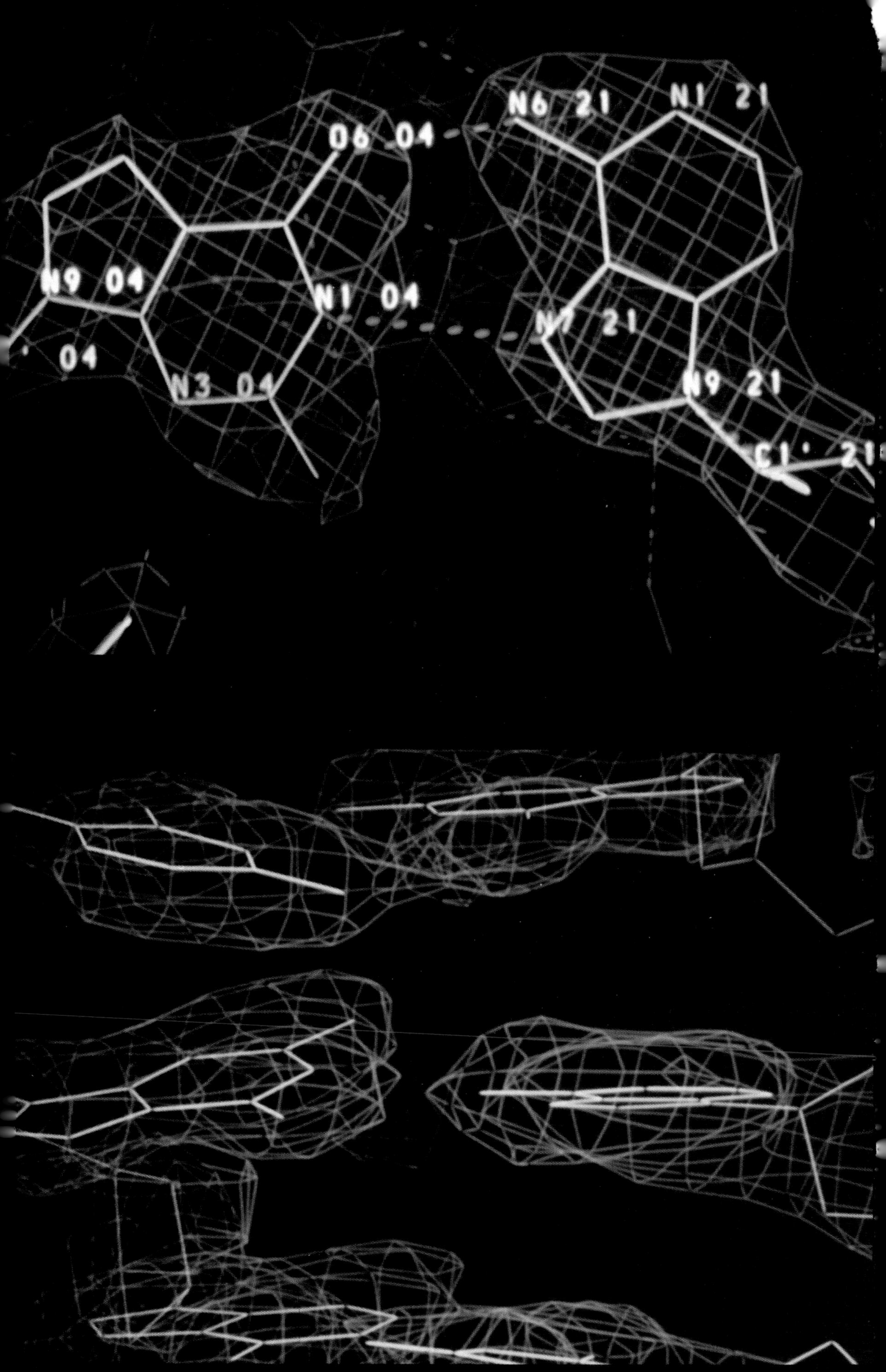

N6 21
N1 21
O6 04
N9 04
N1 04
N7 21
04
N3 04
N9 21
C1' 21

These studies have yielded a wealth of information about DNA fragments up to 12 base pairs long. Single crystal X-ray diffraction studies have led to the discovery, by Rich and co-workers, of the left handed Z-form of DNA (21) and to the correlation between base sequence and local helical parameters in B-DNA (22,23) and in A-DNA (24,25). These studies have established the conformational flexibility of the DNA double helix and brought about a reappraisal of our views of the role of DNA conformation in biological systems. The X-ray studies were all carried out on DNA fragments containing Watson-Crick complementary base pairs. It is, however, feasible to synthesize sequences with different potential base pair mismatches and to study them in a variety of environments by the same techniques.

Patel and co-workers using this approach (26,27) have recently investigated the NMR spectra of several B-DNA dodecamers containing various non-complementary base pairs and concluded that both G•T and G•A can form hydrogen bonded pairs with bases as the major tautomers and in the *anti* orientation around the glycosidic bond. Similar results for G•A base pairs were obtained by Kan and co-workers from a proton NMR study of a decamer (28). The effect of single base pair mismatches on the stability of the double helix was studied by Hilbers and co-workers (29).

In parallel with these investigations we have undertaken a series of single crystal X-ray studies of base pair mismatches in deoxyolignucleotides (30). The present paper gives an account of the structure analysis of the G•T mismatch in A, B and Z DNA fragments and the G•A mismatch in a B-DNA dodecamer. The results are discussed in relation to the fidelity of replication.

"Crystal Engineering"

Deoxyoligonucleotides are notoriously difficult to crystallise. When single crystals are formed they tend to adopt packing motifs which appear to be related to the DNA conformation. Table I lists various self-complementary sequences which crystallise as A, B or Z helices. The A-type helices commonly have C−C or G−G sequences at the 5′ terminus and pack with the base pairs of one double helix stacking against the shallow minor groove of an adjacent helix. The packing of B-type oligomers cannot be generalised since, until recently, only one example of a B-DNA helix was known; the dodecamer d(CGCGAATTCGCG) (31). In this crystal structure the molecules are staggered with each molecule overlapping by three base pairs with its neighbours above and below. We have just solved the structure of a hexamer d(Gp(S)-C-p-G-p(S)-C-p-G-p(S)-C) which contains the phosphorothioate group and crystallises in the B-conformation (32). In this structure the molecules form continuous helical stacks. A large number of Z-DNA (21) crystal structures have been analysed, nearly all as hexamers. Most of these structures are isomorphous and pack with hexamer duplexes stacked above each other along the c axis, forming infinite helices.

Numerous attempts to analyse sequences other than those listed in Table I failed

Table I
Deoxyoligonucleotide X-ray Structures with Watson-Crick base pairs

Sequence	Resolution Å	Space Group	R	Helix Type	Reference
d(GGBrUABrUACC)	1.8	$P6_1$	14.0	A	(25)
d(GGGGCCCC)	2.5	$P6_1$	14.0	A	(33)
d(GGCCGGCC)	2.25	$P4_32_12$	15.8	A	(55)
d(^{I}CCGG)	2.0	$P4_32_12$	20.5	A	(56)
d(pATAT)	1.04	$P2_1$	15.3	B	(57)
d(CGCGAATTCGCG)	1.90	$P2_12_12_1$	17.8	B	(31)
d(CGCGCG)	0.90	$P2_12_12_1$	13.0	Z	(21)
d(m^{5}CGm^{5}CGm^{5}CG)	1.3	$P2_12_12_1$	16.0	Z	(58)
d(m^{5}CGTAm^{5}CG)	1.2	$P2_12_12_1$	16.0	Z	(59)
d(Br^{5}CGATBr^{5}CG)	1.5	$P2_12_12_1$	19.3	Z	(60)
d(CGCG)	1.5	$P6_5$	19.0	Z	(61)
d(CGCG)	1.5	$C222_1$	19.9	Z	(62)

This table is not fully comprehensive. Only one example of the crystal structure for each sequence is referenced although in many cases several structure determinations of the same sequence measured at different temperatures, with different counterions etc. were published.

because either the fragments could not be crystallised or the crystals did not diffract X-rays, or they contained more than one molecule in the asymmetric unit and suitable heavy atom derivatives have not yet been obtained.

In view of this experience we undertook the synthesis, in the first instance, of some of the sequences listed in Table I, but with modifications of the sequences so as to allow the formation of mismatched base pairs. We hoped in this way to maximise the possibility of obtaining single crystals and also to facilitate structure solutions, in the event that the crystals proved to be isomorphous with already known structures.

This strategy of "crystal engineering" led to a higher rate of success in crystallisation. Depending on the nature of the mismatch and the position in which the mismatch was introduced we have, so far, succeeded in crystallising and solving structures with G•T mismatches in A, B and Z helices and a G•A mismatch in a B helix. Two of these structures were modifications of the A-octamer (33): d(GGGGCTCC) and d(GGGGTCCC); two were modifications of the B-dodecamer (31): d(CGCGAATTTGCG) and d(CGCGAATTAGCG) and one a modification of the Z-hexamer (21), d(TGCGCG).

Experimental

All the sequences were prepared by solid phase triester methods (34) scaled to yield about 15 mgs in one synthesis. The deoxyoligomers crystallised from buffered solutions containing various counterions, and in some cases alcohol and spermine. Details will be reported elsewhere. All X-ray measurements were carried out at 4°C using a Syntex $P2_1$ diffractometer and a conventional sealed X-ray tube. Cell constants and other data for the five crystal structures are listed in Table II. The measured intensities were corrected for absorption and for Lorentz and polarisation factors. The intensity data used for the analyses were generally derived from more than one independently measured set. With the exception of the Z-hexamer the compounds diffracted relatively poorly to a resolution of around 2 to 2.5Å. We are planning to extend the resolution using the synchrotron source.

Table II

Crystal Data for Deoxyoligonucleotides containing mismatched base pairs

Mis-match	Oligomer	Helix Type	Space Group	aÅ	bÅ	cÅ	Resolution Å	R Value
G•T	d(GGGGCTCC)	A	$P6_1$	45.20	45.20	42.97	2.2	14%
G•T	d(GGGGTCCC)	A	$P6_1$	44.71	44.71	42.36	2.1	14%
G•T	d(TGCGCG)	Z	$P2_12_12_1$	17.97	30.73	45.11	1.5	18%
G•T	d(CGCGAATTTGCG)	B	$P2_12_12_1$	25.53	41.22	65.36	2.4	24%
G•A	d(CGCGAATTAGCG)	B	$P2_12_12_1$	25.69	41.96	65.19	2.5	18%

Within each conformation type the DNA fragments crystallised isomorphously with the parent compound containing only Watson-Crick complementary bases. The same strategy of isomorphous replacement and refinement was followed in each analysis. A model of the appropriate double helix was constructed using either fibre coordinates (35) or coordinates from the parent compound but with the non-complementary bases substituted at the mismatch positions. The coordinates of the model, after best molecular fit to the parent structure, were the starting point for the refinement. For some of the structures the program ULTIMA (36) was used to position the model in the unit cell. The structure was refined by the constrained-restrained least squares program CORELS (37) beginning with low resolution data. The limit of resolution was gradually increased and more variables introduced until all data were included and the refinement converged. Throughout this procedure the mismatched bases were allowed to move relative to each other, with no hydrogen bond restraints. The other base pairs were restrained to standard Watson-Crick geometry. On convergence of the CORELS refinement the direct and difference electron density maps were calculated and plotted in the plane of the base pairs and the distances between the various atoms in the two bases were calculated. In each of the five structures the technique was powerful enough to shift the bases into positions where acceptable hydrogen bonds could be formed between them. The refinements were continued with the Hendrickson Konnert technique (38). During this refinement solvent molecules were located from difference maps and gradually included in the model. The calculations were terminated when the refinement converged and when an inspection of the difference maps indicated that the remaining peaks were too low to be assigned to specific solvent molecules. Any unusual features were examined during the course of the refinement by calculating fragment difference maps with the appropriate atomic groupings omitted from the structure factor calculation. Full details are being reported elsewhere.

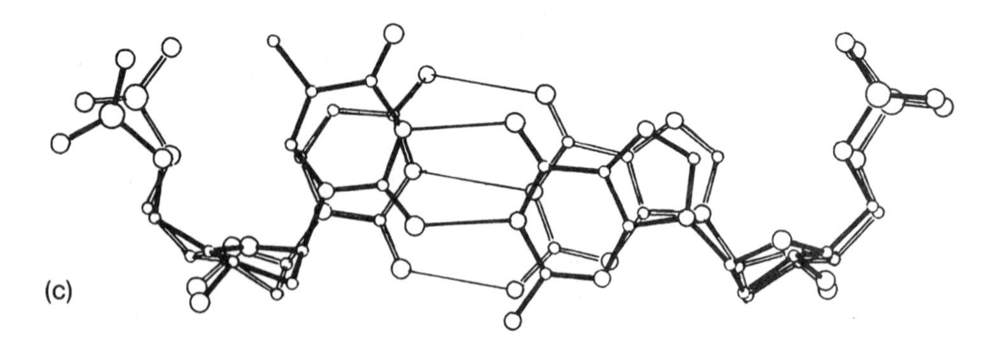

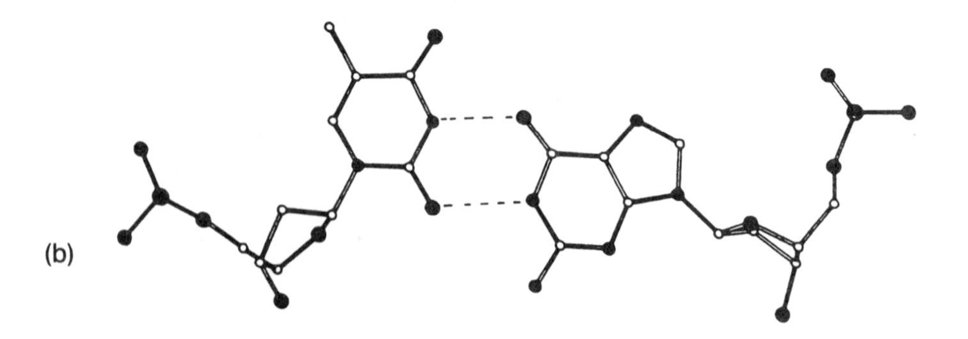

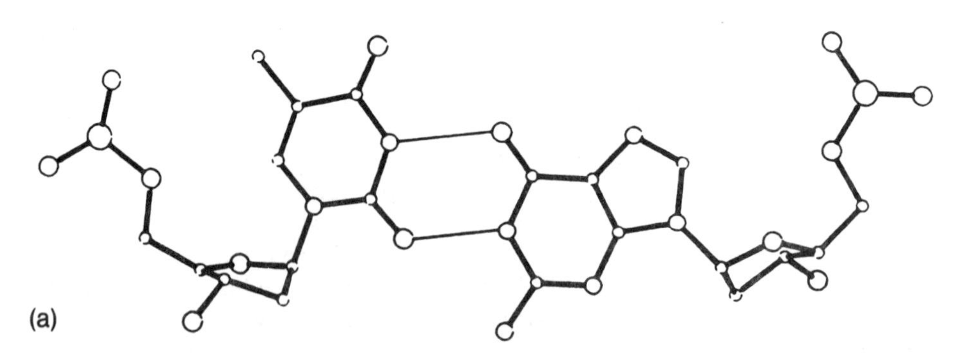

Figure 3. (a) The G•T wobble base pair observed in the crystal structure of the A-DNA octamer d(GGGGCTCC). (b) The G•T wobble base pair observed in the crystal structure of the A-DNA octamer d(GGGGTCCC). (c) Superposition of the G•T wobble base pairs at the two positions observed in the crystal structure of d(GGGGCTCC) on an ideal G•C pair. The figure was constructed by calculating the best molecular fit of the refined structures with a model of d(GGGGCCCC) (30) derived from A-DNA fibre coordinates.

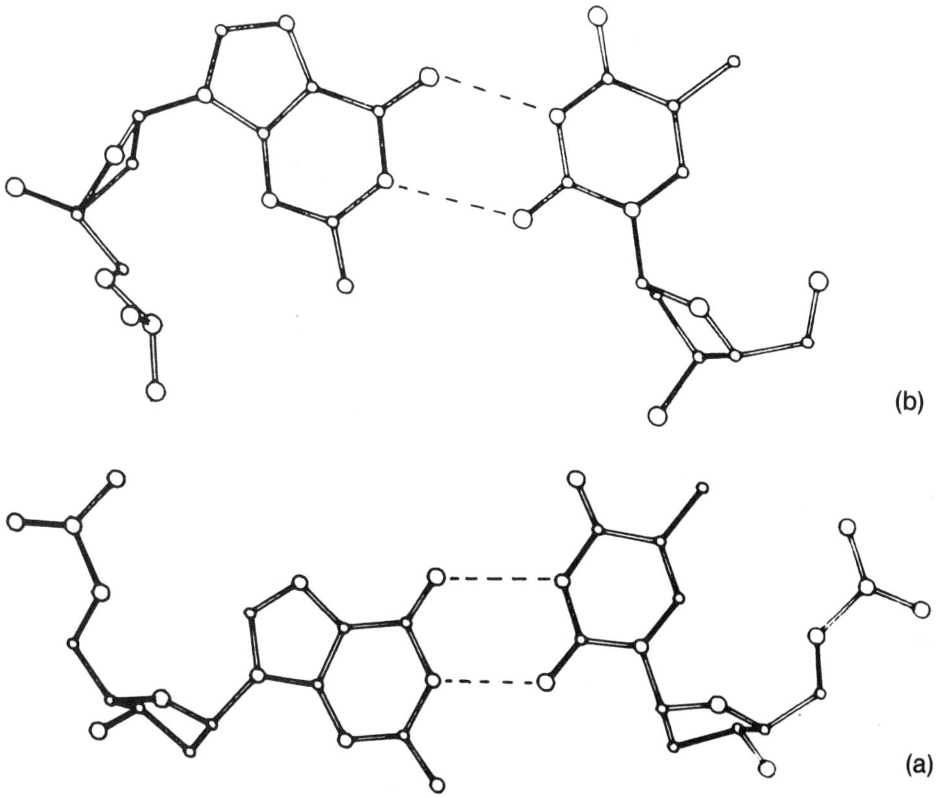

(b)

(a)

Figure 4. (a) The G•T wobble base pair in the B-DNA dodecamer d(CGCGAATTTGCG). (b) The G•T wobble base pair in the Z-DNA hexamer d(TGCGCG).

The G•T base pair in A, B and Z DNA helices

In all four structures containing G•T base pairs (Table II) the mismatched bases form wobble base pairs. These were established at the end of the CORELS refinement from molecular plots, electron density maps and from the hydrogen bonding geometry.

Fig. 3 illustrates the wobble base pairs as found in the two A-DNA octamers. The hydrogen bonding geometry is excellent with NH . . . O values in the range of 2.7 to 3.0Å. Individual values are given in Table III. These values were obtained from the fully refined structure at the convergence of the refinement. Fig. 4a illustrates the G•T mismatch in the B-DNA dodecamer at the end of the CORELS refinement and Fig. 4b the mismatch base pair in Z-DNA at the end of the CORELS refinement. In this diagram the *anti* and *syn* orientations of the nucleosides, characteristic of the left handed Z-DNA conformation, is clearly evident. In all four structures the asymmetric unit contains one duplex which has almost perfect two-fold molecular (non-crystallographic) symmetry, reflecting the symmetry of the base sequence. Only one illustration of a base pair mismatch is thus given for each structure, but two values are listed in Table III for the hydrogen bond distances. In all four

Table III
Hydrogen Bond Distances in G•T Wobble Base Pairs (Å)

	N1(G) .... O2(T)*	O6(G) .... N3(T)*
d(GGGGCTCC)	2.7 2.8	2.9 2.8
d(GGGGTCCC)	2.8 3.1	2.7 2.8
d(TGCGCG)	2.6 2.8	2.8 2.8
d(CGCGAATTTGCG)	2.8 3.0	2.6 2.8

*In each of the structures the bond distance values in the two independent wobble base pairs are listed.

structures there is a marked displacement of the two mismatched bases relative to standard Watson-Crick base pairs. In the two A-DNA structures the thymines move into the major groove and guanines towards the minor groove. The magnitude of the movements in the two structures is slightly different. The displacements of the bases from the standard Watson-Crick base pairs lead to an asymmetry in the disposition of the glycosidic bonds. The angles between these bonds and the C1′ ... C1′ vectors, which are symmetrical in G•C base pairs (54°-58°) (39) become about 70° for T and 40° for G, averaged over the four G•T mismatches in the two A-DNA structures. There are similar movements and consequent asymmetry in the mismatch base pairs of the B-dodecamer and the Z-hexamer. The C1′ ... C1′ distance is in the range of 10.2 to 10.6Å in the four structures.

The observations that DNA fragments containing the G•T wobble base pairs can be crystallised isomorphously with parent fragments containing only Watson-Crick complementary base pairs is a strong indication that this type of mismatch can be accommodated in the DNA double helix without appreciably affecting the global conformation, be this of the A B or Z type. That this is indeed the case is confirmed by the close agreement between the average helical parameters and torsion angles for the mismatch structures and the appropriate parent compounds (Table IV). The isomorphism also implies close similarity in the packing of the molecules in the crystal structures which are virtually indistinguishable, except for details of the solvent structures, from the parent compounds.

The G•A mismatch in B-DNA

A number of different deoxyoligonucleotides with potential for G•A base pairing were synthesised. These included modifications of the A, B and Z type sequences and of the decamer d(CCAAGCTTGG), studied by Kan and co-workers (28). The first of the sequences which yielded sufficiently large crystals is the dodecamer d(CGCGAATTAGCG).

The crystals were isomorphous with the B-dodecamer d(CGCGAATTCGCG) (31). A model of this sequence was prepared using DNA fibre coordinates (35). Adenine was substituted for cytosine with a best molecular fit of the sugar and phosphate groups. Both bases were in the *anti* orientation as found in various tRNA structures (11-15) and in the NMR studies of deoxyoligonucleotides (26-28). The coordinates of the model, after best molecular fit to the parent dodecamer (using fractional coordinates from a refinement by Westhoff et al. (40)), were the starting point for the rigid body refinement with program SHELX (41). This was followed by the CORELS procedure as described earlier. Throughout these calculations the two purine bases were allowed to move relative to each other and no constraints were placed on the hydrogen bonding geometry between them.

The model fitted the difference electron density maps at the end of the CORELS refinement and converged using the Hendrickson Konnert method to R=20% with 50 solvent molecules located. However, an examination of the Fourier difference maps at the end of this refinement indicated electron density in the region of the *syn* position of the adenine. A new model was constructed with the mismatched base pairs in the G(*anti*)•A(*syn*) orientation and the analysis repeated. This model converged at R=17.8% (72 solvents located). There was no indication in the final $2F_o-F_c$ or F_o-F_c maps of any residual electron density which could be assigned to the adenine in the *anti* orientation. We conclude therefore that in the present structure only G•A(*syn*) mispairs are found, in contrast to the tRNA structures (11-15), and the NMR solution studies (26-28).

An examination of $2F_o-F_c$ and F_o-F_c difference maps, using an Evans and Sutherland PS300 colour graphics system showed good agreement between the electron densities of the base, sugar and phosphate groups throughout the entire molecule. A view of the G•A syn base pair in the B-DNA helix is shown in Fig. 5. The C1′ ... C1′ distance at the current stage of refinement is 11.07Å in one G•A pair and 10.52Å in the other—the average of 10.8Å differing by about 0.5Å from the corresponding distance in the G•T mismatches and the parent B-dodecamer (10.3Å av.). It is of interest to note that the average C1′ ... C1′ distance in the G(*anti*)•A(*anti*) base pair at the end of the Hendrickson Konnert refinement was 11.6Å.

Base Stacking

One of the principal differences between DNA duplexes containing mismatched base pairs and those with only Watson-Crick base pairs is at the level of the local environment, particularly stacking interactions between adjacent bases. Such interactions have a major influence on the conformational stability of double stranded DNA and have been shown to give rise to sequence dependent conformational effects. The stacking interactions most affected are those in the vicinity of the mismatched base pairs.

In four of the structures investigated we have examples of the same G•T wobble base pair in different environments of base sequence and helix conformation. Fig. 6 compares the base stacking on either side of the mismatch in the two A-DNA structures with the stacking at the same site in the parent d(GGGGCCCC) helix. In d(GGGGCTCC) (Fig. 6a) we observe a marked improvement in the stacking of the pyrimidines on the 5′ side of the thymine at the C−T(=G−G) step while on the 3′ side, at the T−C (=G−G) step, there is a destacking of pyrimidines, characteristic

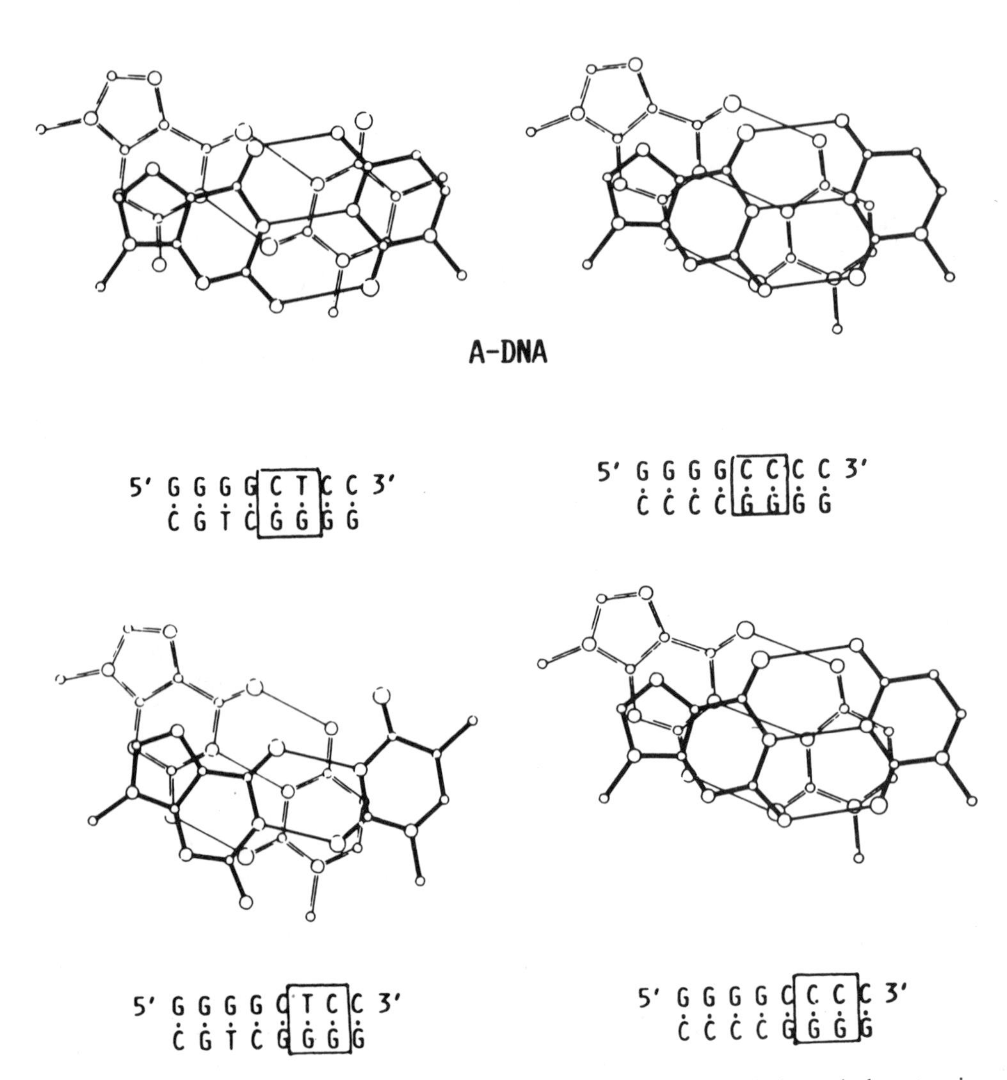

Figure 6. Comparison of base stacking in the A-DNA mismatch octamers with the equivalent steps in d(GGGGCCCC) (30) (a) d(GGGGCTCC) (b) d(GGGGTCCC).

of A-type helices. The purine-purine stacking on either side appears unaffected. In the second A-type mismatch d(GGGGTCCC) (fig. 6b) the two wobble base pairs are adjacent to each other. In this case there is purine−pyrimidine (=purine−pyrimidine) stacking on the 5′ side of T at the G−T (=G−T) step which is excellent and a pyrimidine−pyrimidine (=purine−purine) step at the T−C (=G−G) step on the 3′ side of T which is poor, as in the parent compound and in fibre models.

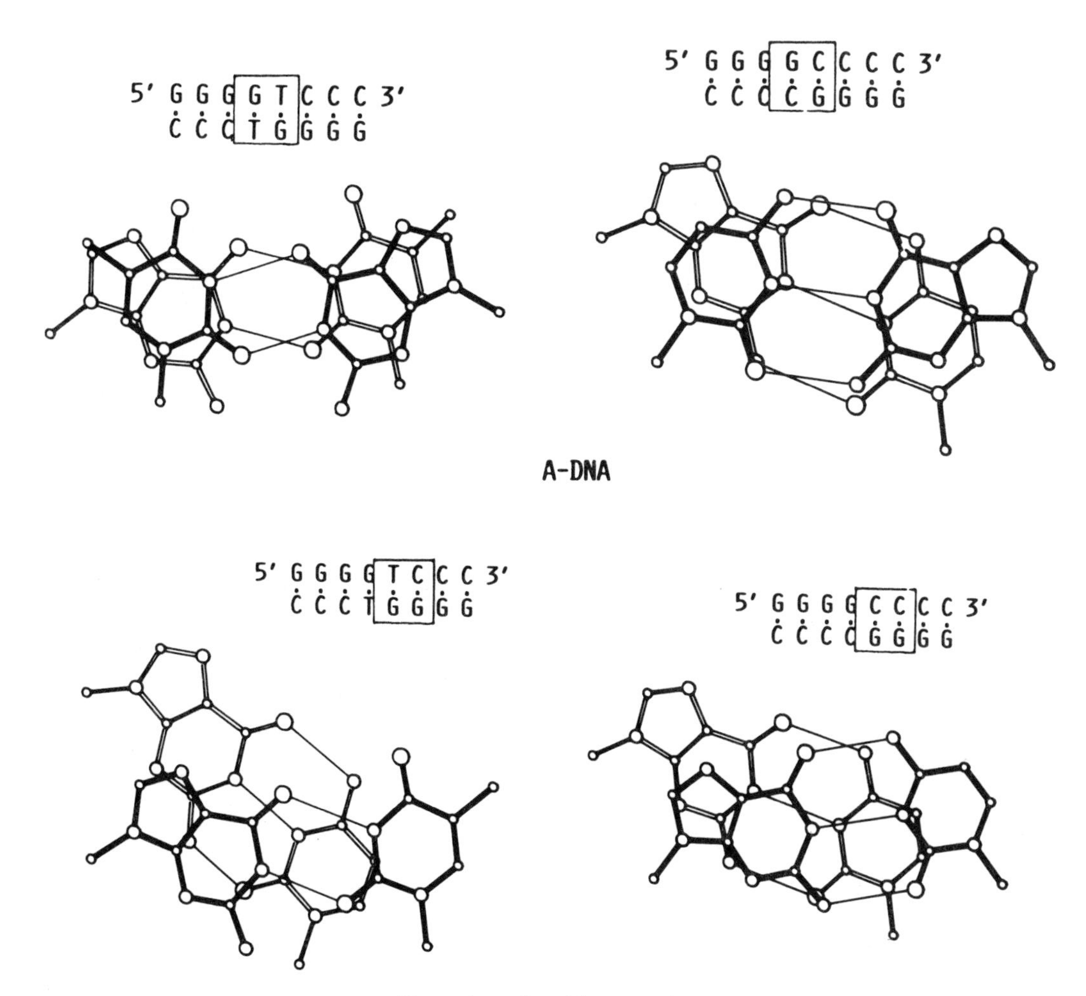

Figure 6 continued from previous page.

The correlation between base sequence and base stacking in B-DNA has been explored extensively by Dickerson and co-workers (42). A comparison of the stacking patterns in the parent dodecamer d(CGCGAATTCGCG) and in the dodecamer d(CGCGAATTTGCG) is shown in Fig. 7. There is improvement in the overall base stacking, particularly on the 5′ side of the mismatched thymines.

In the A and B-DNA structures the mismatched bases are sited in the interior of the double helix whereas in the Z-DNA structure, d(TGCGCG), the mismatch is at the terminal position. In this crystal structure the hexamer duplexes are stacked one atop the other forming continuous helices. Consequently base stacking within the helix occurs only on the 3′ side of T at the T−G (=C−G) step while on the 5′ side of T there is intermolecular stacking between adjacent purine−pyrimidine (=purine−pyrimidine) bases. As can be seen from Fig. 8 the intermolecular stacking

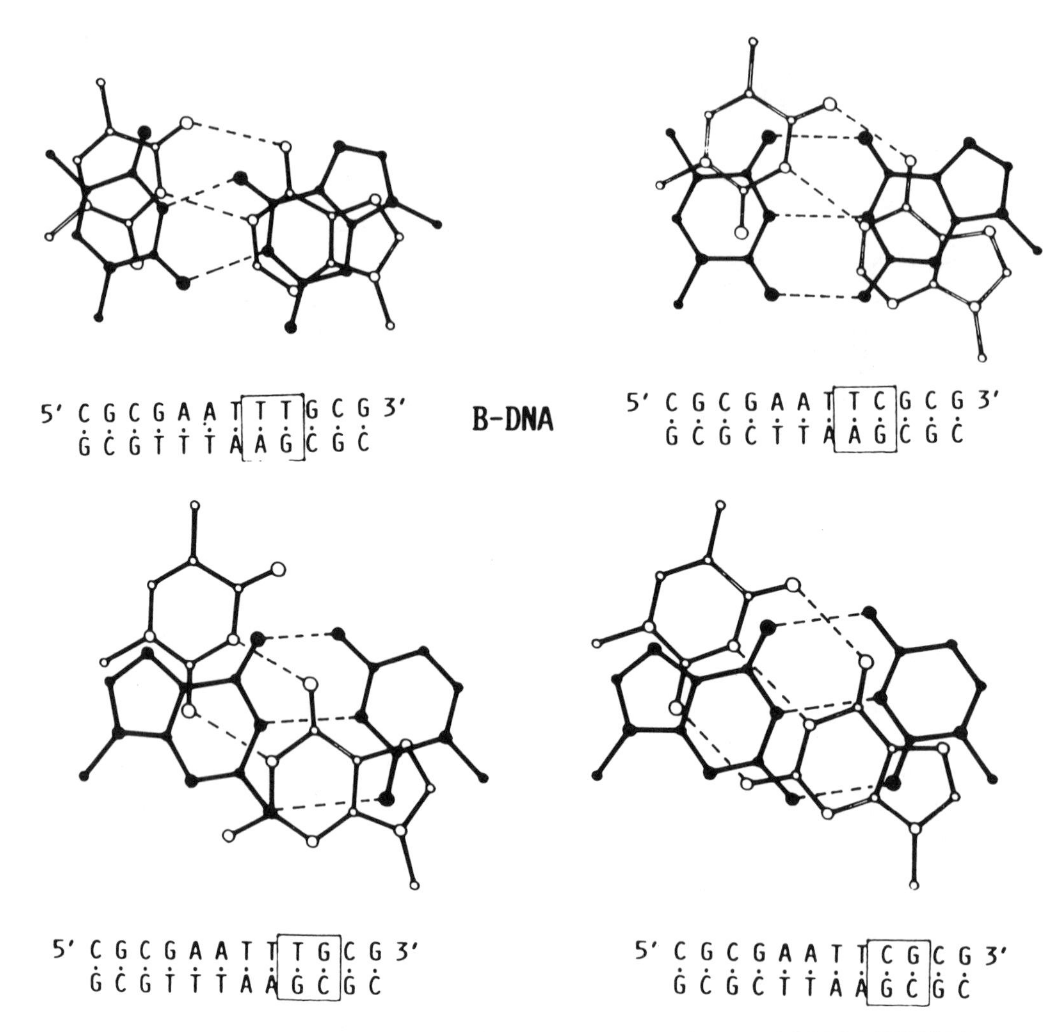

Figure 7. Comparison of base stacking in the B-DNA mismatch dodecamer d(CGCGAATTTGCG) with the equivalent steps in d(CGGGAATTCGCG) (31).

is better than in the parent compound as a result of the mutual rotation of the mismatched base pairs. In the intra helical step the stacking of pyrimidine on purine is less good than in the parent compound. The stacking of bases in the G•A mismatch is currently being analysed.

Solvent structure

DNA is a polymorphic compound with a number of stable conformations. The particular conformation adopted is dependent on the sequence and on environmental conditions such as the nature and concentration of counterions, the presence or absence of alcohol and, in fibres, the degree of relative humidity.

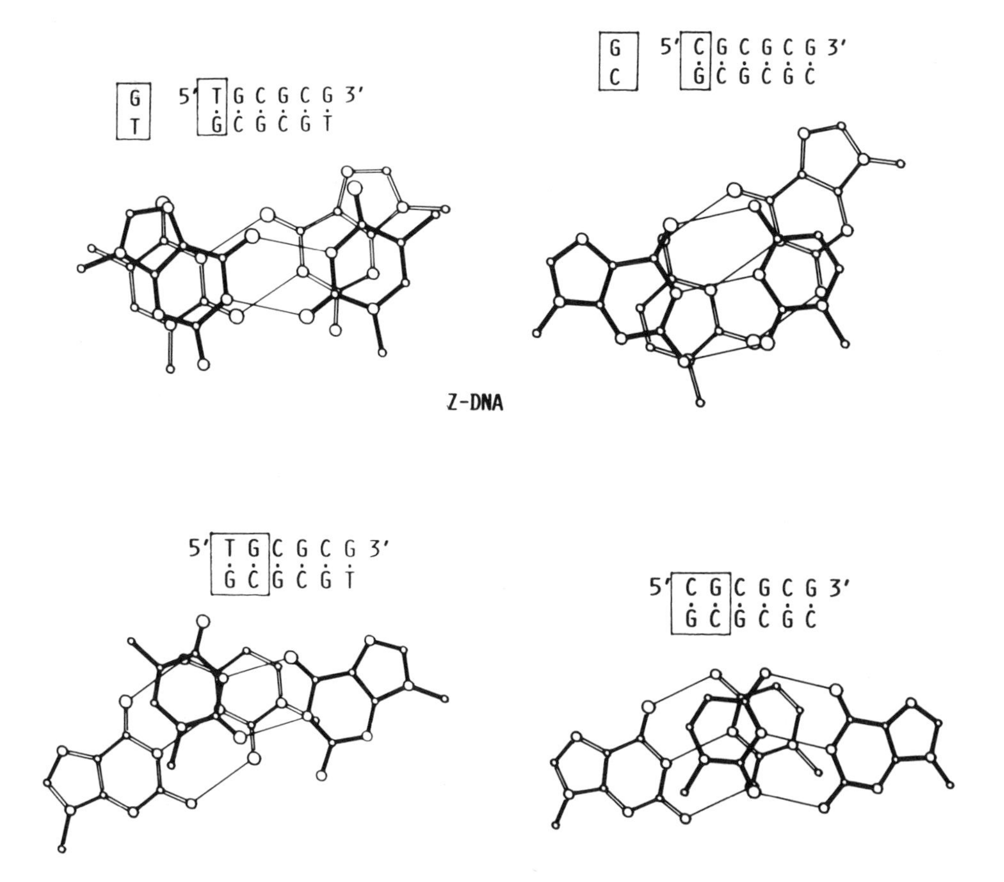

Figure 8. Comparison of base stacking in the Z-DNA mismatch hexamer d(TGCGCG) with the equivalent step in d(CGCGCG) (21).

In crystal structures the correlation between conformation and environment is less evident. All the deoxyoligonucleotides so far analysed contain about 50% solvent (including water), counterions in various concentrations, possibly spermine, alcohol or other compounds used in the crystallisation mixture. All structures are highly hydrated, a typical illustration being the parent A-DNA octamer d(GGGGCCCC) (33) (Fig. 9), where the asymmetric unit contains some 300 water molecules (or ions) for each DNA duplex molecule. Over 100 of these were sufficiently ordered to be located by the X-ray analysis.

The potential number of solvent molecules which can be located in a particular analysis is determined by the degree to which the solvent (and the DNA) are ordered in the crystal structure. This ordering may be related to the difference between the temperature at which the intensities were measured and the melting point of the crystal, as well as other factors. In the more stable octamer d(GGGGTCCC) 104 waters were located whereas in the less stable octamer d(GGGGCTCC), which was refined to the same R factor at the same resolution, only 56 ordered solvents

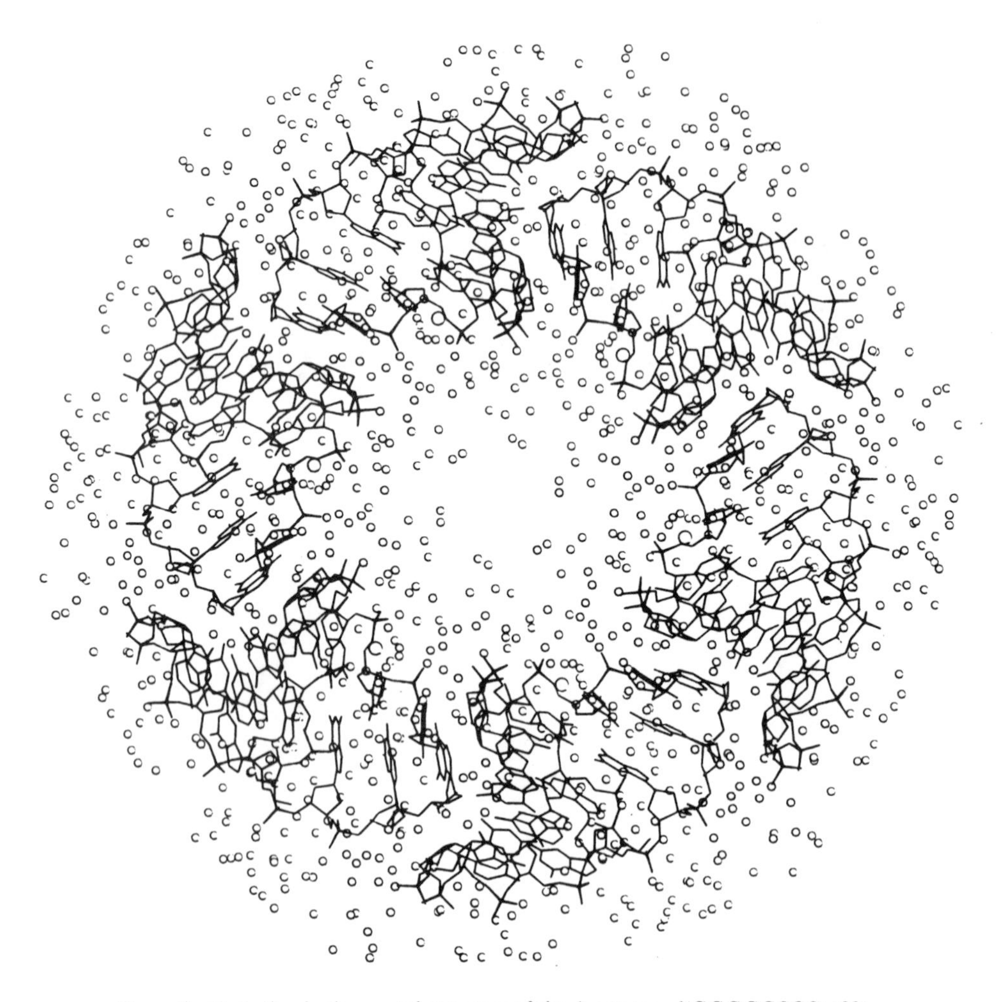

Figure 9. Hydration in the crystal structure of the A-octamer d(GGGGCCCC) (33).

were found. Intensities for both structures were measured at the same temperature of 4°C.

Oligomers which crystallise in the Z-conformation generally have the most ordered solvent structures. In the Z-mismatch hexamer which diffracts to 1.5Å, at the present stage of refinement 88 solvent molecules have been located and there are good indications of the presence of a spermine molecule and identifiable counter ions.

The first hydration shell is an integral part of the DNA double helix and contributes to its stabilisation in crystal structures. A highly ordered solvent structure was found in the parent B-dodecamer, as a spine of hydration linking the bases in the minor groove (43). In the A-octamer d(GGTATACC) a pentagonal network of water molecules fills the centre of the major groove (44) as illustrated in Fig. 10.

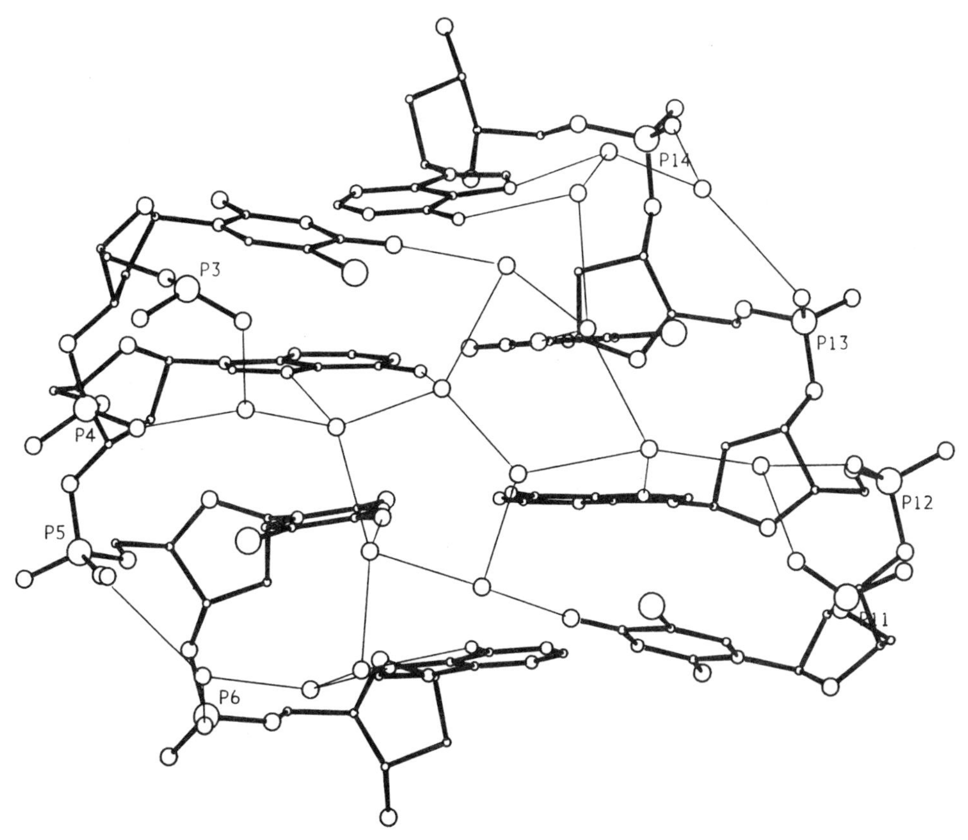

Figure 10. Network of solvent molecules in central TATA region of the A octamer d(GGTATACC) (25) showing the pentagonal arrangement of the water molecules linking the functional groups of the bases and the phosphate O1 atoms.

In the context of the present work we are particularly concerned with water molecules associated with mismatched base pairs. In the G•T wobble pair the keto O4 atom of thymine, which has moved into the major groove, and the N2 amino group of guanine in the minor groove are each free to take part in two hydrogen bonded interactions.

In the two A-octamer structures water molecules were indeed found in close association with the mismatched bases. In d(GGGGTCCC) a water oxygen is bridging the keto O atoms of thymine and guanine on the major groove side, and another water molecule is found in the plane of the bases on the minor groove side, hydrogen bonded to the N2 amino group of guanine and the keto O2 atom of thymine. A second water molecule is bound to the N2 amino group of guanine. This pattern is observed for both mismatched bases. The water molecules in this region are well defined and have some of the lowest temperature factors for solvents in the whole structure, indicating close association with the functional groups of the mismatched bases. In the second A-octamer, d(GGGGCTCC), similar water molecules were found in the plane of the G•T base pairs. Water molecules bound to the mis-

Table IVa

Average Helical Parameter for A-DNA Structures

	twist/ b.p. (°)	b.p./ turn	rise/ b.p. (Å)	base tile (°)	Propellor twist (°)
d(GGGGCCCC)	32.5	11.1	2.9	12	10
d(GGTATACC)	32.2	11.1	2.9	13	15
d(GGGGCTCC)	31.9	11.3	2.8	12	13
d(GGGGTCCC)	32.8	11.0	2.9	12	12

Average Torsional Angles for A-DNA Structures

	alpha	beta	gamma	delta	epsilon	zeta	chi
d(GGGGCCCC)	−76	178	62	84	−156	−70	−158
d(GGTATACC)	−62	173	52	88	−152	−78	−160
d(GGGGCTCC)	−78	171	68	82	−154	−68	−160
d(GGGGTCCC)	−67	172	59	81	−150	−74	−161

Table IVb

Averaged Helical Parameters for B-DNA Structures

	twist/ b.p. (°)	b.p./ turn	rise/ b.p. (Å)	base tilt (°)	Propellor twist (°)
d(CGCGAATTCGCG)	38.8	9.3	3.2	1	13
d(CGCGAATTTGCG)	35.5	10.2	3.3	1	13

Average Torsional Angles (0°) for B-DNA Structures)

	alpha	beta	gamma	delta	epsilon	zeta	chi
d(CGCGAATTCGCG)	63	164	59	123	−170	−99	−117
d(CGCGAATTTGCG)	−29	162	24	150	180	−134	−98

Table IVc

Averaged Helical Parameters for Z-DNA Structures

	twist dimer (°)	dimer turn	rise dimer (Å)	base tilt (°)	Propellor twist (°)
d(CGCGCG)	60	6	7.6	−6.2	4.6
d(TGCGCG)	61.4	5.9	7.6	−8.2	5.4

Average Torsional Angles for Z-DNA Structures

	alpha	beta	gamma	delta	epsilon	zeta	chi
d(CGCGCG)	−137 48	−139 179	56 −170	138 100	−94 −104	80 −69	−159 72
d(TGCGCG)	−149 71	−138 177	65 179	138 101	−92 −114	74 −56	−151 65

matched base pairs were found in the Z-hexamer, and in the B-dodecamer d(CGCGAATTAGCG), where the spine of hydration, in the minor groove, is also evident. Full details are being reported elsewhere.

Discussion

The five X-ray analyses discussed in this paper establish unequivocally that base pairs in major tautomeric form can substitute for standard Watson-Crick base pairs in DNA helices without appreciable perturbation of the global conformations. It appears from the present investigations that the sugar-phosphate backbone is sufficiently flexible to accommodate either the G•T wobble base pair or the relatively large purine-purine base pair, with G in the *anti* and A in the *syn* orientation. These base pairs thus appear to be good models for mismatched base pairs where, as in the experiments reported by Fersht and co-workers, a mismatched base is incorporated in the double helix during replication in the major tautomer form (18).

Other studies by Fersht and co-workers suggest that *in vitro* misinsertion frequencies vary with the nature of the template and may be related to base stacking effects (45). Base stacking may also be important for mutational hotspots where certain mismatches occur with greatly increased frequency. The recognition and deletion of mismatches is influenced by neighbouring base sequences (4,46-48) and here again local structural changes may be pertinent factors. The investigations summarised in this paper give a qualitative indication of changes in base stacking and show a trend towards improved stacking on the 5′ side of thymine in the various structures containing G•T base pairs. We intend to extend the X-ray studies to other structures where the mismatches are substituted at different positions in the same DNA fragment. We plan to use the structural parameters from all these X-ray analyses to calculate the stacking interactions at the various steps in mismatched oligomers and compare them with the equivalent steps in the parent compounds. We hope that such calculations might lead to a more quantitative assessment of the changes related to base sequences in the neighbourhood of mismatches, and to correlate these with those observed in the biological experiments.

The results of the X-ray studies may also be pertinent to an understanding of the way base pair mismatches are recognised during proof-reading or by post replicative repair systems and may throw some light on the difference in the efficiency of repairing the various mismatched base pairs (4,46-48).

Hydrogen bonded interactions are closely involved in protein-nucleic acid recognition and the disposition of the donor and acceptor groups of the bases may be important in discriminating between different sequences. The normal Watson-Crick base pairs are approximately isosteric. The sugars of nucleosides are related by a pseudo two fold symmetry axis in the plane of the base pairs. The two acceptor groups in the minor groove, N3 (of the purines) and O2 (of the pyrimidines), are in symmetrically related positions and approximately equivalent. The disposition of the donor-acceptor groups in the major groove, on the other hand, is characteristic of the particular base pair A•T, T•A, G•C or C•G and can be distinguished by directed hydrogen bonds. Rich and co-workers, in their study of the recognition of base sequences (49), have concluded that two hydrogen bonds would be sufficient for discriminating between these base pairs.

In the purine-pyrimidine wobble base pair G•T the symmetry about the glycosidic bond, characteristic of Watson-Crick base pairs, is lost and the functional groups of the bases are no longer disposed symmetrically. In addition the O4 oxygen atom of the thymine protrudes substantially into the major and the N2 amino group of guanine into the minor groove. The latter has two, instead of one hydrogen available for bonding. It may be significant that in all the G•T mismatch structures well ordered solvent molecules, with relatively low temperature factors, were found, bonded to these functional groups. Such an arrangement might well result in steric hindrance at the active site and may be an important feature in discriminating mismatched base pairs from Watson-Crick base pairs.

The difference between a G•A(*syn*) pair and normal Watson-Crick pairs is even more marked. The G•A(*syn*) base pair has four donor/acceptor groups instead of three only on the major groove side and lacks the keto O2 atom, which is a common feature of all purine-pyrimidine base pairs, in the minor groove. Such differences could readily be detected by repair enzymes. Crystallographic information about protein-DNA interactions is becoming available through several studies of DNA binding proteins (50-53) and their complexes with deoxyoligonucleotides (54). In the latter hydrogen bonded interactions have already been observed between the amino acid side chains and the functional groups of the bases, in line with our suggestion of enzyme recognition of base pair mismatches.

The technique of modifying some of the base sequences in DNA fragments known to crystallise in well defined packing patterns discussed in this paper is proving to be a practical way of investigating errors in DNA and we are now extending the studies to other mismatches and to the deoxyoligomers with various modified bases. Such studies will, we hope, help towards a better understanding of some aspects of mutagenesis and carcinogenesis at a structural level.

Acknowledgements

The work surveyed in this paper was carried out by members of the Crystallographic Chemistry Group at the University Chemical Laboratory, Cambridge: Drs. Tom Brown, Bill Hunter, Geoff Kneale, and Mr. Naveen Anand and by Prof. D. Rabinovich of the Weizmann Institute of Science, Rehovot, Israel. Financial support was provided by the Medical Research Council, EMBL (Visiting Fellowship to D.R) and the Royal Society Guest Research Fellowship. I thank other members of the Group: Drs. Bill Cruse, Maxine McCall, Steve Salisbury and Mr. Joseph Nachman for numerous discussions and help with technical problems, and Mrs. Susan Berry for the preparation of the manuscript. I am grateful to the Head of Department, Professor R.A. Raphael FRS, for his advice and support, the MRC Laboratory for

Molecular Biology for the use of their graphics system, and Mr. Neil Cray for the color photographs.

References and Footnotes

1. Radding, C.M., *A. Rev. Biochem. 47,* 847-880 (1978).
2. Hotchkiss, R.D. *A. Rev. Microbiol. 28,* 445 (1974).
3. Benzev, S. *Proc. Natl. Acad. Sci. USA 47,* 403-416 (1961).
4. Clavery, S.J.P., Mejan, V., Gasc, A.M., & Siccard, A.M. *Proc. Natl. Acad. Sci. USA 80,* 5956-5960 (1983).
5. Watson, J.D. & Crick, F.H.C. *Nature 171,* 964-967 (1953).
6. Topal, M.D. & Fresco, J.R. *Nature 263,* 285-293 (1976).
7. Crick, F.H.C. *J. Molec. Biol. 19,* 548-555 (1976).
8. Churprina, V.P. & Poltev, V.I. *Nucleic Acids Res. 11,* 5205-5222 (1983).
9. Poltev, V.I. & Bruskov, V.I. *J. Theor. Biol. 70,* 69-83 (1978).
10. Rein, R., Shibata, M., Gardino-Juarez, R. & Keiber-Emmons, T. in *Structure and Dynamics of Nucleic Acids and Proteins* (eds. Clementi, E. & Sarma, R.) 269-288, Adenine Press, New York 1983).
11. Ladner, J.E., Jack, A., Robertus, J.D., Brown, R.S., Rhodes, D., Clark, B.F.C. & Klug, A. *Nucleic Acids Res. 2,* 1629-1637 (1975).
12. Quigley, G.J., Seeman, N.C., Wang, A.H.-J., Suddath, F.L. & Rich, A. *Nucleic Acids Res. 2,* 2329-2335, (1975).
13. Sussmann, J.L. & Kim, S.H. *Biochem. Biophys. Res. Comm. 68,* 89-96 (1976).
14. Stout, C.D., Mizuno, H., Rubin, J., Brennan, T., Rao, S.T. & Sundaralingam, M. *Nucleic Acids Res. 3.,* 1111-1123 (1976).
15. Mizuno, H. & Sundarlingam, M. *Nucleic Acids Res. 5,* 4451-4461 (1978).
16. Traub, W. & Sussman, J.L. *Nucleic Acids Res. 10,* 2701 (1982).
17. Early, T.A., Olmstead III, J., Kearns, D.R. & Lezius, A.G. *Nucleic Acids Res. 5,* 1955-1965 (1978).
18. Fersht, A.R., Shi, J.-P. & Tsui, W.-C. *J. Molec. Biol. 165,* 655-667 (1983).
19. Stravinsky, J., Hozumi, T., Narang, S.B., Bahl, C.P., & Wu, R. *Nucleic Acids Res. 4,* 353-371 (1981).
20. Beaucage, S.L., Caruthers, M.H. *Tet. Lett. 22.,* 1859-1862 (1981).
21. Wang, A.H.J., Quigley, G.J. Kolpack, F.J., Crawford, L., van Boom, J.H., van der Marel, & G., Rich, A. *Nature (London) 282,* 680-686 (1979).
22. Calladine, C.R. *J. Molec. Biol. 161,* 343-352 (1982).
23. Dickerson, R.E. *J. Molec. Biol. 166,* 419-441 (1983).
24. Shakked, Z. & Kennard, O. in *Biological Macromolecules and Assemblies,* Volume 2 (eds. McPherson, A. & Jurnak, F.) 1-36, John Wiley & Sons, New York (1985).
25. Shakked, Z., Rabinovich, D., Kennard, O., Cruse, W.B.T., Salisbury, S.A. & Viswamitra, M.A. *J. Molec. Biol. 166,* 183-201 (1983).
26. Patel, D.J., Kozlowski, S.A., Marky, L.A., Rice, J.A., Broka, C-., Dallas, J., Itakura, K. & Breslauer, K.J. *Biochemistry 21,* 437-444 (1982).
27. Patel, D.J., Kozlowski, S.A., Ikuta, S & Itakura, K. *Fedn. Proc. 43,* 2663-2670 (1984).
28. Kan, L.S., Chandrasegaran, S., Pulford, S.M., Miller, P.S., *Proc. Natl. Acad. Sci. U.S.A. 80,* 4263-4265 (1983).
29. Tibanyenda, N., De Bruin, S.H., Haasnoot, C.A., Van der Marel, G.A., van Boom, J.H., Hilbers, C.W., *Eur. J. Biochem. 139,* 19-25 (1983).
30. Brown T., Kennard, O., Kneale, G., Rabinovich, D., *Nature 315,* 604-606 (1985).
31. Wing, R., Drew, H., Takano, T., Broka, C., Tanaka, S., Itakura, K. & Dickerson, R.E. *Nature 287,* 755-758 (1980).
32. Cruse, W.B.T. et al. to be published (1985).
33. McCall M., Brown, T., & Kennard, O. *J. Mol. Biol. 183,* 385-396 (1985).
34. Gait, M.J., Matthes, H.W.D., Singh, M., Sproat, B.S. & Titmus, R.C. *Nucleic Acids Res. 10,* 6243-6248 (1982).
35. Arnott, S. & Hukins, D.W.I. *Biochem. Biophys. Res. Comm. 47,* 1504-1509 (1972).
36. Rabinovich, D., Shakked, Z., *Acta Cryst. A40,* 195-200 (1984).
37. Sussman, J.L., Holbrook, S.R., Church, G.M. & Kim, S.-H *Acta Cryst. 33,* 800-804 (1977).

38. Hendrickson, W.A. & Konnert, J.H. in *Biomolecular Structure, Conformation, Function and Evolution,* Vol. I (ed. Srinivasan, R.) 43-57, Pergamon, Oxford (1981).
39. Saenger, W. *Principles of Nucleic Acid Structure,* Springer Verlag, New York (1984).
40. Westhoff, E., et al. private communication (1985).
41. Sheldrick, G.M., SHELX 76 System of Computing Programs University of Cambridge, England (1976).
42. Dickerson, R.E., Kopka, M.L., Drew, H.R., *Conformation in Biology* (eds. Srinivasan, R. and Sarma, R.H.), Adenine Press 227-257 (1983).
43. Fratini, A.V., Kopka, M.L., Drew, H.R. & Dickerson, R.E. *J. Biol. Chem. 257,* 14686-14707 (1982).
44. Kennard, O., *Pure and Appl. Chem. 56,* 989-1004 (1984).
45. Fersht, A.R., Knill-Jones, J.W. & Tsui, W.C. *J. Molec. Biol. 156,* 37-51 (1982).
46. Benzer, S., Freeze, E., *Proc. Natl. Acad. Sci. U.S.A. 44,* 112-119 (1958).
47. Dohet, C., Wagner, R and Radman, M. *Proc. Natl. Acad. Sci. U.S.A. 82,* 503-505 (1985).
48. Lieb, M. *Mol. Gen. Genet. 191,* 118-125 (1983).
49. Seeman, N.C., Rosenberg, J.M., Rich, A. *Proc. Natl. Acad. Sci. U.S.A. 73,* 804-808 (1976).
50. Anderson, W.F., Ohlendorf, D.H., Takeda, Y., Mathews, B.W. *Nature 290,* 754-758 (1981).
51. Richmond, T.J., Finch, J.T., Rushton, B., Rhodes, D., Klug, A. *Nature 311,* 532-537 (1984).
52. Suck, D., Oefner, C., Kabsch, W. *EMBO Journal 3,* 2423-2430 (1984).
53. Ollis, D.L., Brick, P., Hamlin, R., Xuong, N.G., Steitz, T.A. *Nature 313,* 762-766 (1985).
54. Fredrick, C.A., Grable, J. Melia, M., Samudzi, C., Jen-Jacobson, L., Wang, B.-C., Greene, P., Boyer, H.W. & Rosenberg, J.M. *Nature 309,* 327-331 (1984).
55. Wang, A.H.J., Fujii, S., van Boom, J. and Rich, A. *Proc. Natl. Acad. Sci. U.S.A., 79,* 3968-1972 (1982).
56. Connert, B.N., Takano, T., Tanaka, S., Itakura, K. and Dickerson R.E., *Nature, 295,* 294-299, (1982).
57. Viswamitra, M.A., Shakked, Z., Jones, P.G., Sheldrick, G.M., Salisbury, S.A. and Kennard, O. *Biopolymers, 21,* 513-532 (1982).
58. Fujii, S., Wang, A.H.-J., van der Marel, G., van Boom, J.H. and Rich, A. *Nucleic Acids Res., 10,* 7879-7892 (1982).
59. Wang, A.H.-J., Hakoshima, T. van der Marel, G., van Boom, J.H. and Rich A. *Cell 36,* 321-331 (1984).
60. Wang, A.H.-J., Gessner, R.V., van der Marel, G.A., van Boom, J.H. and Rich, A. *Proc. Natl. Acad. Sci. U.S.A., 82,* 3611-3615 (1985).
61. Kolpak, F.J., Wang, A.H.-J., Quigley, G.J., van Boom, J.H., van der Marel, G., Rich, A. *Proc. Natl. Acad. Sci., U.S.A., 77* 4016-4020 (1980).
62. Drew, H., Takano, T., Tanaka, S., Itakura, K. and Dickerson, R.E. *Nature, 286,* 567-573 (1980).
63. Reprinted from the *Journal of Biomolecular Structure* & Dynamics 3, 205-226 (1985).

Biomolecular Stereodynamics IV, Proceedings of the Fourth Conversation in the Discipline Biomolecular Stereodynamics, State University of New York, Albany, NY, June 04-09, 1985, Eds., Ramaswamy H. Sarma & Mukti H. Sarma, ISBN 0-940030-18-7, Adenine Press, ©Adenine Press 1986.

Determination of the Three-Dimensional Structures of Oligonucleotides in Solution: Combined Use of Nuclear Overhauser Enhancement Measurements and Restrained Least Squares Refinement

G. Marius Clore and Angela M. Gronenborn
Max-Planck-Institut für Biochemie
D-8033 Martinsried bei München
West Germany

Abstract

Obtaining the three-dimensional structures of proteins and oligonucleotides in solution involves three stages: (i) assignment of proton resonances, (ii) determination of interproton distances from nuclear Overhauser enhancement measurements, and (iii) determination of the three-dimensional structure of the macromolecule on the basis of the interproton distance data. The approach we have chosen for the third stage in the case of oligonucleotides involves restrained least squares refinement of an initial model structure based solely on distance and planarity restraints. The application of this approach is illustrated by the refinement of the solution structures of two DNA oligonucleotides, 5′d(CGTACG)$_2$ and 5′d(AAGTGTGACAT)•5′d(ATGTCACACTT), the latter comprising the specific target site of the cAMP receptor protein in the *gal* operon, and one RNA oligonucleotide 5′r(CACAG)•5′r(CUGUG) comprising the stem of the TψC loop of yeast tRNA^{Phe}. In each case the RMS difference between the calculated and experimental interproton distances is $\lesssim$ 0.2 A. Although the refined structures of the two DNA oligonucleotides have an overall B-type conformation and that of the RNA oligonucleotide an A-type conformation, there are large variations in many of the local conformational parameters including backbone and glycosidic bond torsion angles, and helical and propellor twist angles.

Introduction

It has long been a goal of NMR spectroscopists to obtain three-dimensional structures of small proteins and oligonucleotides at a resolution comparable to X-ray crystallography. This involves three stages: (i) assignment of proton resonances, (ii) determination of interproton distances from nuclear Overhauser enhancement (NOE) measurements, and (iii) determination of the three-dimensional structure of the macromolecule on the basis of the interproton distance data. It is only over the past few years, however, with the advent of very high field (500 MHz) NMR spectrometers, and new experimental (viz. the whole array of two-dimensional NMR experiments) as well as computational techniques that this has become feasible. At the present time there are four examples of proteins (1-4) and three of oligonucleotides (5-7) where complete three-dimensional solution structures have been obtained. In this paper we will outline the various steps involved in obtaining three-dimensional solution structures of oligonucleotides and illustrate this with three examples from our laboratory, namely the self complementary DNA hexamer 5′d(CGTACG)$_2$, the double stranded DNA undecamer 5′d(AAGTGTGACAT)•5′d(ATGTCACACTT) comprising a portion of the specific DNA target site for the cAMP receptor protein in the gal operon, and the double stranded RNA undecamer 5′r(CACAG)•5′r(CUGUG) comprising the stem of the TψC loop of yeast tRNA^{Phe}.

Proton resonance assignment

The general approach for obtaining sequential resonance assignments in ^{1}H-NMR spectra of proteins has been pioneered by Wüthrich and his collaborators (8-13). In a similar vein, comprehensive sequential resonance assignment strategies have recently been put forward for nucleic acids by a number of groups simultaneously (14-24). These strategies involve two types of NMR experiments.

First it is helpful to delineate through bond connectivities by either decoupling or two-dimensional J correlated spectroscopy (COSY). In this manner sugar resonances belonging to the same network of coupled spins via the intranucleotide H1′⇔H2′/H2″⇔H3′⇔H4′⇔H5′/H5″ pathway, H5 and H6 resonances of cytosine, and CH_3 and H6 resonances of thymine can be grouped together. Because of the limited chemical shift dispersion of the H3′ resonances, the intranucleotide connectivity between the H3′ and H4′ resonances cannot be established unambiguously from the COSY spectrum alone. In this respect, two-dimensional relayed and double relayed coherence transfer (RCT) spectroscopy can potentially resolve this ambiguity by establishing connectivities between two remote nuclei separated by one and two nuclei, respectively, within a given spin system. However, it should be borne in mind that the line widths in oligonucleotides of six base pairs or longer are sufficiently broad to render both COSY and RCT spectroscopy of limited value.

Fortunately, complete or virtually complete resonance assignments in the case of oligonucleotides can be obtained entirely from the second type of NMR experiment, namely the establishment of connectivities between all protons separated by short distances ($\lesssim$ 5 A) in space by means of nuclear Overhauser enhancement (NOE) measurements. This can be accomplished either by one-dimensional pre-steady state NOE measurements or by two-dimensional NOE spectroscopy (NOESY). In this manner, neighboring bases can be identified as well as bases belonging to two different strands which are involved in base pairing. Fig. 1 summarizes a comprehensive NOE strategy for the assignment of all proton resonances in right handed oligonucleotides (24). In the case of left-handed Z-DNA, the intranucleotide distance relationships are the same as those in right handed DNA (although the relative magnitudes of the sugar-base NOEs are significantly different for the purine residues which adopt a syn conformation in contrast to the anti conformation in right handed nucleic acids). The internucleotide distance relationships, however, are entirely different for Z-DNA and these are summarized in Fig. 2 (24).

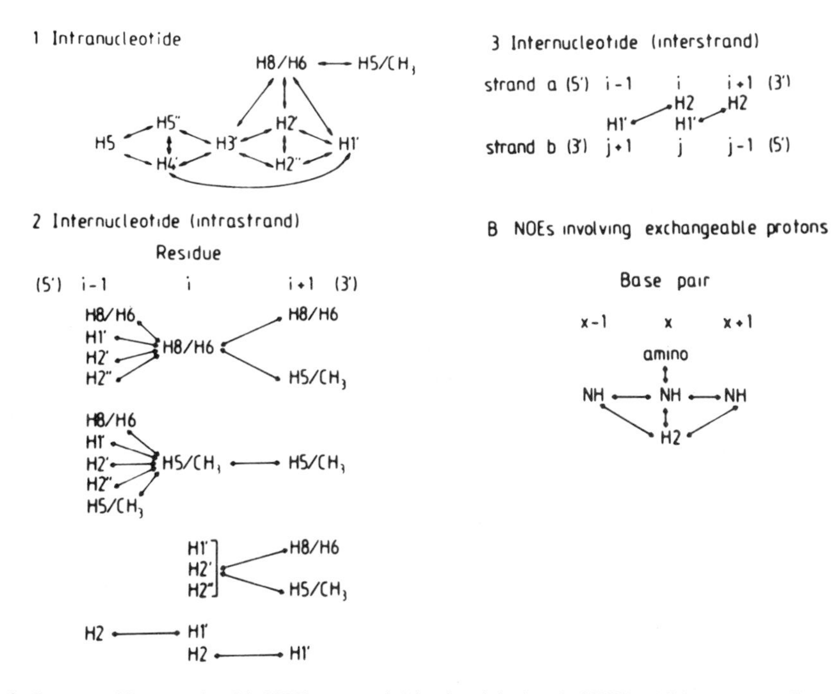

Figure 1. Intra- and internucleotide NOE connectivities for right handed DNA. All interproton distances < 5 A are listed and are applicable to both A- and B-DNA.

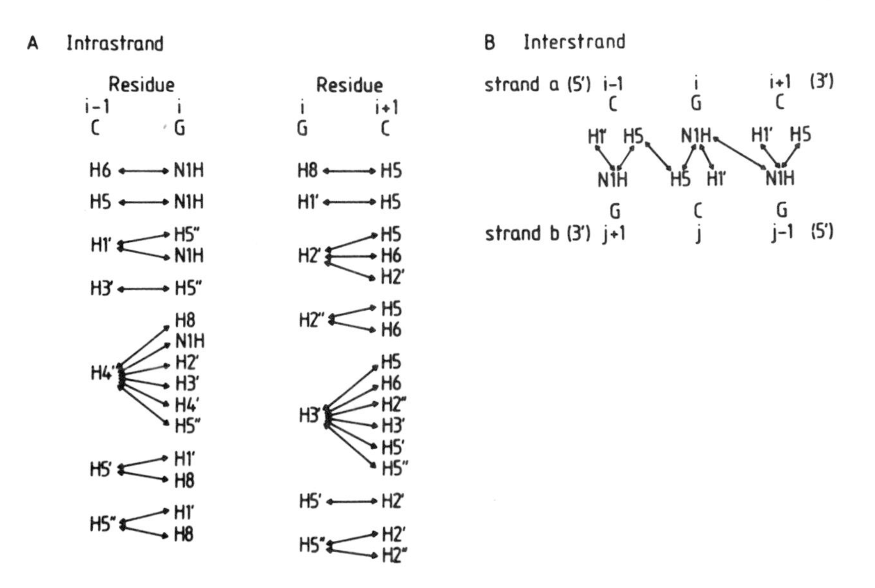

Figure 2. Internucleotide NOE connectivities for left-handed Z-DNA [Poly d(GC)]. All interproton distances < 5 A are listed.

Low resolution structure

Because of the r^{-6} dependence of the pre-steady state NOE, the relative magnitudes of the NOEs provide a sensitive probe of conformation and enable one to readily distinguish between the A, B and Z families of nucleic acids.

The glycosidic bond and sugar pucker conformations can be assessed by examining the relative magnitudes of the intranucleotide sugar-base NOEs. In B DNA, the pattern of NOE intensities observed will be $N_{H2'-H8/H6} \gg N_{H1'-H8/H6} \gtrsim N_{H3'-H8/H6}$ corresponding to a glycosidic bond conformation in the conventional anti range ($\chi \sim -115\pm30°$; IUPAC Nomenclature) and a sugar pucker extending from O1′-endo (δ(C4′-C3′)~100°) to C2′-endo (δ~140°) (25). In contrast, in A DNA the pattern of NOE intensities observed is $N_{H3'-H8/H6} \gg N_{H1'-H8/H6} \sim N_{H2'-H8/H6}$ corresponding to a low anti glycosidic bond conformation (χ −160±10°) and a 3′-endo sugar pucker (δ~80°) (26,27). In Z DNA, the pattern of sugar-base NOEs for the pyrimidine residues is the same as that for B DNA but that of the purine residues is quite different with $N_{H1'-H8/H6} \gg N_{H2''-H8/H6} \sim N_{H2'-H8/H6} \gg N_{H3'-H8/H6}$ on account of the syn glycosidic bond conformation (χ ~ 60-70°) and the 3′-endo sugar pucker (28,29).

The handedness and overall helical conformation can be assessed from the internucleotide NOEs with directional specificity (see Fig. 1). Thus, for right handed helices, internucleotide NOEs are observed between the H1′, H2′ and H2″ sugar proton of a given residue and the H8/H6 base proton of the adjacent 3′ but not 5′ residue, and between the $T(CH_3)$ protons of a given residue and the H8/H6 base proton of the adjacent 5′ but not 3′-residue. The relative magnitudes of intra- and internucleotide sugar-base NOEs involving the H2′ and H2″ protons enable one to readily distinguish between A and B DNA. Thus in B DNA we find that $N_{H2'(i)-H8/H6(i)} \gg N_{H2''(i-1)-H8/H6(i)} > N_{H2'(i-1)-H8/H6(i)}$ whereas in A DNA $N_{H2'(i-1)-H8/H6(i)} \gg N_{H2''(i-1)-H8/H6(i)} > N_{H2'(i)-H8/H6(i)}$ (30).

Interproton distance determination

The initial slope of the time dependence of NOE, N_{ij}, between two protons i and j is given by (31,32)

$$\left.\frac{dN_{ij}}{dt}\right|_{t=0} = \sigma_{ij} \qquad (1)$$

where σ_{ij} is the cross-relation rate between protons i and j given by (33,34)

$$\sigma_{ij} = \frac{\gamma^4\hbar^2}{10r_{ij}^6}\left(\tau_{eff} - \frac{6\tau_{eff}}{1+4\omega^2\tau_{eff}^2}\right) \qquad (2)$$

where γ and $\hbar$ have their usual meanings, τ_{eff} is the effective correlation time of the i-j interproton vector, and r_{ij} is the distance between protons i and j. Thus distance ratios, or distances, if one distance is already known can be obtained either from the exact equation

$$r_{ij}/r_{kl} = (\sigma_{kl}/\sigma_{ij})^{1/6} \quad (3)$$

or the approximate relationship

$$r_{ij}/r_{kl} \sim [N_{kl}(t)/N_{ij}(t)]^{1/6} \quad (4)$$

at short values of t, providing the correlation time of the i-j and k-l interproton vectors are the same. Note that the approximation (4) introduces only small errors ($\lesssim$ 0.2 A) in the estimate of the unknown distance r_{ij} up to relatively long times (viz. values of t up to 4-8 times longer than that at which the initial rate approximation $N_{ij} \sim \sigma_{ij}t$ breaks down) providing either $\sigma_{ij} \gtrsim \sigma_{ki}$ or $\sigma_{ij} \gtrsim \sigma_{kj}$ or $\sigma_{ij} \gtrsim \sigma_{li}$ or $\sigma_{ij} \gtrsim \sigma_{lj}$ (35).

A certain degree of caution is required in using either Eqs (3) or (4) to determine unknown interproton distances. Namely, σ_{ij} is not only proportional to $\langle r_{ij}^{-6} \rangle$ but also to $\tau_{eff}(ij)$ when $\omega\tau_c \gg 1$. Hence differences in effective correlation time between the different interproton vectors may result in errors in the estimation of the unknown distances. Fortunately, because of the $\langle r_{ij}^{-6} \rangle$ dependence of σ_{ij}, quite substantial variations in $\tau_{eff}(ij)$ relative to the effective correlation time τ_{eff} (ref) for the reference interproton distance vector only lead to relatively small errors in the estimation of r_{ij}.

Considering DNA there are three internal reference distances, $r_{H2'-H2''}$, $r_{C(H6)-C(H5)}$ and $r_{T(H6)-T(CH_3)}$ which have values of 1.78, 2.46 and 2.70 A fixed by the geometry of the sugars and bases themselves (note that the latter distance is a $(\langle r^{-6} \rangle)^{-1/6}$ mean calculated on the assumption of free rotation of the methyl group). Cross relaxation rate measurements on a double stranded DNA hexamer and undecamer (36) have shown that although there is no residue to residue variation in the effective correlation times of these interproton vectors, the effective correlation time of the H2′-H2″ vector is significantly shorter than that of the C(H5)-C(H6) and T(H6)-T(CH_3) vectors, both of which have the same effective correlation times (the contribution from free rotation of the methyl group of the T(H6)-T(CH_3) vector being negligible). Thus, it becomes necessary to make an appropriate choice of reference distance. Fortunately, this choice can be made easily on the basis of stereochemical considerations taking into account the expected ranges of the various interproton distances and the types of motion of different portions of the oligonucleotide as deduced from the analysis of X-ray thermal factors (37), molecular dynamics calculations (38-40) and NMR relaxation studies (41-43). Taking these facts into account, it seems reasonable to assume in the case of DNA oligonucleotides that (i) the effective correlation times for the sugar-sugar and sugar-base (with the exception of the H1′sugar-base) interproton vectors are the same as those of the intranucleotide H2′-H2″ vector, and (ii) the effective correlation times for the base-base and H1′ sugar-base interproton vectors are the same as those of the intranucleotide C(H5)-C(H6) and T(CH_3)-T(H6) vectors, as the contribution from internal motion to the effective correlation times for (i) will be dominated by motions within the sugar ring and for (ii) by motions about the glycosidic bond (17,44). A check on the validity of these assumptions has been carried out by calculating various interproton distances whose values are affected by $\lesssim \pm 0.2$ A by alterations in conformation: in each case the calculated values are all within 0.2 A of the predicted ones (17,44).

In the case of RNA oligonucleotides the situation is somewhat simpler as there appears to be no difference in effective correlation time of the base and sugar moieties (7). This is not surprising as the mobility of the ribose ring in RNA would be expected to be considerably reduced relative to that of the deoxyribose ring in DNA because of steric hindrance arising from the presence of the bulky O2′H hydroxyl group on the ribose and electrostatic interactions between the O2′H group of residue i and the O4′ atom of residue i+1, thereby immobilizing one ribose ring with respect to the neighbouring ones.

Bearing all these considerations in mind, we estimate that, in the case of oligonucleotides, the errors in the values of the calculated interproton distances are in general $\lesssim$ 0.2 A except for distances > 3.3 A where the errors may be somewhat larger ($\lesssim$ 0.3 A). It is worthwhile pointing out that at the present time such accuracy is not attainable in the case of proteins as there are very few accessible (from the spectroscopic point of view) fixed internal interproton reference distances and stereospecific assignments (e.g. the distinction between the two β methylene protons of a particular residue) are difficult to make. As a result, interproton distances in proteins can usually only be estimated within an accuracy of $\sim\pm$0.5 A (4,45,46).

High resolution structures – Structure refinement

The determination of high resolution solution structures on the basis of interproton distances determined by NOE measurements is characterized by a poor observation-to-parameter ratio, a situation similar to that found in protein crystallography. Thus, a satisfactory structure determination method must call upon other sources of information, in particular the stereochemical restraints on a molecule which arise from energy considerations (i.e. bond lengths, bond angles, van der Waals radii). Two different approaches can be applied to tackle this problem. One approach, adopted by Wuthrich and his collaborators, involves the *ab initio* computation of the three-dimensional structure using triangulation. This makes use of distance-geometry algorithms based on the interconvertability of intermolecular distances, torsion angles and cartesian coordinates providing the chirality of the structure is known (46-50). This has been applied successively to a number of small proteins, namely lipid bound glucagon (1), Insectotoxin I_5A (2), BUSY IIA (4) and bovine pancreatic trypsin inhibitor (46). (Note the latter involved testing the approach with distance data derived from the crystal structure rather than NMR data). Although this is possibly the most elegant approach as no initial trial model is required, it suffers from a potential false minimum problem. The second approach involves refining an initial trial model by either restrained least squares refinement or restrained molecular dynamics. In our view this second path appears more promising as for both oligonucleotides and proteins low resolution structures can be obtained by model building on the basis of a qualitative interpretation of the NOE data alone (3,4,6,7,51), thereby reducing the false minimum problem. We have applied restrained least squares refinement to two DNA oligonucleotides (5,6) and one RNA (7) oligonucleotide, and are beginning to apply restrained molecular dynamics to refine oligonucleotide (52) and polypeptide (53) structures. Restrained molecular dynamics has also been applied successfully by Kaptein et al. (3) to refine the solution structure of *lac* repressor head piece.

Of the two refinement methods, there is little doubt that restrained molecular dynamics can search a wider range of conformational space than restrained least squares refinement (52,53). However, restrained molecular dynamics is considerably more expensive in terms of computing time. Thus, for example, whereas a 10 ps dynamics run on a 700 atom system such as the duplex DNA undecamer 5'd(AAGTGTGACAT)•5'd(ATGTCACACTT) requires 1hr of cpu time on a CRAY 1A supercomputer, the complete restrained least squares refinement requires only 1 min. Thus, on small computers such as a VAX 11/780, restrained least squares refinements can be easily performed whereas restrained molecular dynamics calculations become prohibitively expensive in terms of time. In the case of oligonucleotides, it is easy to choose an initial model within an RMS difference of ~ 1A from the final structure, as the three classes of DNA, A, B and Z, are easily distinguished. Consequently restrained least squares refinement, despite its smaller range of accessible conformational space, is perfectly adequate for oligonucleotides.

In the case of the restrained least squares refinement procedure we have used, which is based on the crystallographic refinement program RESTRAIN (54,55), the function minimized in cartesian coordinate space is given by

$$C = \Sigma W_d(d_r - d_c)^2 + \Sigma W_v|V| + \Sigma W_b(b_o - b_{min})^2 \qquad (b_o < b_{min}) \qquad (5)$$

where W_d, W_v and W_b are weighting coefficients, d_r and d_c are the target and calculated interatomic distances respectively, $|V|$ is the determinant of the product-moment matrix for planar groups of atoms (the necessary and sufficient condition for a set of atoms to be planar being that $|V|$ is zero), and b_o and b_{min} are the calculated and minimum allowed distances between two non-bonded atoms. The interatomic distances include all distances between covalently bonded atoms, between atoms defining fixed bond angles and between atoms defining base pair hydrogen bonding, as well as the experimental interproton distances determined from the NOE measurements. It should be noted that restrained least squares refinement is quite distinct from a full scale energy refinement. This is because the change in conformation on restrained least squares refinement, in contrast to that on energy refinement, arises solely from the interproton distance restraints, as all other restraints are well satisfied in the initial and final structures as well as in the intermediate structures sampled during the entire course of the refinement.

Examples of the application of restrained least squares refinement to oligonucleotides

To date, we have refined three oligonucleotide structures on the basis of interproton distance data derived from NOE measurements using restrained least squares refinement. These are the self-complementary DNA hexamer 5'd$(CGTACG)_2$ (6), the duplex DNA undecamer 5'd(AAGTGTGACAT)•5'(ATGTCACACTT) (5) and the duplex RNA pentamer 5'r(CACAG)•5'r(CUGUG) (7). In all three cases, two refinements were carried out. For the DNA hexamer two different initial models were used, namely classical B-DNA derived from fibre diffraction data (56) and a B-DNA structure derived by energy minimization of the classical structure; similarly for the RNA pentamer where the first initial structure was that of classical A-RNA (57) with the second initial structure derived from it by applying 10 cycles of regularization including non-bonded contact restraints (but excluding, of course, the interproton distances). For these four refinements the weightings for the three distance ranges $r < 2.12$ A, 2.12 A $< r < 2.6$ A and $r > 2.6$ A, and the planes were applied throughout in the ratio 5:4:3:4. (Note these three distance ranges correspond to distances between directly bonded atoms, between atoms separated by two bonds and between atoms separated by three or more bonds, respectively). The weights were chosen to represent approximately the gradation of error as a function of distance for the experimental interproton distances. In the case of the undecamer only a single initial structure was used, namely that of classical B DNA, but the refinement was taken through two different pathways. In the first, the weights were applied in the ratio 10:10:7:7 for 60 cycles and then, following 5 cycles of regularization, in the ratio 5:4:3:4 for the remaining 15 cycles. In all 6 cases, the RMS difference between the experimental and calculated interproton distances for the refined structures was $\lesssim 0.2$ A compared to > 0.6 A for the initial structures. At the same time, all other restraints (viz. bond length and bond angle restraints) were well satisfied in the refined structures and no undesirably close non-bonded contacts

Table I

Features of the least squares refined structures of the DNA hexamer 5'$(dCGTACG)_2$ (6), the DNA undecamer 5'd(AAGTGTGACAT)•5'd(ATGTCACACTT) (5) and the RNA pentamer 5'r(CACAG)•5'r(CUGUG) (7)

	Number of restraints			RMS
	DNA 6mer	DNA 11mer	RNA 5mer	difference (A)
Covalent and bond angle restraints[a]				
r<2.12 A	522	1044	444	<0.04
2.12A<r<2.6 A	584	1027	468	<0.07
r>2.6 A	12	29	10	<0.04
Planes[b]	12	22	10	<0.02
Base pairing restraints	32	59	31	<0.03
Interproton distances[c]	190	150	55	<0.2
Total number of atoms	374	700	311	
Total number of restraints	1352	2331	1018	
Total number of refinement cycles	30	80	35	

[a]It should be noted that bond angles are defined by interatomic distances

[b]For each residue the C1'-atom of the ribose and all the atoms of the base (with the exception of the methyl protons of thymine) are constrainted to lie in the same plane.

[c]These interproton distances do not include interproton distances which are fixed by the geometry of the sugar and bases themselves.

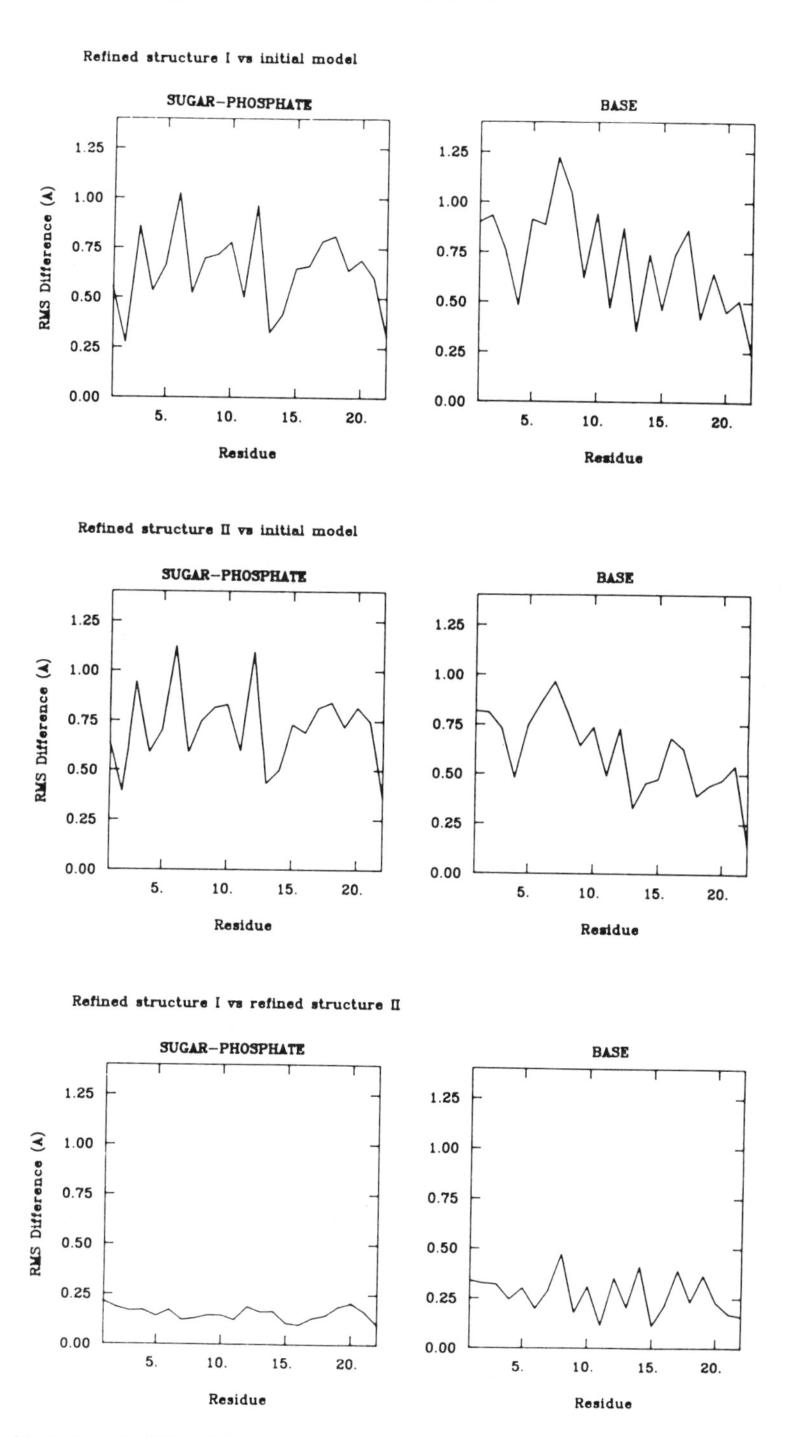

Figure 3. Variations in RMS difference for the sugar-phosphate and base moieties between the initial classical B-DNA model and the refined structures I and II of the DNA undecamer 5′d(AAGTGTGACAT)·5′d(ATGTCACACTT) (5).

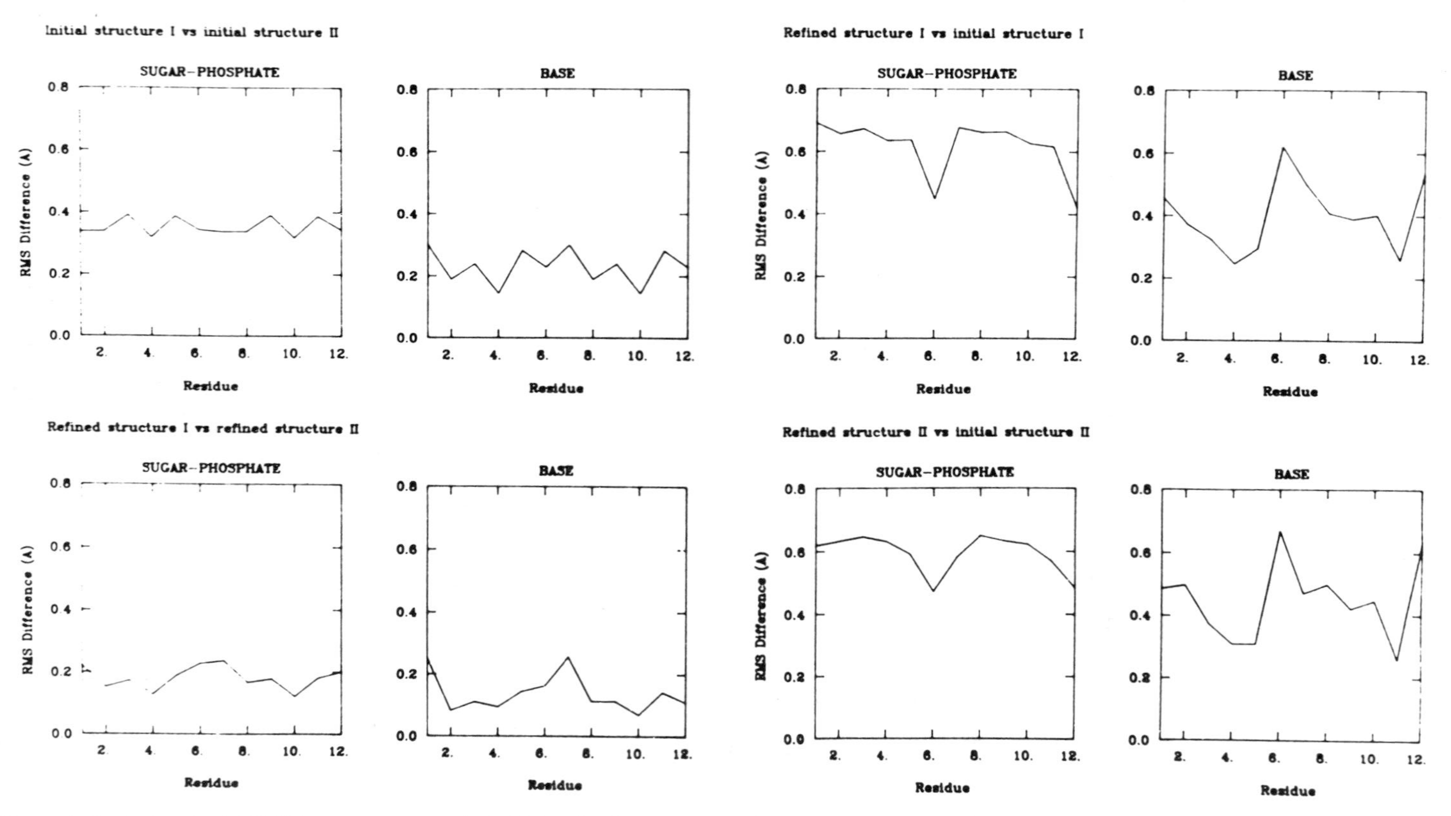

Figure 4. Variations in RMS difference between the initial and refined structures of the DNA hexamer 5′d(CGTACG)$_2$. Initial model I is classical B-DNA and initial model II was derived from initial model I by energy minimization (6).

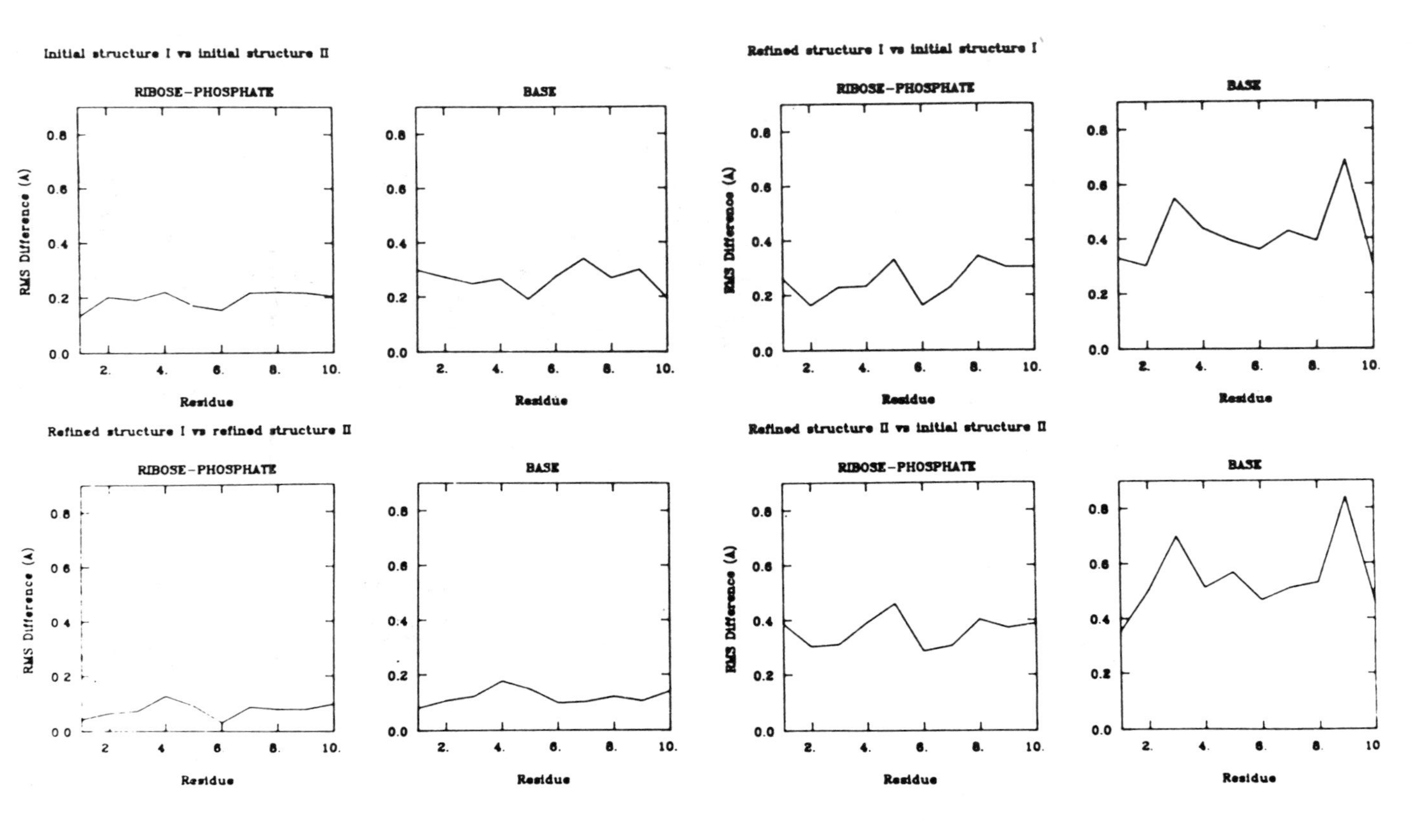

Figure 5. Variations in RMS difference between the initial and refined structures of the RNA pentamer 5′r(CACAG)·5′r(CUGUC). Initial model I is classical A-RNA and initial model II was derived from initial model I by subjecting it to 10 cycles of regularization including non-bonded contract restraints but excluding the interproton distance restraints (7).

were exhibited. In addition, the overall RMS difference between the pairs of refined structures was insignificant ($\lesssim$ 0.2 A) providing good evidence for the convergence properties of the restrained least squares refinement. The essential features of the refined structures are given in Table I and the RMS differences between the initial and refined structures are shown in Figs. 3 to 5.

In considering the refined structures, it is essential to bear in mind that the measured interproton distances are $(\langle r^{-6}\rangle)^{-1/6}$ means. As a result, the measured interproton distances will invariably be weighted towards fluctuations with the shortest interproton distances in a dynamic structure as is the case for molecules in solution. Thus, the average structure represented by the refinements is an $(\langle r^{-6}\rangle)^{-1/6}$ time average in cartesian coordinate space in contrast to the average structure obtained by X-ray crystallography which can be considered as a linear super-position of structures. Consequently, the refined structures could potentially be distorted relative to an arithmetic time averaged structure or a crystal structure. Fortunately, the refinement

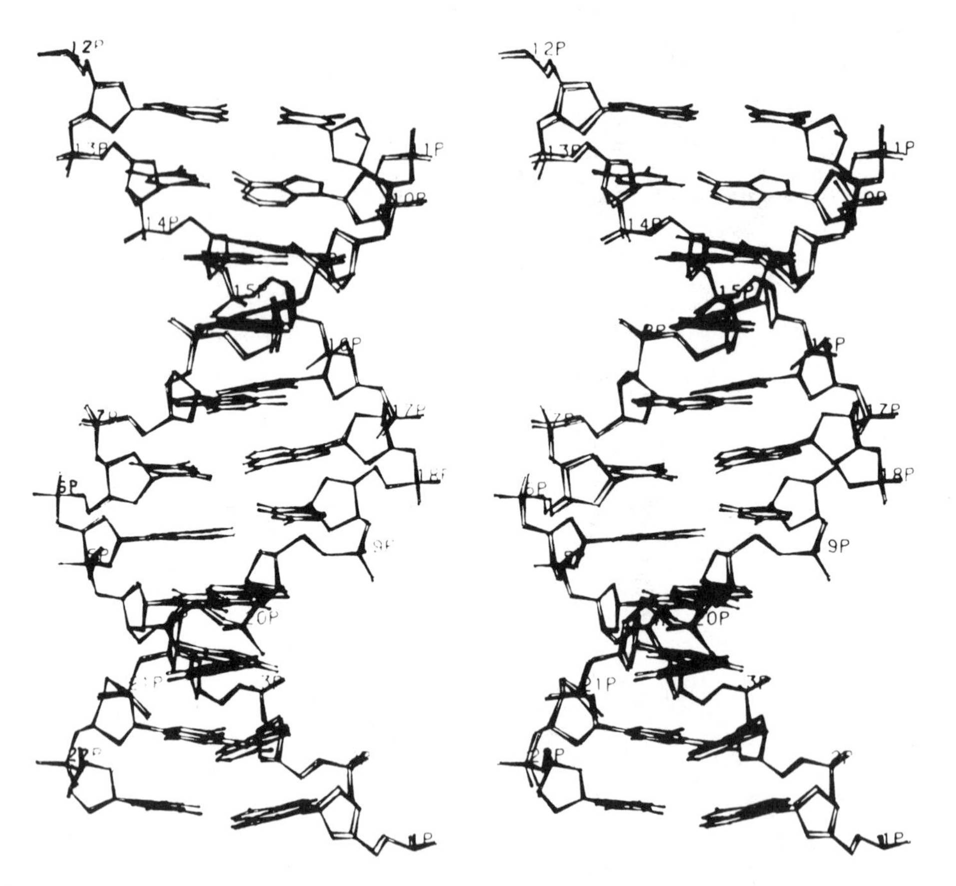

Figure 6. Stereoview of refined structures I and II of the DNA undecamer 5′d(AAGTGTGACAT)· 5′d(TAGTCACACTT) superimposed (5). The overall RMS difference between the two structures is 0.23 A. For the sake of clarity protons have been omitted.

itself provides information both on the magnitudes of the internal motions as well as on the relationship between the refined $(\langle r^{-6} \rangle)^{-1/6}$ time averaged structure and the arithmetic time averaged solution structure. This is due to the fact that a large number of interproton distances are used in the refinement and that many of these distances are correlated so that a single structure would not be able to provide an adequate fit to the experimental data if the magnitude of the internal motions is large (5-7). In the three cases considered here, the magnitude of the internal motions must be small in order to accommodate an RMS difference of $\lesssim$ 0.2 A between experimental and calculated interproton distances (a difference comparable to the error in the experimental interproton distances). Hence, it can be concluded that the $(\langle r^{-6} \rangle)^{-1/6}$ time averages represented by the refined structures are close to the respective arithmetic time average structures, and that the small RMS difference between the pairs of refined structures ($\lesssim$ 0.2 A) provides a measure of the error in the refined coordinates.

Structural features of the three refined oligonucleotides

Stereoviews of the refined structures of the DNA undecamer, the DNA hexamer and the RNA pentamer are shown in Figs 6 to 8 respectively, and the average values of the conformational parameters describing their structures are given in Table II. It is clear from the stereoviews as well as the data in Table II that the overall B-type conformation for the two DNA oligonucleotides and the overall A-type conformation for the RNA oligonucleotide is preserved in the refined structures. Like the crystal structures of B (25) and A (26,27)-DNA oligonucleotides, however, the refined structures are no longer regular helices but exhibit local structural variations in glycosidic and backbone torsion angles as well as in helical and propellor twist

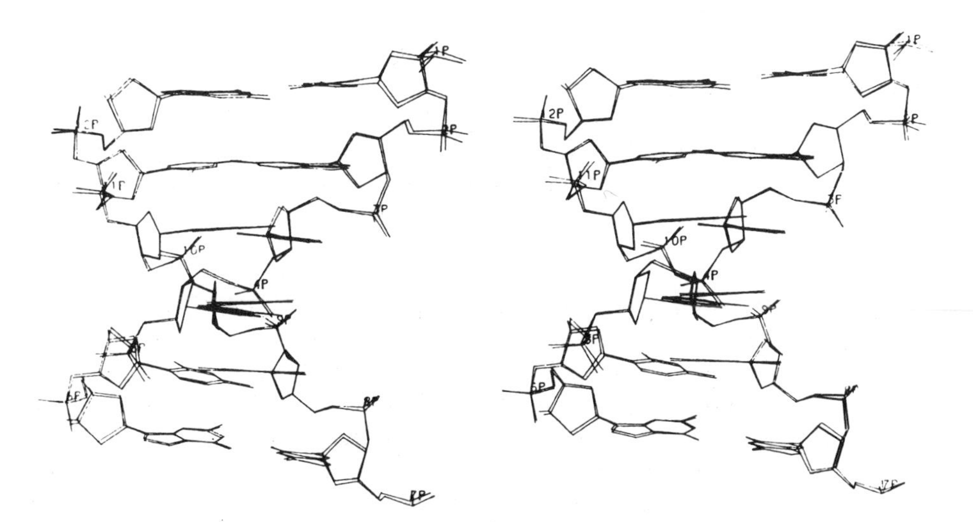

Figure 7. Stereoview of refined structures I and II of the DNA hexamer 5′d(CGTACG)$_2$ superimposed (6). The overall RMS difference between the two structures is 0.16 A. For the sake of clarity protons have been omitted.

angles. For example, in the case of the DNA undecamer propellor twist angles range from 20° to 0°, helical twist angles from 39° to 30°, and so on. These sorts of local structural variations should have important consequences for specific DNA-protein interactions if, as seems likely, they involve recognition not only of base sequence but also of local structure.

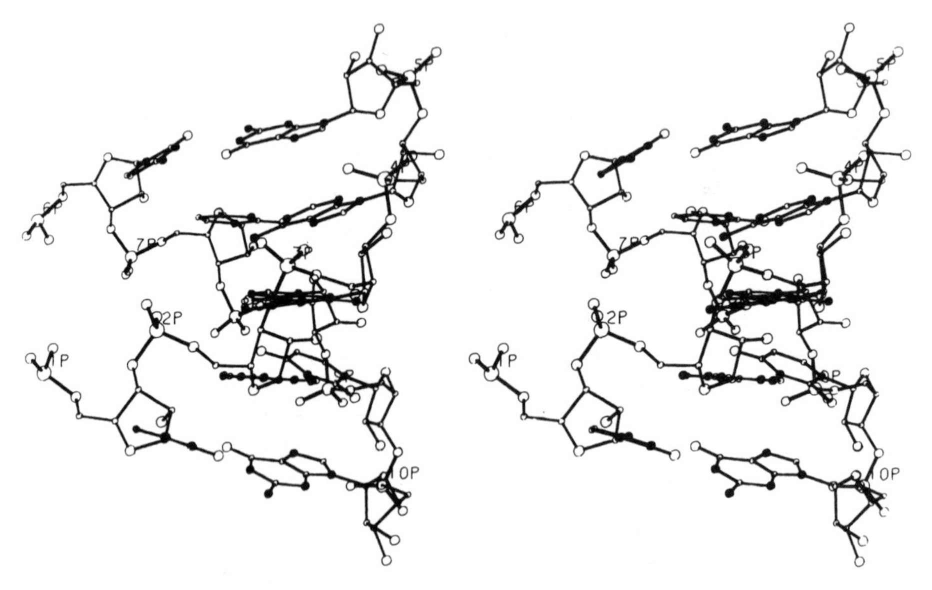

Figure 8. Stereoview of one of the refined structures of the RNA pentamer 5′d(CACAG)•5′d(CUGUG) (7). Because the two refined structures are virtually indistinguishable with an overall RMS difference of 0.11 A, only one refined structure is shown. For the sake of clarity protons have been omitted.

The most obvious correlation in conformational parameters we observe is between the C4′-C3′ (δ) and glycosidic (χ) bond torsion angles, the smaller the value of δ, the more negative that of χ. This correlation is also observed in the crystal structures. In addition, we find that for a given value of χ, the value of δ is likely to be higher for a pyrimidine than a purine residue.

Correlations of local structural variation with base sequence, however, are less obvious for the three refined solution structures and there appear to be no generalized sets of rules governing the sequence dependence of these variations. This is in contrast to the findings for the small number of A- and B-DNA fragment crystallized to date where, with the exception of a DNA-RNA hybrid (58), the local structural variations in helical twist, propellor twist, base roll and C4′-C3′ (δ) torsion angles appear to be reasonably well predicted by a set of simple sum functions (59) based on the avoidance of steric clash between purines on opposite strands of the helix at adjacent base pairs (60). For example, the variations in helical twist for the DNA undecamer are almost exactly the opposite to those predicted by Dickerson's sum functions (5). Similarly, in the case of the RNA pentamer all the base pairs (with the

Table II

Average values of the conformational parameters describing the refined solution structures of the DNA hexamer, the DNA undecamer and the RNA pentamer

	Glycosyl (°)	Main chain torsion angles (°)						Adjacent phosphorus atom atom separation conformation (A)	Observed sugar twist (°)	Base Pair propellor (°)	Helical twist (°)
	χ	α	β	γ	δ	ϵ	ξ				
DNA 6mer	−114±9	−87±6	−147±7	65±2	113±5	142±7	−65±3	6.45±0.1	C1′-exo	8.5±3	34.6±2
DNA 11mer	−113±14	−61±16	−153±11	38±14	129±7	155±13	−71±23	6.42±0.2	C1′-exo	10±7	35±3
RNA 5mer	−156±10	−83±6	−162±7	49±4	84±11	179±6	−57±13	5.7±0.1	C3′-endo	14±8	30.5±0.5
B_c-DNA[a]	−117±14	−63±8	171±14	54±8	123±25	−169±25	−108±34	6.68±0.23	C1′-exo	13±5	37.3±3.8
A_c-DNA[b]	−160±8	−62±12	173±8	52±14	88±3	−152±8	−78±7		C3′-endo	13±4	33.7
A_c-DNA[c]	−157±14	−73±4	−179±9	73±30	89±25	−161±7	−66±5		C3′-endo	16±4	

[a]From the crystal data of Dickerson and Drew (25)
[b]From the crystal data of Shakked et al. (26)
[c]From the crystal data of Conner et al. (27)

exception of the fifth base pair G_5C_6) are highly propellor twisted (7) whereas Dickerson's sum functions predict that the first four base pairs should hardly be propellor twisted while the fifth base pair should show maximal propellor twist. Interestingly, a recent molecular dynamics calculation with and without hydrated cations presented on a DNA pentamer also showed a variation in propellor twist opposite to that predicted by the sum functions (40). On reflection, these observations may not be all too surprising. First, the data base from which the predictions were derived is small; second, the contributions to the sum functions arise only from Pyr-Pur and Pur-Pyr steps and no distinction is made between pyrimidine and purine residue types; third, only nearest neighbour interactions are considered; and fourth, and possibly most important of all, the balance of competing non-bonded interactions may be different in the solution and crystalline states. For example, the sum functions are based on increased steric clash arising from an increase in propellor twisting which is particularly marked in alternating pyr-pur sequences. On the other hand, an increase in propellor twist also leads to improved stacking interactions (61). Thus, in solution the energetic advantage of increased stacking may outweigh the disadvantage of some steric clash.

Clearly at the present time there are too few refined solution structures of oligonucleotides available to deduce any generalized set of rules governing sequence dependence of local structural variations in solution. Whether such rules do emerge will no doubt become clear in the fullness of time when the solution structures of other oligonucleotides, refined on the basis of interproton distance data, become available.

References and Footnotes

1. Braun, W., Wider, G., Lee, K.H. and Wuthrich, K., *J. Mol. Biol. 169,* 921 (1983).
2. Arseniev, S.A., Kondakov, V.I., Maiorov, V.N. and Bystrov, V.F., *FEBS Lett. 165,* 57 (1984).
3. Kaptein, R., Zuiderweg, E.R.P., Scheek, R.M., Boelens, R. and van Gunsteren, W.F., *J. Mol. Biol. 182,* 179 (1985).
4. Williamson, M.P., Havel, T.F. and Wuthrich, K., *J. Mol. Biol. 182,* 295 (1985).
5. Clore, G.M. and Gronenborn, A.M., *EMBO J. 4,* 829 (1985).
6. Clore, G.M., Gronenborn, A.M., Moss, D.S. and Tickle, I.J., *J. Mol. Biol., 185,* 219 (1985).
7. Clore, G.M., Gronenborn, A.M. and McLaughlin, L.W., *Eur. J. Biochem., 151,* 153 (1985).
8. Wagner, G., Kumar, A. and Wuthrich, K., *Eur. J. Biochem. 114,* 375 (1981).
9. Wagner, G. and Wuthrich K., *J. Mol. Biol. 159,* 347 (1982).
10. Wagner, G. and Wuthrich, K., *J. Mol. Biol. 160,* 334 (1982).
11. Strop, P., Wider, G. and Wuthrich, K., *J. Mol. Biol. 166,* 641 (1983).
12. Williamson, M.P., Marion, D. and Wuthrich, K., *J. Mol. Biol. 173,* 341 (1984).
13. Zuiderweg, E.R.P., Kaptein, R. and Wuthrich, K., *Eur. J. Biochem. 137,* 279 (1983).
14. Reid, A.G., Salisbury, S.A., Bellard, S., Shakked, Z. and Williams, D.H., *Biochemistry 22,* 2012 (1983).
15. Clore, G.M. and Gronenborn, A.M., *EMBO J. 2,* 2109 (1983).
16. Clore, G.M. and Gronenborn, A.M., *Eur. J. Biochem. 141,* 119 (1984).
17. Gronenborn, A.M., Clore, G.M. and Kimber, B.J., *Biochem. J. 221,* 723 (1984).
18. Clore, G.M., Lauble, H.P., Frenkiel, T.A. and Gronenborn, A.M., *Eur. J. Biochem. 145,* 629 (1984).
19. Clore, G.M., Gronenborn, A.M., Piper, E.A., McLaughlin, L.W., Graeser, E. and van Boom, J.H., *Biochem. J. 221,* 735 (1984).
20. Scheek, R.M., Boelens, R., Russo, N., van Boom, J.H. and Kaptein, R., *Biochemistry 23,* 1371 (1984).
21. Hare, D.R., Wemmer, D.E., Chou, S.H., Drobny, G. and Reid, B.R., *J. Mol. Biol. 171,* 319 (1983).

22. Weiss, M.A., Patel, D.J., Sauer, R.T. and Karplus, M., *Proc. Natl. Acad. Sci. U.S.A. 81,* 130 (1984).
23. Feigon, J., Leupin, W., Denny, W.A. and Kearns, D.R., *Biochemistry 22,* 5943 (1983).
24. Clore, G.M. and Gronenborn, A.M., *FEBS Lett. 179,* 187 (1985).
25. Dickerson, R.E. and Drew, H.R., *J. Mol. Biol. 149,* 751 (1981).
26. Shakked, Z., Rabinovich, D., Kennard, O., Cruse, W.B.T., Salisbury, S.A. and Viswamitra, A., *J. Mol. Biol. 166,* 183 (1983).
27. Conner, G.N., Yoon, C., Dickerson, J.L. and Dickerson, R.E., *J. Mol. Biol. 174,* 663 (1984).
28. Wang, A.H.J., Quigley, G.J., Kolpak, F.J., Crawford, J.L., van Boom, J.H., van der Marel, G.A. and Rich, A., *Nature 282,* 680 (1979).
29. Drew, H.R. and Dickerson, R.E., *J. Mol. Biol. 152,* 723 (1981).
30. Gronenborn, A.M. and Clore, G.M., *Progr. Nucl. Magn. Reson. Spectroscopy 17,* 1 (1985).
31. Wagner, G. and Wuthrich, K., *J. Magn. Reson. 33,* 675 (1979).
32. Dobson, C.M., Olejniczak, E.T., Pousen, F.M. and Ratcliffe, R.G., *J. Magn. Reson: 48,* 87 (1982).
33. Solomon, I., *Phys. Rev. 90,* 559 (1955).
34. Kalk, A. and Berendsen, H.C.J., *J. Magn. Reson. 24,* 343 (1976).
35. Clore, G.M. and Gronenborn, A.M., *J. Magn. Reson. 61,* 158 (1985).
36. Clore, G.M. and Gronenborn, A.M., *FEBS Lett. 172,* 219 (1984).
37. Holbrook, S.R. and Kim, S.H., *J. Mol. Biol. 173,* 361 (1984).
38. Levitt, M., *Cold Spring Harbor Symp. Quant. Biol. 47,* 251 (1983).
39. Tidor, B., Irikura, K.K., Brooks, B.R. and Karplus, M., *J. Biomolec. Struc. Dyn. 1,* 231 (1983).
40. Singh, U.C., Weiner, S.J. and Kollman, P., *Proc. Natl. Acad. Sci. U.S.A. 82,* 755 (1985).
41. Hogan, M.E. and Jardetzky, O., *Proc. Natl. Acad. Sci. U.S.A. 76,* 6341 (1979).
42. Bolton, P.H. and James, T.L., *J. Chem. Phys. 83,* 3359 (1979).
43. Lipari, G. and Szabo, A., *Biochemistry 20,* 6250 (1981).
44. Clore, G.M. and Gronenborn, A.M., *FEBS Lett. 175,* 117 (1984).
45. Keepers, J.W. and James, T.L., *J. Magn. Reson. 57,* 404 (1984).
46. Havel, T.F. and Wuthrich K., *J. Mol. Biol. 182,* 281 (1985).
47. McKay, A.L., *Acta Crystallogr. A30,* 440 (1974).
48. Crippens, G.M. and Havel, T.F., *Acta Crystallogr. A34,* 282-284.
49. Cohen, F.E. and Sternberg, M.J.E., *J. Mol. Biol. 138,* 321 (1980).
50. Wuthrich, K., Wider, G., Wagner, G. and Braun, W., *J. Mol. Biol. 155,* 311 (1982).
51. Zuiderweg, E.R.P., Billeter, M., Boelens, R., Scheek, R.M., Wuthrich, K. and Kaptein, R., *FEBS Lett. 174,* 243 (1984).
52. Nilsson, L., Brunger, A.T., Clore, G.M., Gronenborn, A.M. and Karplus, M., *J. Mol. Biol. submitted.*
53. Clore, G.M., Gronenborn, A.M., Brunger, A.T. and Karplus, M., *J. Mol. Biol. 186,* in the press (1985).
54. Moss, D.S. and Morffew, A.J., *Comput. Chem. 6,* 1 (1982).
55. Haneef, I., Moss, D.S., Stanford, M.J. and Borkakoti, *Acta Crystallogr. A., in press* (1985).
56. Arnott, S. and Hukins, D.W.L., *Biochem. Biophys. Res. Commun. 47,* 1504 (1972).
57. Arnott, S., Hukins, D.W.L. and Dover, S.D., *Biochem. Biophys. Res. Commun. 48,* 1342 (1972).
58. Wang, A.H.J., Fjuii, S., van Boom, J.H., van der Marel, G.A., van Boeckel, S.A.A. and Rich, A., *Nature 299,* 601 (1982).
59. Dickerson, R.E., *J. Mol. Biol. 166,* 419 (1984).
60. Calladine, C.R., *J. Mol. Biol. 161,* 343 (1982).
61. Levitt, M., *Proc. Natl. Acad. Sci. U.S.A. 75,* 640 (1978).

Biomolecular Stereodynamics IV, Proceedings of the Fourth Conversation in the Discipline Biomolecular Stereodynamics, State University of New York, Albany, NY, June 04-09, 1985, Eds., Ramaswamy H. Sarma & Mukti H. Sarma, ISBN 0-940030-18-7, Adenine Press, ©Adenine Press (1986).

Structural Studies on Nucleic Acid Double Helices In Solution: Principle and Application of One-Dimensional Nuclear Overhauser Effect Measurements

R.H. Sarma, M.H. Sarma and Goutam Gupta
Institute of Biomolecular Stereodynamics
State University of New York at Albany
Albany, New York 12222 USA

Abstract

Polymorphism of DNA essentially reflects the inherent conformational preference of the constituent nucleotides to adopt one structural form or another. While single crystal data of oligonucleotides reveal that different nucleotide conformations lead to different structural morphologies, 1D NOE measurements for DNA molecules in solution provides us with a tool by which the average conformation of the constituent nucleotides could be determined and hence the polymorphous form and the handedness. Using 1D NOE measurements, it has been shown that (i) poly(dG)•poly(dC) adopts the A-form in solution with dG/dC residues in (C3′-*endo*, low *anti*) conformation, (ii) poly(dG)•poly(dC^{5Me}) takes up the B-form with dG/dC^{5Me} in (C2′-*endo*, *anti*) conformation, (iii) poly(dA)•poly(dT) exhibits the B-form with both dA/dT residues in (C2′-*endo*, *anti*) conformation while the hybrid poly(rA)•poly(dT) adopts a slightly altered B-form with C1′-*exo* sugars for rA/dT residues [C1′-*exo* being a minor variant of C2′-*endo*]. Also, a brief discussion on Z-DNA and B⇄Z transition is included.

Introduction

Under appropriate physico-chemical conditions DNA undergoes structural changes—the phenomenon is known as polymorphism. The polymorphous forms of the same DNA molecule could be of opposite handedness viz salt/alcohol/drug induced B⇄Z transition in the DNA copolymer poly(dG-dC)•poly(dG-dC) [See Table I]. The structural changes in a natural DNA molecule are normally less drastic-viz calf thymus/sperm head DNA exhibits humidity/alcohol induced A⇄B transition in which only helical parameters change but not the handedness (Table I). There are examples, as well, in which the DNA molecule retains the same polymorphous form under all physico-chemical conditions viz. poly(dG)•poly(dC) showing only the A-form while poly(dA)•poly(dT) the B-form (Table I). For some time it has been pointed out that the polymorphism of DNA is a natural expression of inherent conformational preference of the constituent nucleotide units under the given conditions (22-23). This hypothesis is vindicated by the discoveries of the A-, B- and Z-forms of DNA in the single crystals of oligonucleotides (Table II). The single crystal data reveal the structural differences of the A-, B- and Z-forms in terms of

Table I
Conditions Under Which DNAs Adopt Different Forms

DNA	Physical-Chemical Conditions	Form (n,h)* Handedness	References
Calf-thymus	Fiber; 98% r.h,Na-Salt-. Li-Salt- al-all r.h	B-form(n=10,h=3.40Å) Right-handed	1-4
Calf-thymus	Fiber; 50% r.h Na-Salt	A-form(n=11,h=2.56Å) Right-handed	5-7
Calf-thymus	Solution Alcohol (v/V<30%)	B-form	8-9
Calf-thymus	Solution Alcohol(v/V>70%)	A-form	8-9
Poly(dG-dC)• Poly(dG-dC)	Fiber; Low salt & no alcohol	B-form	10-11
Poly(dG-dC)• Poly(dG-dC)	Fiber; High salt/alcohol (v/V~50%)	Z-form (n=6,h=7.2Å) Left-handed	10-11
Poly(dG-dC)• Poly(dG-dC)	Solution Low salt/in presence of ethidium, actinomycin D	B-form	12-13
Poly(dG-dC)• Poly(dG-dC)	Solution High salt/alcohol	Z-form	12-14
Poly(dG)• Poly(dC)	Fiber; at all r.h	A-form	15
Poly(dG)• Poly(dC)	Solution in presence/ absence of alcohol	A-form	16-17
Poly(dA)• Poly(dT)	Fiber; at all r.h	B-form	18-19
Poly(dA)• Poly(dT)	Solution; under all conditions	B-form	20-21

*n = no. of repeating units per turn of the helix.

h = average separation between the successive repeating units along the helix-axis.

For the A- and B-form the repeating unit is a nucleotide while for the Z-form it is a dinucleotide—(n,h) values could be precisely obtained only from the fiber-diffraction studies. Inference about the form of DNA in solution is made by comparing the Circular Dichroism/Raman Spectra (as the case may be) of the DNA in the fiber (with the predetermined structural form) and the DNA in solution (8-9).

constituent nucleotide conformations (Table I); the data clearly show how different nucleotide prototypes represent different structural morphologies (i.e. A, B, Z-DNA). The data also indicate that there could be subtle variations within the same polymorphous form i.e. B_I and B_{II} within the B-DNA and Z_I and Z_{II} within the Z-DNA (Table II). As seen from Table II, the nucleotides in the A-, B- or Z-DNA belong to very different conformational domains. This leads us to the questions

Table II

Conformational Parameters for A, B & Z-DNA Structures as Observed in the Single Crystals

	Reference	Average (n,h)	Average Torsion Angles (deg) α	β	γ	δ	ϵ	ξ	χ
A-DNA									
Sequence									
d(CCGG)	24	11,2.87Å	291	174	60	84	207	287	200
d(GGTATACC)	25								
d(GGTACC)	26								
B-DNA									
Sequence									
d(CGCGAATTCGCG)	27-29	~10,3.30Å	297	164	59	123	190	261	243
d(CGTACG)									$-B_I$
+daunomycin	30		310	149	42	139	221	212	253
									$-B_{II}$*
Z-DNA									
Sequence									
d(CGCG)	31-32								
d(CGCGCG)	32-33	6,~7.5Å	C223	221	56	138	266	80	201
d(^{Met}CG^{Met}CG^{Met}CG)	34		G47	179	191	99	256	291	78
									$-Z_I$
			C146	164	66	147	260	74	212
			G92	193	157	94	181	55	62
									$-Z_{II}$

*An example of B_{II} conformation of nucleotides in the crystal which is also the structural motif of B-DNA in the fiber (35). The torsion angles are defined as follows:

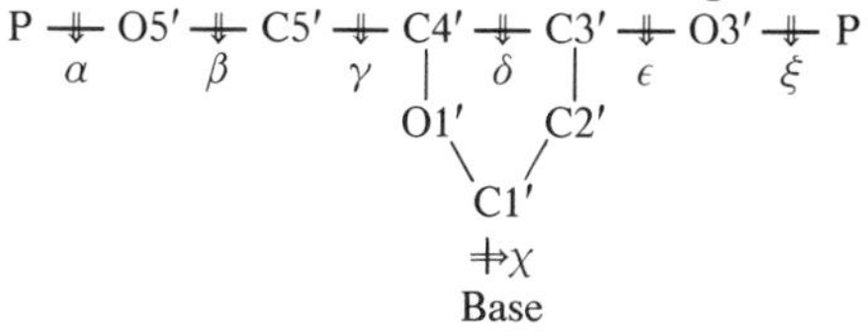

```
P -+- O5' -+- C5' -+- C4' -+- C3' -+- O3' -+- P
   α      β      γ  |  δ  |  ε      ξ
                  O1'    C2'
                     \  /
                     C1'
                     +χ
                    Base
```

whether by using any spectroscopic technique in solution, nucleotides belonging to different conformational domains could be identified. It turns out that NMR spectroscopy using NOE measurements provides us with such a tool. This article is a brief account of the *principle and application of NOE spectroscopy for identification of polymorphous forms of DNA in terms of nucleotide conformations.* For details of the results as obtained for different DNA polymers, see refs. 36-40.

Identification of Nucleotide Geometry In a DNA Polymer From NOE Measurements: The Principle

Sugar pucker (defined by δ) and glycosyl torsion χ can be used to assign different nucleotide geometries as present in the A-, B- and the Z-forms of DNA. For example, in the A-form the nucleotides belong to (C3′-*endo*, low *anti*) domain with $70° \leq \delta \leq 260°$ and $180° \leq \chi \geq 210°$; in the B-form the nucleotides belong to (C2′-*endo*, *anti*) domain with $120° \leq \delta \leq 160°$ and $240 \leq \chi \leq 260°$. For the Z-form, the repeat unit is a dinucleotide: pyrimidine nucleotides in (C2′-*endo*, low *anti*) domain i.e. $120 \leq \delta \leq 160°$ and $180 \leq \chi \leq 210°$ while the purine nucleotides in (C3′-*endo*, *syn*) domain i.e. $70° \leq \delta \leq 100°$ and $30° \leq \chi \leq 80°$. The nucleotide units belonging to different conformational domains (as present in different forms of DNA) are distinguishable from each other in terms of inter-proton distances, especially the ones involving the base proton H8/H6 and the sugar protons H1′,H2′/H2″ and H3′ (See Table III-V).

A-DNA

The relevant inter-proton distances for (C3′-*endo*, low *anti*) conformation in the A-DNA are: H8/H6---H3′ = 2.8 ± 0.2Å in the same nucleotide and H8/H6---H2′ of the 5′ neighbor = 2.2 ± 0.3Å (Tables III-IV). Thus, when H8/H6 in a DNA polymer is pre-saturated, if the primary sites of NOE are H3′ of the same nucleotide and H2′ of the 5′ neighbor (the magnitude of NOE at the latter site being stronger than the former) we can conclude that the nucleotides in the polymer exhibit (C3′-*endo*, low *anti*) conformation as in the A-DNA.

B-DNA

The relevant inter-proton distances for (C2′-*endo*, *anti*) conformation in the B-DNA are: H8/H6---H2′ = 2.2 ± 0.2Å in the same nucleotide and H8/H6---H2″ of the 5′ neighbor = 2.2 ± 0.2Å and H8/H6---H3′ of the same nucleotide >4.5Å (Tables III-IV). Therefore, when H8/H6 in a DNA polymer is presaturated, two almost equally strong sites of NOE at H2′/H2″ region and absence of NOE at H3′ are the signatures of (C2′-*endo*, *anti*) conformation of nucleotides as in the B-DNA.

Z-DNA

In this polymorphous form, the primary NOE patterns are different when H8 and H6 are presaturated. In this DNA, purines adopt (C3′-*endo*, *syn*) conformation and thus the primary site of NOE from H8 is H1′ of the same nucleotide (H8---H1′ = 2.3 ± 0.2Å in the same nucleotide, Table V). For the pyrimidines with (C2′-*endo*, low *anti*) conformation, only weak NOEs are expected at sugar protons when H6 is irradiated (Table V).

Identification of the Nucleotide Geometry vis-a-vis the Polymorphous Form of a DNA Polymer in Solution from NOE Measurements: The Applications

I. A-Form for Poly(dG)•Poly(dC)

Table VI lists the chemical shift values of the base and sugar protons of poly(dG)•poly(dC) at 30°C; for complete assignment see ref. 37. As seen from Table VI, GH8 and CH6 of poly(dG)•poly(dC) do overlap in frequency at 30°C. Hence, it is not

Table III
Ranges of Inter Proton Distances (Å) in the Same Nucleotide

	A-DNA (n=11, h=2.56Å) nucleotide geometry: (C3′-*endo*, low *anti*)					
	H6/H8	H1′	H2′	H2″	H3′	CH_3/H5
H6/H8	—	3.8±0.2	3.8±0.2	>4.5	2.8±0.2	3.0*/2.5*
H1′		—	2.9±0.2	2.3±0.2	3.8±0.2	>4.2
H2′			—	1.8*	2.4±0.2	>4.2
H2″				—	3.1±0.2	>4.2
H3′					—	±4.2
CH_3/H5						—
	B-DNA (n=10, h=3.40Å) nucleotide geometry: (C2′-*endo*, *anti*)					
	H6/H8	H1′	H2′	H2″	H3′	CH_3/H5
H6/H8	—	3.9±0.2	2.2±0.2	3.9±0.2	>4.2	3.0*/2.5*
H1′		—	3.1±0.2	2.4±0.2	3.9±0.2	>4.2
H2′			—	1.8*	2.6±0.2	>4.2
H2″				—	2.7±0.2	>4.2
H3′					—	>4.2
CH_3/H5						—

*Fixed distance

Table IV
Ranges of Inter Proton Distances (Å) Between Neighbouring (5′⇒3′) Nucleotides*

	A-DNA (n=11, h=2.56Å) nucleotide geometry: (C3′-*endo*, low *anti*)					
5′⇒3′	H6/H8	H1′	H2′	H2″	H3′	CH_3/H5
H6/H8	—	—	—	—	—	3.8±0.4
H1′	4.4±0.3	—	—	—	—	—
H2′	2.4±0.3	—	—	—	4.1±0.3	3.3±0.3
H2″	3.5±0.3	—	—	—	—	—
H3′	3.5±0.3	—	—	—	—	3.7±0.3
CH_3/H5	—	—	—	—	—	4.0±0.3
	B−DNA (n=10, h=3.40Å) nucleotide geometry: (C2′-*endo*, *anti*)					
5′⇒3′	H6/H8	H1′	H2′	H2″	H3′	CH_3/H5
H6/H8	—	—	—	—	—	3.7±0.4
H1′	3.3±0.3	—	—	—	—	4.3±0.3
H2′	3.9±0.2	—	—	—	—	3.2±0.3
H2″	2.1±0.2	—	4.1±0.2	—	—	2.7±0.3
H3′	—	—	—	—	—	—
CH_3/H5	—	—	—	—	—	—

*Distances over 4.5Å are marked (—)

possible to assign the nucleotide conformation of dG and dC separately by simultaneously irradiating GH8 and CH6. This apparent problem has been circumvented by selectively deuterating GH8 and then conducting the NOE experiments by

Table V
Ranges of Intra Proton Distances Å In Z-DNA (n=6, h=7.5 Å)

Proton Pair	In the Same Nucleotide H8G(H6C)	H1'G(H1'C)	H2'G(H2'C)	H2"G(H2"C)	H3'G(H3'C)	H5C
H8G	—	2.3±0.2	4.1±0.2	>4.5	>4.5	—
(H6C)	(—)	(3.7±0.2)	(3.4±0.2)	(4.4±0.2)	>4.5	(2.4.)
H1'G	—	—	2.8±0.2	2.5±0.2	3.9±0.2	—
(H1'C)	(—)	(—)	(3.2±0.2)	(2.3±0.2)	(3.9±0.2)	(>4.5)
H2'G	—	—	—	1.8*	2.6±0.2	—
(H2'C)	(—)	(—)	(—)	(1.8*)	(2.6±0.2)	(>4.5)
H2"G	—	—	—	—	3.2±0.2	—
(H2"C)	(—)	(—)	(—)	(—)	(2.9±0.2)	(>4.5)
H3'G	—	—	—	—	—	—
(H3'C)	(—)	(—)	(—)	(—)	(—)	(>4.5)
H5C						

Proton Pair 5'-end	With the Neighboring Nucleotide 3'-end H6C	H1'C	H2'C	H2"C	H3'C	H5C
H8G	—	—	—	—	—	—
H1'G	—	—	—	—	—	—
H2'G	3.8±0.2	—	—	—	—	3.9±0.2
H2"G	—	—	—	—	—	—
H3'G	4.1±0.2	—	3.3±2	—	—	—

*Fixed Distance

irradiating CH6 alone (37). Figure 1 shows the result of such a NOE experiment with a presaturation time of 10 msec (found to be optimum for observing essentially the primary NOE pattern). For comparison, the NOE spectrum for 50 msec of presaturation time is also included; notice that non-specific spin-diffusion sites disappear at 10 msec—and the primary sites of NOE are retained (37). From CH6

Table VI
Chemical Shifts of Various Protons (ppm) in Different DNA Duplexes*

DNA/ Temp(°)C	Nucleotide	Chemical Shifts(ppm) for the protons H8/H6	H2	H1'	H2',H2"	H3'	Me/H5
Poly(dG)•Poly(dC)	dG	7.56	—	5.62	2.60,2.44	4.61	—
30°-40°C	dC	7.56	—	5.84	2.44,2.17	4.61	5.51
Poly(dG)•Poly(dC^{5Me})	dG	7.36	—	5.74	2.58,2.24	4.64	—
50°	dC^{5Me}	7.36	—	5.84	2.58,2.24	4.64	1.65
Poly(dA)•Poly(dT)	dA	7.70	6.82	5.64	2.68,2.28	4.88	—
30°-40°C	dT	7.54	—	6.00	2.56,2.10	4.82	1.50
Poly(rA)•Poly(dT)	rA	7.98	7.11	5.68	4.35	4.63	—
30°-40°C	dT	7.54	—	6.18	2.18,1.94	4.60	1.80

*Duplex nature of the DNA polymers was confirmed by monitoring the Watson-Crick protons involved in base-pairing (37-40).

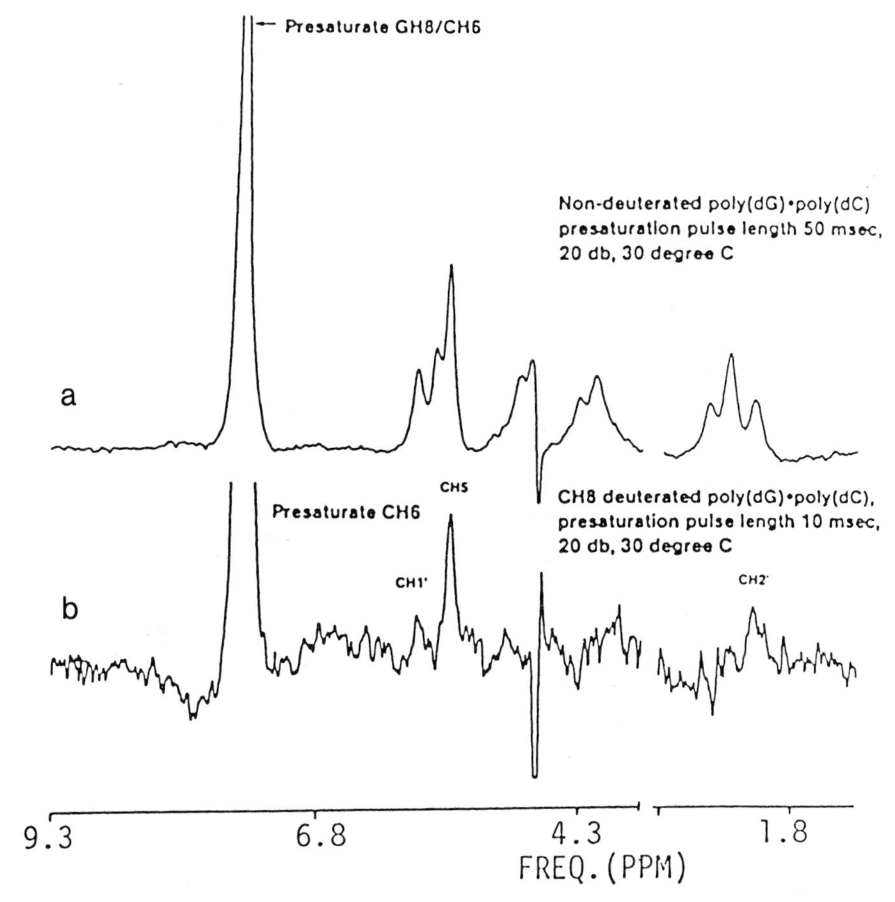

Figure 1. Difference NOE spectrum for presaturation time τ_m=10 msec of GH8-deuterated poly(dG)•poly(dC) at 30° in which CH6 is irradiated; the concentration of DNA was 16 mM in phosphates in 2.5 mM sodium phosphate buffer pH 7.0, 100 mM Nacl, 2.5 mM EDTA. A difference NOE spectrum of non-deuterated poly(dG)•poly(dC), in which GH8/CH6 combined was irradiated for τ_m=50 msec, is also shown for comparison. It is clearly seen how upon reduction of τ_m the primary sites of NOE are alone retained. From CH6 the primary sites of NOE are at CH1′ and only one site in the H2′,H2″ region i.e. CH2′. The NOE pattern is consistent with (C3′-*endo*, low *anti*) geometry for dC (Tables III-IV).

the primary sites of NOE are at CH1′, CH5 and *only one site of NOE in the H2′,H2″ region.* Appearance of only one NOE site in the H2′,H2″ region is consistent with the (C3′-*endo*, low *anti*) conformation of C as in the A-DNA (Tables III-IV). For such a conformation of C, H3′ of C should also show NOE from CH6. However, at 30°C, GH3′/CH3′ of poly(dG)•poly(dC) are hidden under HDO signal and hence, the NOE spectrum of Figure 1 does not permit us to ascertain the presence of NOE at CH3′ from CH6. Therefore, we raised the temperature to 40° such that both GH3′/CH3′ of poly(dG)•poly(dC) came clear of HDO signal. At 40°C, we conducted a NOE experiment in which GH3′/CH3′ combined signal of non-deuterated poly(dG)•poly(dC) was irradiated (Figure 2). At 10 msec of presaturation time, a

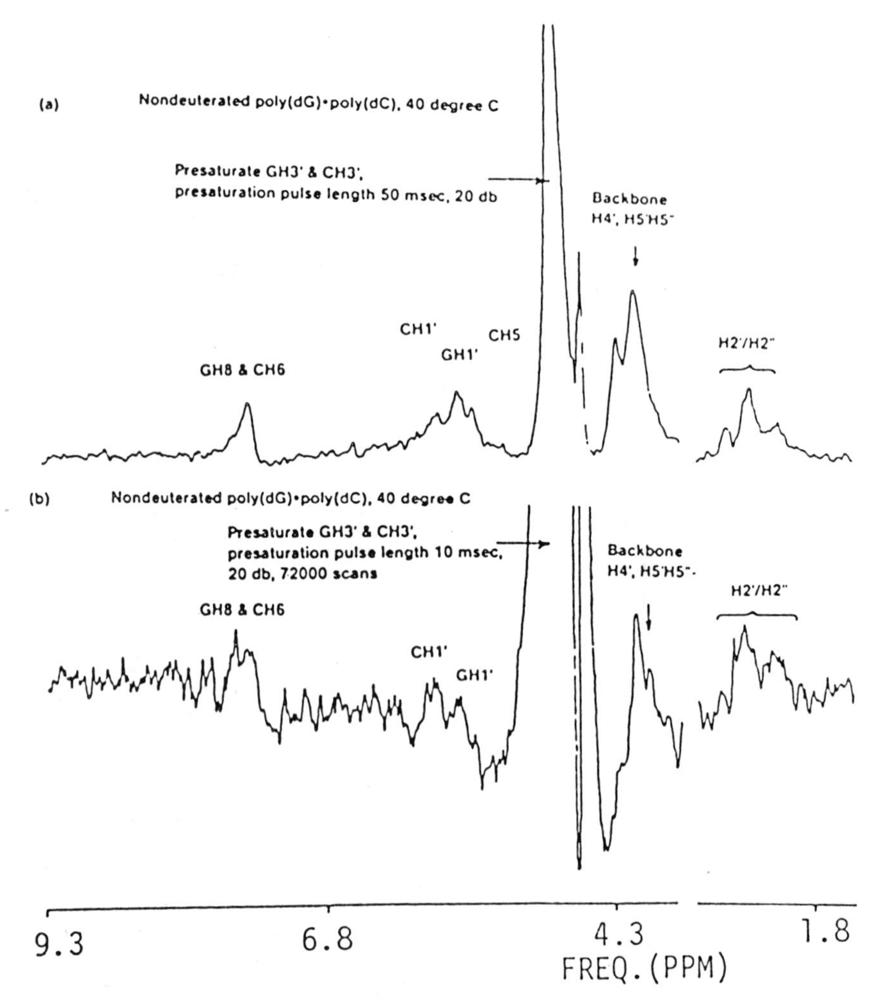

Figure 2. *Spectrum a:* NOE difference spectrum of poly(dG)•poly(dC) at 40°C in which GH3′ and CH3′ are presaturated at 50 msec, 20 db (other conditions being same as in Figure 1). Note that the temperature has to be raised from 30° to 40°C so that H3's could be irradiated without simultaneously presaturating HDO peak. However, GH3′ and CH3′ at 40°C were too close in chemical shift to be individually presaturated and hence GH3′/CH3′ combined was presaturated. *Spectrum b:* NOE difference spectrum of poly(dG)•poly(dC) at 40°C in which GH3′/CH3′ combined is presaturated at 10 msec, 20 db, number of scans were 72,000. The primary sites of NOE from GH3′/CH3′ are GH8/CH6, CH1′, GH1′ and other sugar protons. NOEs from GH3′/CH3′ to the base protons GH8/CH6 clearly suggest that both G and C nucleotides adopt (C3′-*endo*, χ=200°-220°) conformation, as in the A-DNA.

primary site of NOE from H3′ was located at GH8/CH6. Thus, the combined results of Figures 1, 2 proved that in poly(dG)•poly(dC) both dG and dC residues adopt (C3′-*endo*, low *anti*) conformation as in the A-DNA. CD spectra of poly(dG)•poly(dC) within a temperature range of 30°-60°C revealed the gross morphology of the A-form and alcohol had no effect upon the CD spectra (37). Molecular models were then constructed subject to the following constraints: (i) helical parameters as in the A-form (Table I), (ii) dG/dC nucleotides in (C3′-*endo*, low *anti*) domain as required by the NOE data, (iii) models being free of steric compression. Table VII lists the conformational parameters of one such model which also compares very well with the average conformations of nucleotides in the A-DNA crystals (Table II).

Table VII

Conformational Parameters for A,B,Z-models Proposed by us from our NOE data

	DNA/RNA	Form	Nucl	Torsion Angles (Deg) α	β	γ	δ	ϵ	ξ	χ
I.	Poly dG)• Poly(dC)	A	dG dC	288	185	50	79	193	300	207
II.	Poly(dG)• Poly(d^{5Me}C)	B(B_{II})	dG dC^{5Me}	310	149	42	139	221	212	253
III.	Poly(dA)• Poly(dT)	B(B_{II})	dA dT	310	149	42	139	221	212	253
IV.	Poly(rA)• Poly(dT)	B(B_I)	rA dT	286	161	72	120	198	260	243
V.	Poly(dG-dC)• Poly(dG-dC)	Z(Z_I)	dG dC	86 218	169 213	154 55	96 128	248 260	314 90	55 200

A⇒B transition in Poly(dG)•Poly(dC): Effect of 5 Methyl Cytosine

As mentioned above physico-chemical factor viz. temperature, salt or presence of alcohol in the solvent, cause no change in poly(dG)•poly(dC) in solution; however, replacement of C by ^{5Me}C induced A⇒B transition in poly(dG)•poly(dC). Figure 3 shows the difference NOE spectra of poly(dG)•poly(dC^{5Me}) at 50°C in which combined GH8/CH6 was irradiated, notice that the spectrum at 25 msec of presaturation time reveals the primary NOE sites from GH8/CH6: they are H1′, H2′, H2″ and methyl group of ^{5Me}C. Absence of NOE at H3′ and strong NOEs at H2′,H2″ region suggest dG/dC^{5Me} adopt (C2′-*endo*, *anti*) conformation as in the B-DNA. Figure 4 shows the difference NOE spectrum of poly(dG)•poly(dC^{5Me}) in which GH3′/CH3′ combined were presaturated for 25 msec at 70°C and the absence of NOE at GH8/CH6 reaffirmed that dG/dC^{5Me} residues do not take up (C3′-*endo*, low *anti*) conformation as in the native A-form of poly(dG)•poly(dC) (compare Figs. 2 and 4). CD spectra of poly(dG)•poly(dC^{5Me}) within 50°-70° also reveal the gross-morphology of the B-form (37). The conformational parameters of the model generated for poly(dG)•poly(dC^{5Me}) in the B-form are listed in Table VII. Figure 5 illustrates the difference in the stacking arrangements in poly(dG)•poly(dC) [A-form] and poly(dG)•poly(dC^{5Me}) [B-form].

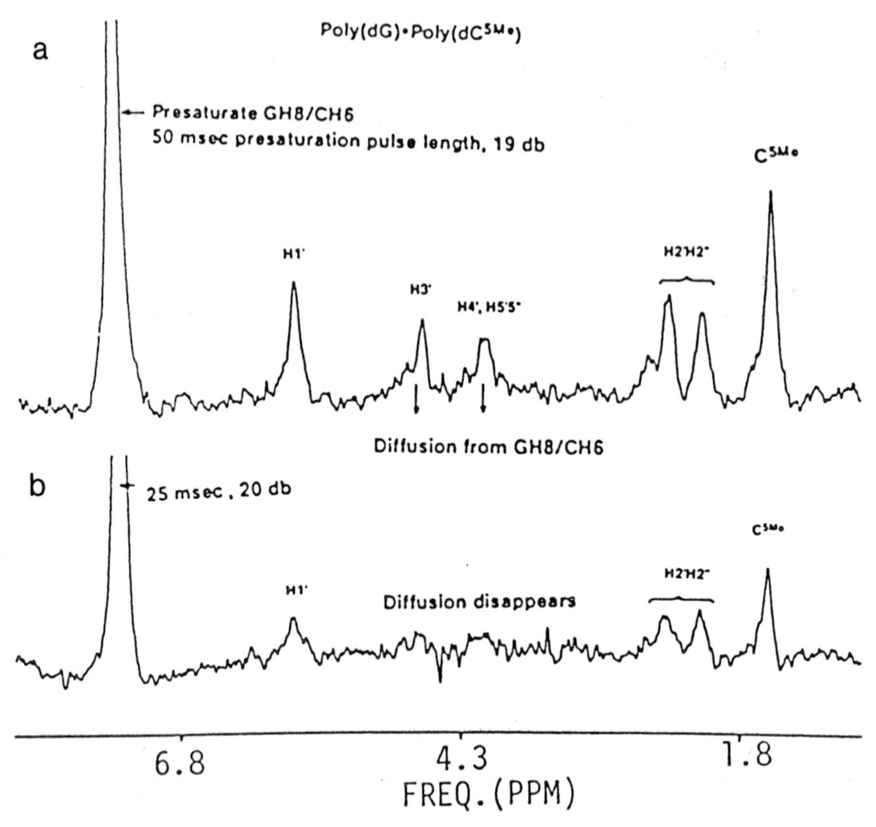

Figure 3. *Spectrum a:* NOE difference spectrum of poly(dG)•poly(dC^{5Me}) at 50°C when GH8/CH6 were presaturated at 50 msec, 19 db, number of transients were 32,000. It should be noted that at 70°C the H3′ protons of G and C residues resolve to separate positions at 4.74 and 4.60 ppm, this is not the case at 50°C. *Spectrum b:* NOE difference spectrum of poly(dG)•poly(dC^{5Me}) at 50°C when GH8/CH6 were presaturated at 25 msec, 20 db, number of scans were 56,680; it took 49 hrs to complete the data accumulation. Notice that the sites of spin diffusion present in spectrum (a) have disappeared here. The primary sites of NOE from GH8/CH6 are H1′, H2′H2″ and C^{5Me} as expected from the interproton distances for (C2′-*endo*, χ=240°-260°) as in B-DNA. Note that there is no NOE at H3′-region thus ruling out the possibility of (C3′-*endo*, χ=200°-220°) conformation as in A-DNA (Tables III-IV).

II. B-form for Poly(dA)•Poly(dT)

Recently it has been proposed from the fiber diffraction data that the B-form of poly(dA)•poly(dT) is heteronomous in nature (19) i.e. two chains of the duplex are conformationally very different with (C3′-*endo*, low *anti*) conformation for dA residues and (C2′-*endo*, *anti*) conformation for the dT residues. Following exhaustive NOE studies we find that poly(dA)•poly(dT) in solution (100 mM NaCl, 30°C) adopts a conventional B-form with two chains being conformationally very similar (38).

Figures 6, 7 show the difference NOE spectra of poly(dA)•poly(dT) in solution in which AH8 and TH6 are presaturated for 10 msec. Notice the primary NOE pattern

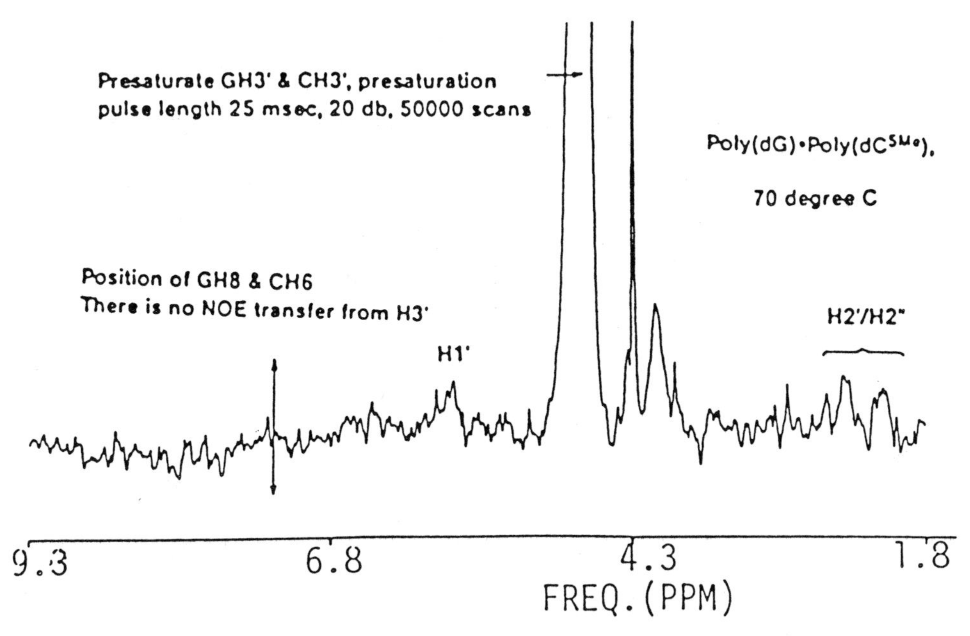

Figure 4. NOE difference spectrum of poly(dG)•poly(dC^{5Me}) when H3′ region has been irradiated at 25 msec, 20 db;—conditions under which there is very little spin-diffusion; 50,000 transients were recorded (for 49 hours) for this spectrum. Notice that there is no NOE from H3′ to the base protons GH8/CH6 thereby ruling out the possibility of (C3′-*endo*, χ=200°-220°) conformation for both G and C. This NOE difference spectrum of poly(dG)•poly(dC^{5Me}) provides a contrast to that of poly(dG)•poly(dC) in Figure 2.

of NOE from AH8 and TH6 bears the signatures of (C2′-*endo, anti*) conformation for both dA and dT residues i.e. equally strong NOEs at H2′,H2″, relative weak NOE at H1′; and absence of NOE at H3′. Also, the absence of NOE from H3′ to AH8/TH6 reaffirmed that neither dA nor dT residue adopts (C3′-*endo*, low *anti*) conformation. Thus, the results of Figures 6, 7 rule out the possibility of the heteronomous DNA model of poly(dA)•poly(dT) in solution [because in such a model dA residues are in (C3′-*endo*, low *anti*) conformation]. Raman spectroscopic studies conducted under conditions same as in our NMR experiments also indicated that both dA/dT residues are conformationally similar (41). Table VII lists the conformational parameters for the B-DNA model of poly(dA)•poly(dT) in solution; the model agrees with the NOE and raman spectroscopic data (38).

Subtle structural changes in the hybrid poly(rA)•poly(dT) as required due to the introduction of the 2′OH group

Zimmerman and Pheiffer postulated that the hybrid poly(rA)•poly(dT) adopts a heteronomous structure in fiber i.e. the rA chain with residues in C3′-*endo* sugar and the dT chain with residues in C2′-*endo* sugar. It was emphasized that C3′-*endo* sugar for rA residues was a stringent stereochemical requirement for accommo-

a

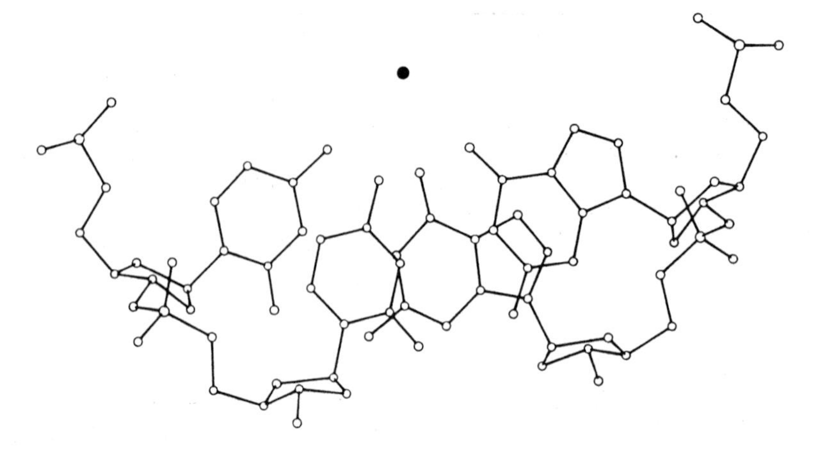

Figure 5. A. (pdGpdG)•(pdCpdC) fragment in the final A-DNA model (n=11, h=2.56Å) of poly(dG)•poly(dC). In this model both G and C nucleotides adopt (C3'-*endo*, χ=200°-220°) conformation. *On top* is the view along the helix-axis which shows that bases do have a considerable tilt (20°); *on bottom* is the view down the helix-axis which shows that bases are displayed away from the helix center (by about 4Å). B. (pdGpdG)•(pdC^{5Me}pdC^{5Me}) fragment of the final B-DNA model (n=10, h=3.40Å) of poly(dG)•poly(dC^{5Me}) viewed along and down the helix-axis. Notice that upon methylation bases are moved from the peripheri towards the inner core of the helix concomitant with A⇒B transition.

dating the 2'-OH group (42). However, we find that the hybrid poly(rA)•poly(dT) adopts a B-DNA like structure in solution with both rA and dT residues having conformation in the (C2'-*endo*/C1'-*exo*, *anti*) domain. Although, subtle structural changes occur in going from poly(dA)•poly(dT) to poly(rA)•poly(dT) structure in dating the 2'-OH group (42). However, we find that the hybrid poly(rA)•poly(dT) adopts a B-DNA like structure in solution with both rA and dT residues having

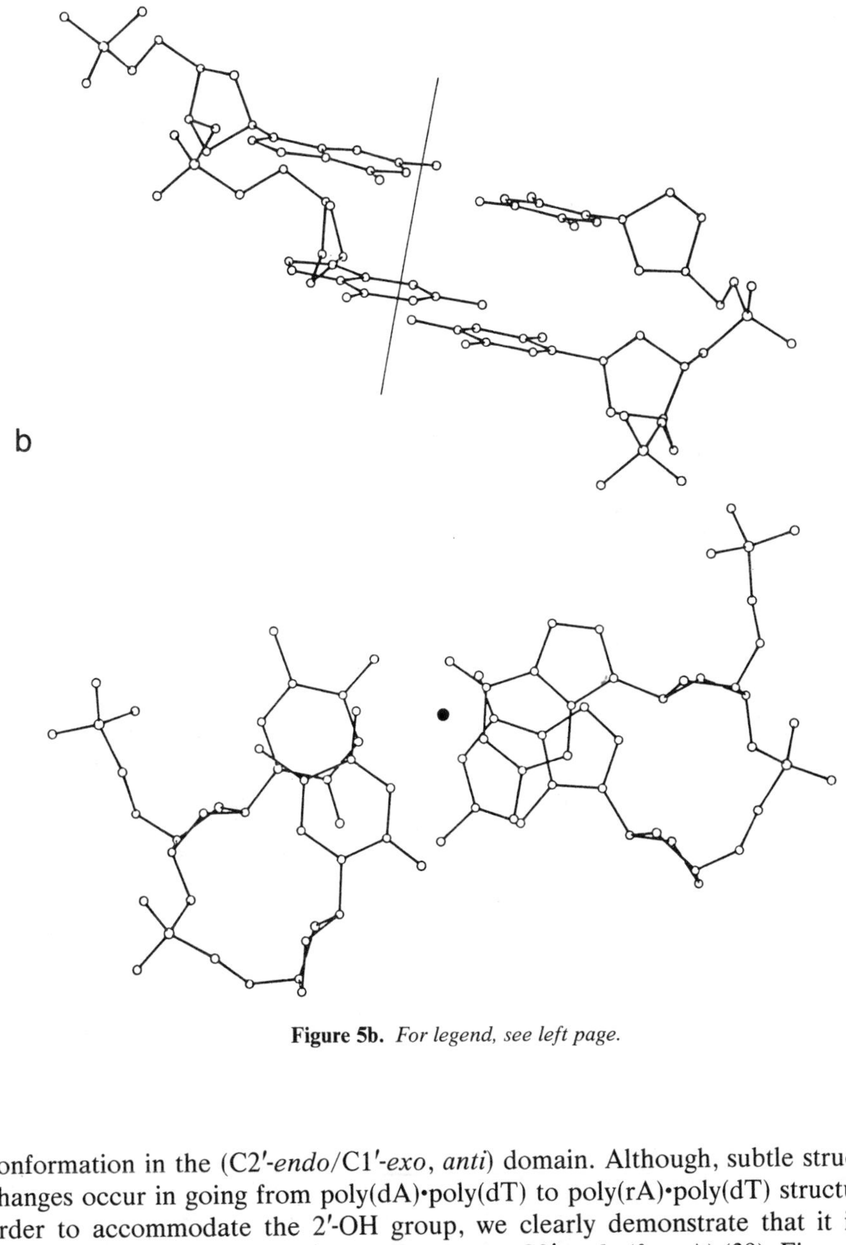

b

Figure 5b. *For legend, see left page.*

conformation in the (C2′-*endo*/C1′-*exo*, *anti*) domain. Although, subtle structural changes occur in going from poly(dA)•poly(dT) to poly(rA)•poly(dT) structure in order to accommodate the 2′-OH group, we clearly demonstrate that it is not obligatory to switch from C2′-*endo* (for dA) to C3′-*endo* (for rA) (39). Figures 8, 9 show the difference NOE spectra of poly(rA)•poly(dT) in which AH8 and TH6 are presaturated. Notice that the primary NOE pattern (for the presaturation time of 25 msec) indicates that both rA/dT residues are conformationally similar and they fall in the (C2′-*endo*/C1′-*exo*, *anti*) domain. Absence of NOE from AH8 to AH3′ rules out the possibility of (C3′-*endo*, low *anti*) conformation for rA residues. Figure 10 shows the stacking arrangement in a regular B-DNA model of poly(dA)•poly(dT) in

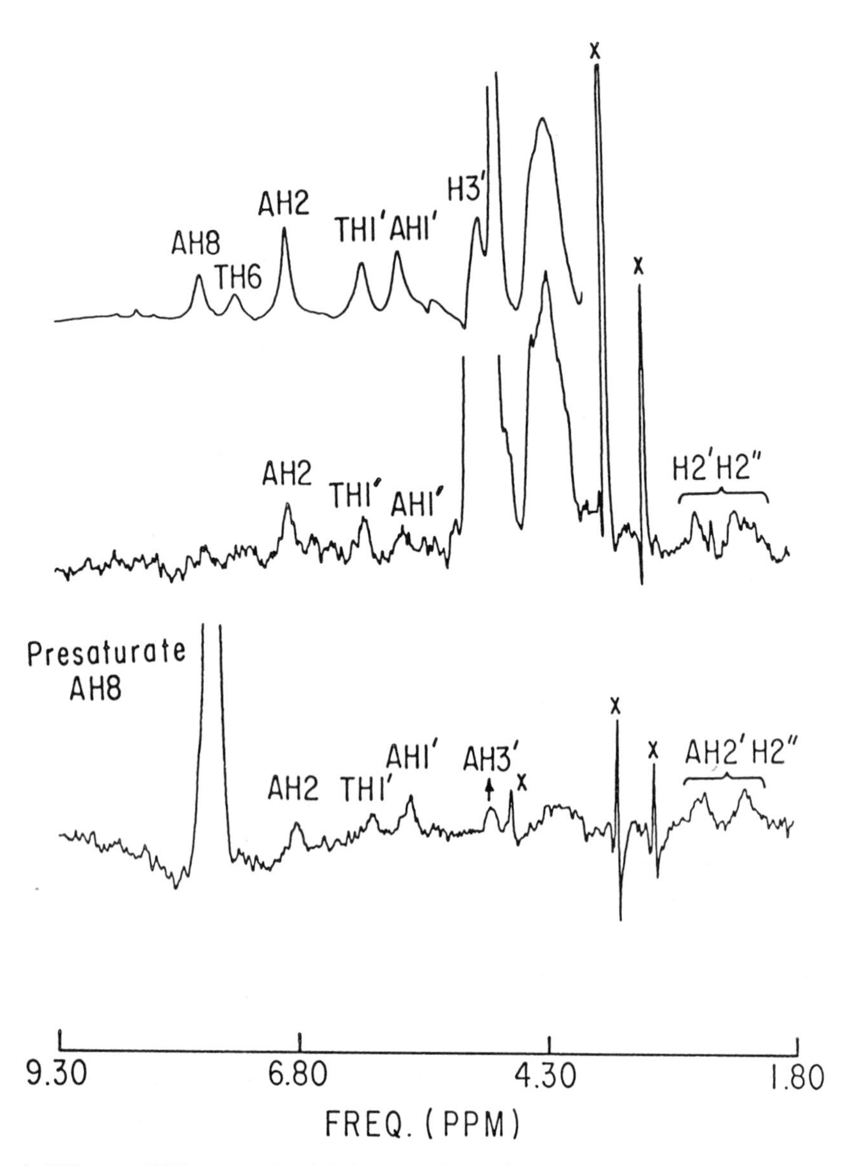

Figure 6. Difference NOE spectra of poly(dA)•poly(dT) at 30°C in which AH8 (bottom) and AH3′ and TH3′ (top) are irradiated for τ_m=10 msec, 20 db. DNA concentration was 30 mM in phosphate in 2.5 mM sodium phosphate buffer, pH 7.0, 100 mM NaCl, 2.5 mM EDTA. For both the spectra, no. of transients was 56,000. The sharp spikes (x) are substration artifacts. The primary sites of NOE from AH8 are: H2′,H2″ (two being equally strong), AH1′; notice that there is a small but noticeable residual NOE at AH3′ which we believe diffused from H2′,H2″. Absence of NOE from AH3′ and TH3′ to base protons AH8/TH6 (top spectrum) confirms that both dA and dT residues adopt (C2′-*endo*, *anti*) conformation as in B-DNA.

which sugar pucker [for both dA and dT] is classical C2′-*endo* (δ=139°) and P-O torsions (ξ,α) are tg$^-$ (212°,310°); incorporation of 2′-OH in such a conformation results in severe short contact between base and sugar. However, by changing the

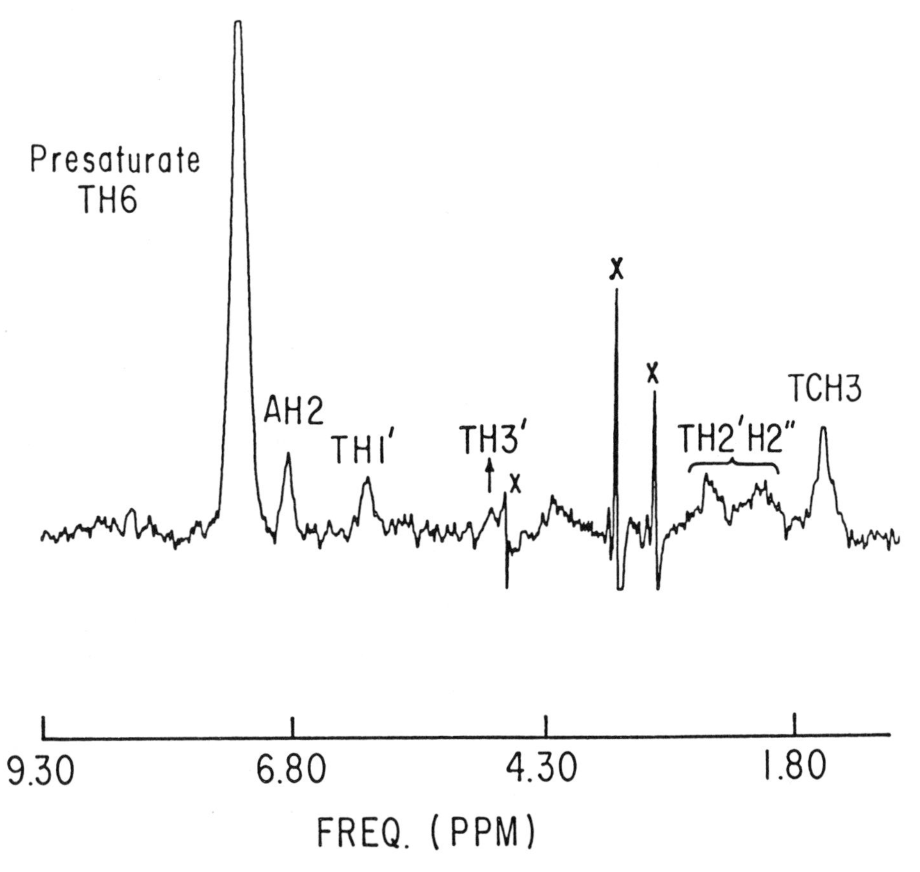

Figure 7. Difference NOE spectrum of poly(dA)•poly(dT) at 30°C in which TH6 is irradiated (τ_m=10 msec, no. of scans = 56,000) other conditions being same as in Figure 6. The primary NOE sites from TH6 are TH2′,H2″ (two being equally strong) TH1′,TCH3 consistent with (C2′-*endo*, *anti*) conformation for dT. For explanation of NOE at AH2 (see ref. 38).

sugar pucker to C1′-exo (δ=120°) (both for rA and dT) and a concomitant change in P-O torsions to g^-g^- (ξ=260°,α=286°), it is possible to generate a B-DNA model for poly(rA)•poly(dT) without switching to C3′-*endo* sugar for rA. In this connection it may be mentioned that Reid et. al. (43) observed that in the hybrid d(TCACAT)•r(AUGUGA) nucleotides on both the strands adopt (C2′-endo/C1′-*exo*, *anti*) conformation. Figure 10 also displays the stacking arrangement in poly(rA)•poly(dT) as suggested by us from our NOE data—notice that the model is free of steric compression.

III. On Z-DNA and B-Z Transition

Based upon the principle outlined above we are able to identify the Z-form of poly(dG-dC) [in high salt] in terms of nucleotide geometry. We were able to show

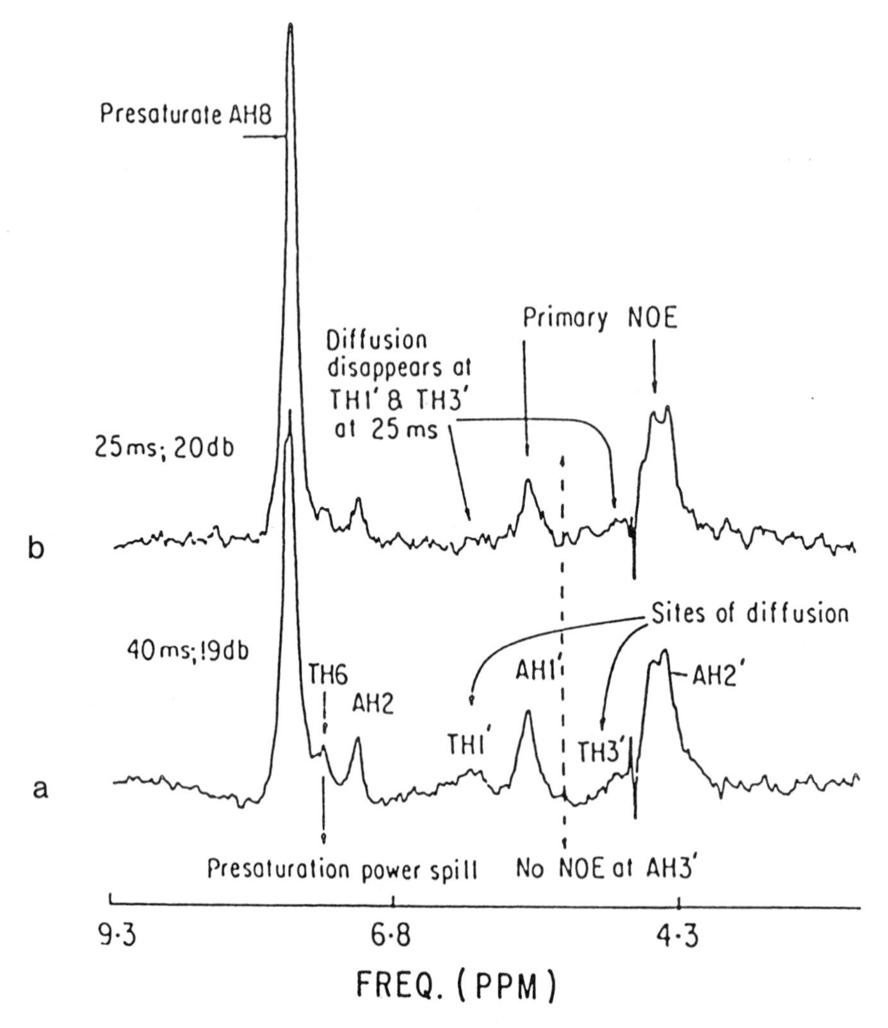

Figure 8. NOE difference 500 MHz ^1H n.m.r. spectra of poly(rA)•poly(dT) in 100% ^2H_2O at 30°C in which AH8 is presaturated using different presaturation pulse lengths and power: (a) 40 ms. 19 db; and (b) 25 ms. 20 db. The concentration of DNA was 20mM in phosphate in 2.5 MM-sodium phosphate buffer (pH 7.0), 100 mM NaCl, 2.5 mM EDTA. The 40 ms spectrum was recorded after 240×40 cycles. Comparison of the two NOE difference spectra indicate that as one lowers the presaturation pulse length and power, the secondary NOEs at TH1′ and TH3′ completely disappear. The primary NOEs from AH8 are at AH1′ and AH2′ of the rA residue. Notice that there is no NOE at the AH3′ position when AH8 is irradiated, i.e. AH3′-AH8 distances are not close enough. This rules out the possibility that the rA nucleotide is in C3′-*endo* sugar pucker and χ low anti (200° to 220°) where the AH3′-AH8 distance in the same nucleotide is 2.9Å (Table III). However, the primary NOE pattern is consistent with a nucleotide geometry for rA in which the sugar pucker is C2′-*endo* or C1′-*exo* and χ is *anti* (240° to 260°), i.e. the AH2′-AH8 distance in the same nucleotide is 2.4Å while AH1′-AH8 distances are within 3.4 to 3.8Å. (Tables III-IV).

that in the Z-DNA dG and dC residues are conformationally dissimilar—dG residues adopt (C3-*endo*, *syn*) conformation while dC residues adopt (C2′-*endo*, low

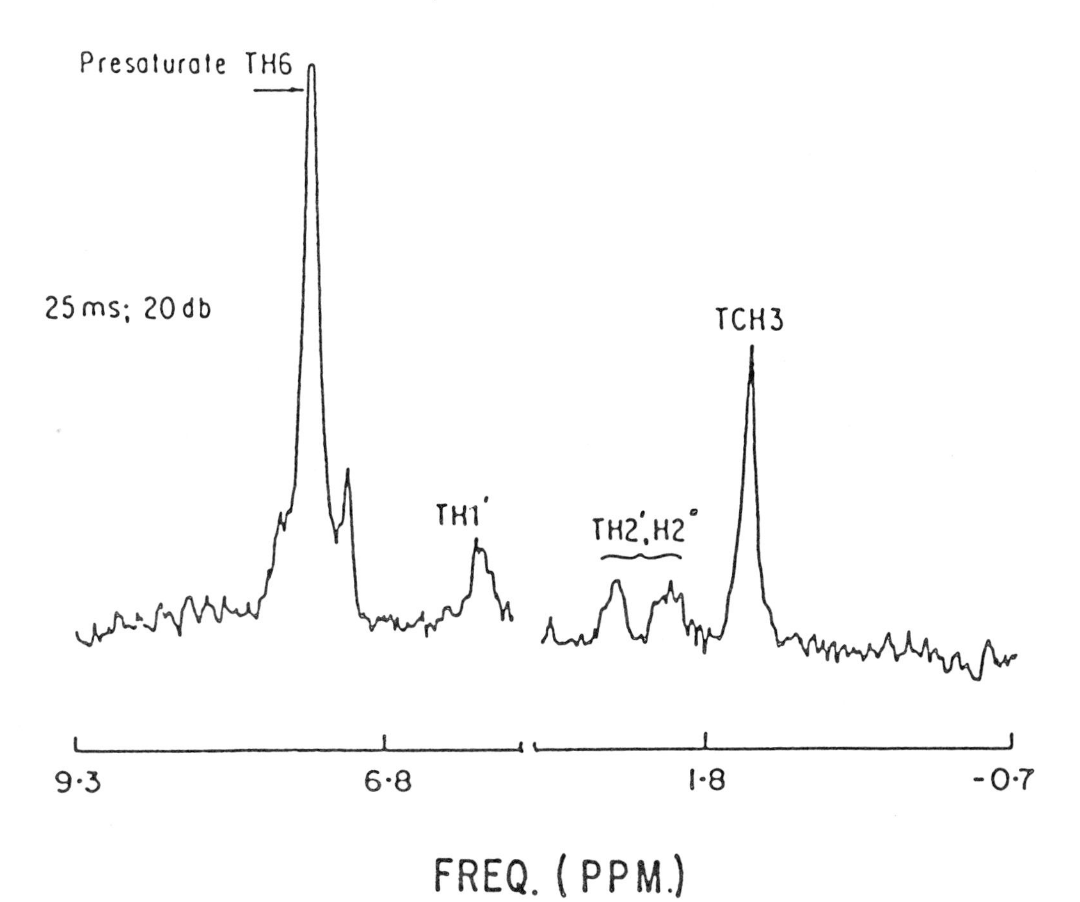

Figure 9. Primary NOE sites (TH1′, TH2′H2″, TCH3) in poly(rA)•poly(dT) when TH6 is presaturated; presaturation pulse length 25 msec., 20 db. Nearly equally strong NOEs at TH2′ and TH2″ from TH6 suggest that dT nucleotides are in (C2′-*endo*, or C1′-*exo*, χ = 240° to 260°) conformation in which the distance TH6-TH2′ of the same nucleotide is 2.4Å, while the distance TH6-TH2″ of the neighboring nucleotide is 2.1Å. For dT nucleotides in (C3′-*endo*, χ=200° to 220°) conformation the distance TH6-TH2′ of the neighboring nucleotide is 1.9Å, while the distances TH6-TH2″ are always greater than 4Å (Tables III-IV). Thus in such a case very dissimilar primary NOEs are expected at TH2′ and TH2″ when TH6 is irradiated contrary to what is seen in the Figure.

anti) conformation (40,44). By monitoring the N1-H proton of the G-C pair, we have been able to describe the stereochemical trajectory of salt induced B⇒Z transition (45): the transition involves coupling of longitudinal and transverse breathing modes which initiates the transition by progressively unwinding the B-form but without breaking the H-bonds, at the transition midpoint the base-pairs are totally unwound but H-bonds are intact and finally the base-pairs wind into the Z-form. During the whole process, there is no large-scale breakage of Watson-Crick G-C pairs as revealed by our experiments (45). Also from formaldehyde exchange studies (46) it has been shown that there is no large-scale breakage of H-bonds during salt-induced B⇒Z

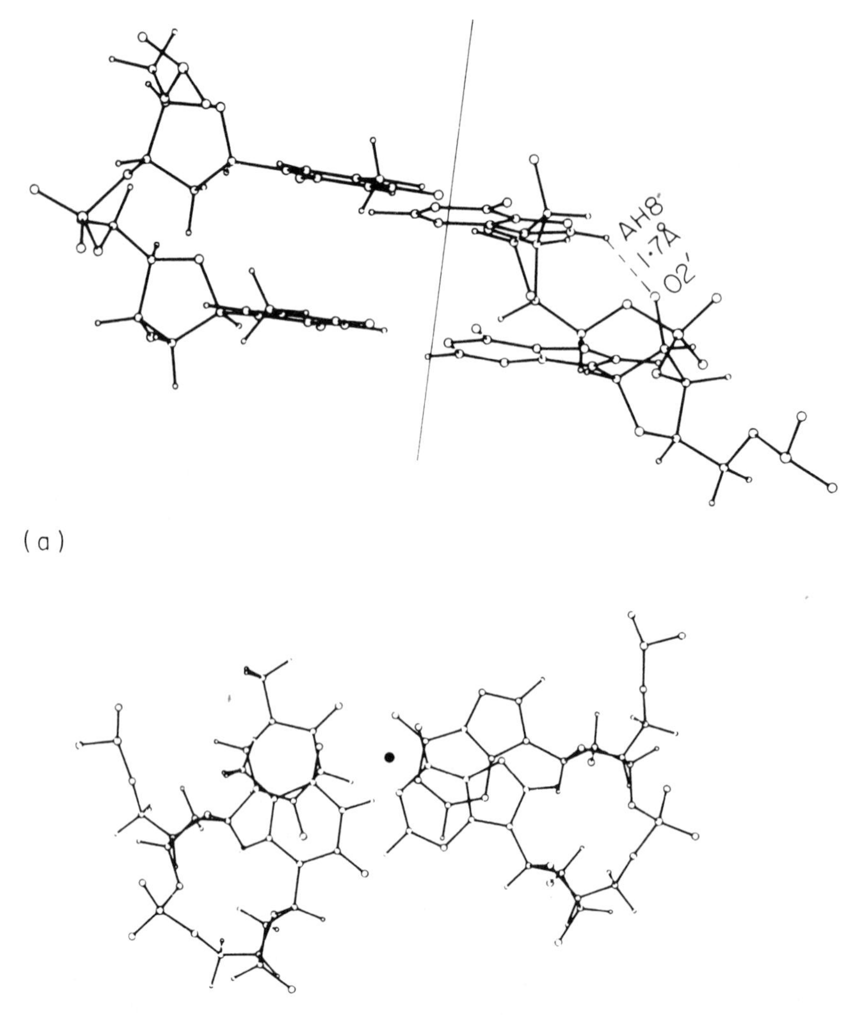

Figure 10. A. d(pApA)•d(pTpT) fragment in the B-form of poly(dA)•poly(dT). Both dA and dT residues adopt the B_{II} conformations (Table VII) i.e. sugar pucker is pure C2′-*endo* (δ=139°), P-O torsions are tg^-. Notice, that in such a structure, the incorporation of a 2′-OH results in a severe sugar-base short contact (shown by a broken line). B. r(pApA)•d(pdTpdT) fragment in the B-form of poly(rA)•poly(dT). Both rA and dT residues adopt the B_I conformation (Table VII) i.e. sugar pucker is C1′-*exo* (δ=120°), a minor variant of C2′-*endo*, P-O torsions are g^-g^-. Notice, that in such a structure bases are displaced away from the helix centre (shown by a dark circle) by about 1.2Å, and also bases are tilted and twisted. Such a structure can easily accommodate the 2′-OH group without any steric compression.

transition. We have also carried out molecular model building studies and showed that Z-DNA can support intercalators like ethidium, actinomycin D etc. (47). We also proposed that such transient intermediate could be the primary initiation process in drug induced Z⇒B transition—which found experimental support by recent studies (48).

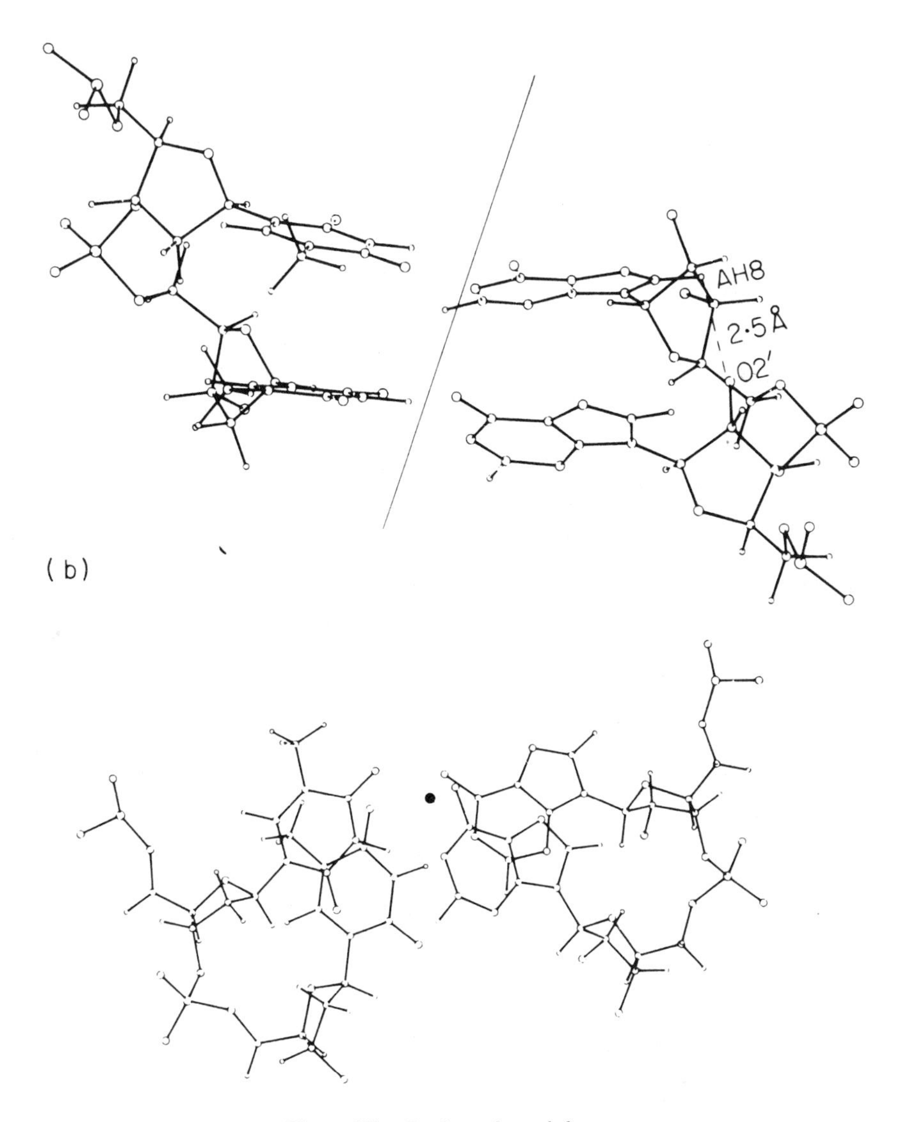

Figure 10b. *For legend, see left page.*

Acknowledgement

This research is supported by a grant from the National Institutes of Health (GM29787) and by a contract from the National Foundation for Cancer Research. The high field NMR experiments were performed at the NMR facility for Biomolecular Research located at the F. Bitter National Magnet Laboratory, M.I.T. The NMR Facility is supported by grant no. RR00995 from Division of Research Resources of the NIH and by the National Science Foundation under contract no. C-670.

References and Footnotes

1. Franklin, R.E. and Gosling, R.G., *Acta Cryst. 6,* 673 (1953).
2. Feughelman, M., Langridge, R., Seeds, W.E., Stokes, A.R., Wilson, H.R., Hooper, C.W., Wilkins, M.H.F., Barclay, R.K. & Hamilton, L.D., *Nature, 175,* 834 (1955).
3. Langridge, R., Seeds, W.E., Wilson, H.R., Hooper, C.W., Wilkins, M.H. and Hamilton, L.D., *J. Biophys. Biochem. Cytol., 3,* 767 (1957).
4. Langridge, R., Wilson, H.R., Hooper, C.W. Wilkins, M.H.F. and Hamilton, L.D., *J. Mol. Biol. 2,* 19 (1960).
5. Franklin, R.E. and Gosling, R.H., *Nature 172,* 156 (1953).
6. Fuller, W., Wilkins, M.H.F., Wilson, H.R. and Hamilton, L.D., *J. Mol. Biol., 12,* 60 (1965).
7. Wilkins, M.H.F., *J. Chem. Phys. 65,* 891 (1961).
8. Tunis-Schneider, M., -J., B. and Maestre, M.F., *J. Mol. Biol., 52,* 521 (1970).
9. Ivanov, V.I., Minchenkova, L.e., Schyolkina, A.K. and A.I. Poletayer, *Biopolymers, 12,* 89 (1973).
10. Arnott, S., Chandrasekaran, R., Birdsall, D.L., Leslie, A.G.W. and Ratliff, R.L., *Nature, 283,* 743 (1980).
11. Sasisekharan, V. and Brahmachari, S.K., *Curr. Sci. 50,* 10 (1981).
12. Pohl, F.M. and Jovin, T.M., *J. Mol. Biol., 67,* 375 (1972).
13. Pohl, F.M., Jovin, T.M., Baehr, W. and Holbrook, J.J., *Proc. Natl. Aca. Sci. USA, 69,* 3805 (1972).
14. Pohl, F.M., *Nature 260,* 365 (1976).
15. Arnott, S. and Selsing, E., *J. Mol. Biol. 88,* 551 (1974).
16. Gray, D.M., Hamilton, F.D. and Vaughn, M.R., *Biopolymers, 17,* 85 (1978).
17. Gray, D.M. and Bollum, F.J., *Biopolymers, 13,* 2087 (1974).
18. Arnott, S. and Selsing, S., *J. Mol. Biol., 88,* 509 (1974).
19. Arnott S., Chandrasekharan, R., Hall, I.H. and Puigjaner, L.C., *Nucl. Acid Res. 11,* 4141 (1983).
20. Allen, F.S., Gray, D.M., Roberts, G.P. and Tinoco, J., Jr., *Biopolymers 11,* 853 (1972).
21. Greeve, J., Maestre, M.F. and Levin, A., *Biopolymers, 16,* 1489 (1977).
22. Gupta, G., Bansal, M. and Sasisekharan, V., *Prot. Natl. Acad. Sci. USA, 77,* 6486 (1980).
23. Kim, S.H., Berman, H.M., Seeman, N.C. and Newton, M.D., *Acta Cryst. B29,* 703 (1973).
24. Conner, B.M., Takano, T., Tanaka, S., Itakura, K. and Dickerson, R.E., *Nature 295,* 294 (1982).
25. Wang, A.H.J., Fujii, S., van Boom, J.H. and Rich, A., *Proc. Natl. Acad. Sci. USA, 79,* 3968 (1982).
26. Shakked, Z., Rabinovich, D., Kennard, O., Cruse, W.B.T., Salisbury, S.A. and Viswamitra, M.A., *J. Mol. Biol., 166* 183 (1983).
27. Wing, R.M., Drew, H.R., Takano, T., Broka, C., Tanaka, S., Itakura, K. and Dickerson, R.E., *Nature 287,* 755 (1980).
28. Dickerson, R.E., Drew, H.R., Conner, B.N., Kopka, M.L. and Pjura, P.E., *Cold Spring Harbor Symp. Quant. Biol. 47* (1983).
29. Fratini, A.V., Kopka, M.L., Drew, H.R. and Dickerson, R.E., *J. Biol. Chem. 257,* 14684 (1982).
30. Quigley, G.J., Wang, A.H.J., Ughetto, G., van der Marel, G., van Boom, J.H. and Rich, A., *Proc. Natl. Acad, Sci. USA, 77,* 7204 (1980).
31. Drew, H.R. and Dickerson, R.E., *J. Mol. Biol., 152,* 723 (1981).
32. Wang, A.H.J., Quigley, G.J., Kolpak, F.J., van der Marel, G., van Boom, J.H. and Rich, A., *Nature, 282,* 680 (1979).
33. Wang, A.H.J., Quigley, G.J., Kolpak, F.J., van der Marel, G., van Boom, J.H. and Rich, A., *Science, 211,* 171 (1981).
34. Wang, A.H.J., Fujii, S., van Boom, J.H. and Rich, A., *Cold Spring Harbor Symp. Quant. Biol. 47* (1983).
35. Gupta, G., Ph.D. Thesis (1981), Indian Institute of Science, Bangalore, India.
36. Gupta, G., Sarma, M.H. and Sarma, R.H., *Int. J. Quantum Chem: Quantum Biol. Symp., 12,* 183 (1986).
37. Sarma, M.H., Gupta, G. and Sarma, R.H., *Biochemistry* (1986) *in press.*
38. Sarma, M.H., Gupta, G. and Sarma, R.H., *J. Biomol. Str. & Dyn., 2,* 1057 (1985).
39. Gupta, G., Sarma, M.H. and Sarma, R.H., *J. Mol. Biol. 186,* 463 (1985).
40. Dhingra, M.M., Sarma, M.H., Gupta, G. and Sarma, R.H., *J. Biomol. Str. & Dyn., 1,* 417 (1983).
41. Thomas, G.A. and Peticolas, W.L., *J. Am. Chem. Soc., 105,* 993 (1983).
42. Zimmerman, S.B. and Pheiffer, B.H., *Proc. Natl. Acad. Sci. USA, 78,* 78 (1981).

43. Reid, D.G., Salisbury, S.A., Brown, T., William, D.H., Vasseur, J., Rayner, B. and Imbach, *J. Eur. J. Biochem., 135,* 307 (1983).
44. Patel, D.J., Kozlowski, S.A., Nordheim, A. and Rich, A., *Proc. Natl. Acad. Sci. USA, 79,* 1413 (1982).
45. Sarma, M.H., Gupta, G., Dhingra, M.M. and Sarma, R.H., *J. Biomol. Str. & Dyn., 1,* 59 (1983).
46. Poverenny, A.M., Kiseleva, V.F., Tyaglov, B.V. and Permogorov, V.I., *FEBS Letters, 186,* 197 (1985).
47. Gupta, G. and Sarma, R.H., In 'Nucleic Acids—The Vectors of Life' (Ed. Pullman, B. and Jortner, J.), Recidel Press, Dodricht (1983) pp. 457.
48. Shafer, R.H., Brown, S.C., Delbarre, A. and Wade, D., *Nucl. Acid. Res., 12,* 4679 (1984).

Biomolecular Stereodynamics IV, Proceedings of the Fourth Conversation in the Discipline Biomolecular Stereodynamics, State University of New York, Albany, NY, June 04-09, 1985, Eds., Ramaswamy H. Sarma & Mukti H. Sarma, ISBN 0-940030-18-7, Adenine Press,

The Left-handed Z-form of Double-Stranded RNA

Phillip Cruz, Kathleen Hall, Joseph Puglisi, Peter Davis, Charles C. Hardin, Mark O. Trulson, Richard A. Mathies, Ignacio Tinoco, Jr.
Department of Chemistry
University of California, Berkeley
Berkeley, CA 94720

W. Curtis Johnson, Jr.
Department of Biochemistry and Biophysics
Oregon State University
Corvallis, OR 97331

and

Thomas Neilson
Departments of Biochemistry and Chemistry
McMaster University
Hamilton, Ontario, L8N 3Z5, Canada

Abstract

Double-stranded RNA containing alternating guanosine-cytidine sequences has been shown to assume a high-salt conformation analogous to the left handed Z-form of DNA. The evidence for Z-RNA in poly[r(G-C)] in concentrated salt solutions originates from several spectroscopic techniques. Phosphorus NMR indicates that there are two very different phosphate environments, consistent with the zig-zag backbone characteristic of Z-form. Proton NMR studies reveal that the guanosine residues adopt a *syn* conformation. The Raman spectrum shows the loss of the phosphodiester symmetric stretch band at 810 wavenumbers. The right to left handed transition for RNA is accompanied by large changes in the circular dichroism spectra. Comparison of the vacuum UV circular dichroism spectra of RNA and DNA in both right and left handed forms indicates that this technique is a sensitive indicator of handedness. The solvent conditions required to effect the transition are more stringent with RNA than DNA. These include high concentrations of specific salts such as sodium perchlorate and sodium bromide, and high concentrations of alcohols. Increasing the temperature also favors formation of Z-form. Bromination of poly r(G-C) facilitates conversion to the Z-form. Base pair opening rates, as monitored by imino proton exchange, are much greater with Z-RNA than Z-DNA. The tetranucleotide r(CGCG) can also assume the Z-form, as shown by CD and NMR methods. NMR experiments to examine the kinetics of the right to left handed transition are described.

Introduction

Recently we have shown that the synthetic RNA polymer, poly[r(G-C)], can undergo a transition from the right-handed A-form to a conformation analogous to the left-handed Z-form observed in DNA (1). This observation is significant in that it demonstrates that double-stranded RNA, like DNA, can display conformational lability, i.e., RNA is not locked into A-type structures, and can even exist in left-handed forms. This is the first known occurrence of RNA in other than A-form. This ability to assume a Z-structure may be used *in vivo* as a source of torsional motion, as well as a means for recognizing specific sequences.

Before discussing Z-RNA further it is useful to illustrate some of the properties of Z-DNA that can be sought in spectroscopic probes of an unknown structure. Several of these are listed in Table I. Z-DNA was originally recognized in an atomic resolution X-ray crystal diffraction study of the alternating pyrimidine-purine oligomer d(CpGpCpGpCpG) (2). The corresponding polymer, poly[d(G-C)], has long been

Table I
Differences Between A-form RNA and Z-DNA that Can Be Probed Spectroscopically

Property	A-form	Z-DNA
Backbone	Regular[1]	Zig-zag
Glycosidic Torsion	All *anti*	Cytidine—*anti* Guanosine—*syn*
Sugar Pucker	All 3′-*endo*	Cytidine—2′-*endo* (S) Guanosine—3′-*endo* (N)
GH8 Position	Inside Major Groove	Out in Solution

[1]The A-form backbone does have some small, sequence-dependent changes.

known to undergo a cooperative structural transition upon increasing salt concentration to a form characterized by an "inversion" in the near UV circular dichroism (CD) spectrum, and a red shift in the UV absorption spectrum (3). There is convincing evidence from Raman spectroscopy that the crystal structure corresponds to the high salt form of the polymer (4). A major feature of this structure is a zig-zag sugar-phosphate backbone: hence the name, Z-DNA. The zig-zag backbone is manifested in a dinucleotide repeat unit with the purines having a different conformation than the pyrimidines. Although the requirement of alternating pyrimidine-purine sequences is not absolute (5), it is these sequences, and especially alternating C-G, that are most closely identified with Z-form.

In addition to the alternation in phosphate conformation, there is also alternation of the glycosidic torsion angle in Z structure. Guanosine residues adopt the *syn* conformation about the glycosidic bond, while cytidines are in the *anti* conformation assumed by all nucleosides in the right-handed forms. The sugar pucker also alternates; a 3′-*endo* sugar pucker is found with guanosine, and a 2′-*endo* sugar pucker is found with cytidine. Another property of Z-form that can be investigated is the increased exchange rate of the slightly acidic purine H8 protons for solvent deuterium relative to B or A-form (6).

In an effort to search for the RNA counterpart to Z-DNA, we have enzymatically synthesized poly[r(G-C)] (7), and used the above criteria to determine the presence of Z-form in various solvent systems.

The Evidence for Z-RNA

The zig-zag backbone in Z-form is largely due to the very different conformations assumed by alternate phosphates. Several studies have shown that this is reflected in the ^{31}P NMR spectra of the Z-form, which shows that half the phosphates are shifted downfield by up to 1.5 ppm. This is in contrast to the right handed forms which show a relatively narrow range (up to about 0.5 ppm) of phosphate chemical shifts. Figure 1 shows the ^{31}P NMR spectrum of poly[r(G-C)] in 3M $NaClO_4$ and 6M $NaClO_4$ at 45 C. There are two resonances in each case, corresponding to the two types of phosphates in the polymer. In low salt, presumably A-form, these peaks are separated by only 0.5 ppm, while in 6M $NaClO_4$ the separation increases to 1.33 ppm. These spectra indicate that raising the salt concentration from 3M to 6M induces a conformational change from a structure with similar phosphate environments, to one with very different phosphate environments. The magnitude of the difference in high salt is similar to that seen in Z-DNA (8), and implies that an alternation in phosphate conformation is present. It is this conformation that we propose is Z-RNA.

The corresponding DNA polymer, poly[d(G-C)], shows only one peak in low salt solution, implying a regular backbone. Poly[d(G-m^5C)], however, has two peaks separated by 0.4 ppm in low salt, similar to what we see with RNA. All these molecules have a dinucleotide repeat unit, so two different ^{31}P peaks are expected although the structural alternation is apparently not nearly as dramatic as in Z-form.

Proton NMR has been used to study the glycosidic bond conformation assumed by poly[r(G-C)], as well as the rate of guanosine H8 exchange. Figure 2a shows the proton NMR spectrum of the polynucleotide in 6M $NaClO_4$ at 45C. The solvent is D_2O. A standard technique for assigning purine H8 protons is to heat the nucleic acid in D_2O and observe the diminution of the peak as these protons exchange with solvent deuterium. An observable decrease in the peak at 7.82 ppm can be seen after heating at 70 C for 2 hours; this peak is thus assigned to the GH8 proton. The similar experiment in 3M perchlorate, where the polymer is in the A-form, results in no observable proton exchange at 70 C, even after 5 hours. These data are consistent with Raman experiments (6) on double stranded nucleic acids which show that purine exchange is fastest with Z-form, and nearly an order of magnitude slower with A-form. In the Z-DNA crystal structure, the bases form a convex outer surface in the location the major groove normally occupies (2). The GH8 proton is much more exposed to solvent in the Z-conformation than in the A or B conformations, where it is buried in the major groove. It has, therefore, a

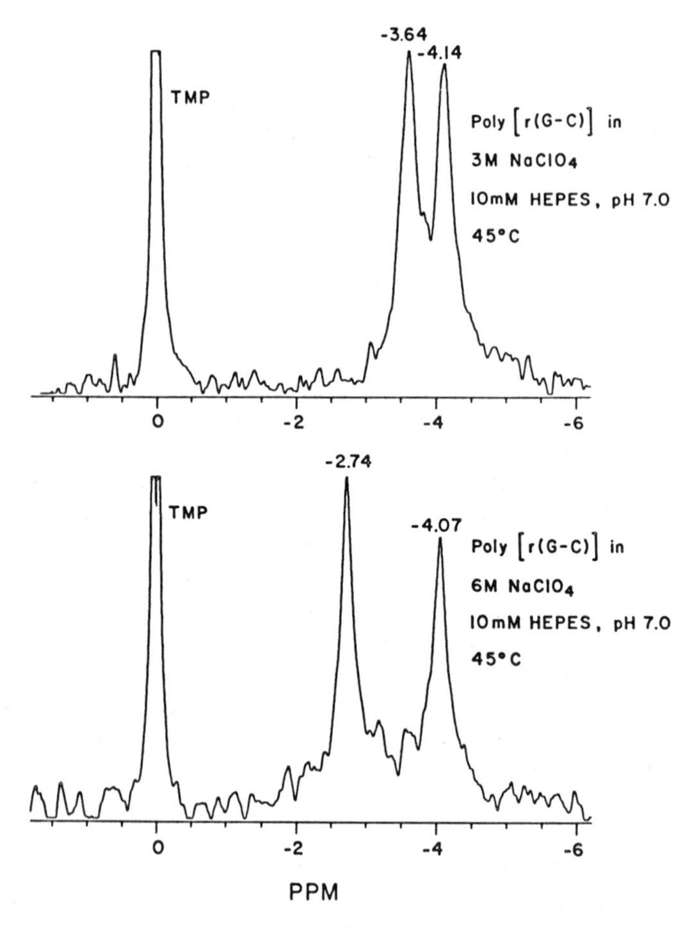

Figure 1. Phosphorus NMR spectra of poly[r(G-C)] (approximately 30mM in nucleotides) at 45 C in 10 mM HEPES, pH 7.0, 0.1 mM EDTA and 3M $NaClO_4$ (top panel) or 6M $NaClO_4$ (bottom panel). Spectra were obtained at 81.75 MHz on the UCB-200 spectrometer, using proton broad-band decoupling. Chemical shifts are referenced to internal trimethyl phosphate (TMP). These spectra are taken from reference (1).

correspondingly greater solvent exchange rate in Z (1,6). In fact, the A-form exchange is so slow that in order to assign the A-form GH8 proton it was necessary to heat the polymer to 80 C for several hours in a solution of 1 mM phosphate, 0.1 mM EDTA (conditions in which the polymer is partly melted to single strands).

Once the GH8 proton is assigned, the other aromatic resonance at 7.16 ppm can be assigned to the CH6 proton by default. We have used transient proton nuclear Overhauser effect (NOE) measurements on these protons in order to determine the conformation about the glycosidic bond. NOE arises from through-space dipolar interactions between spins, and can be observed as a change in one resonance upon irradiation of another. Since the magnitude of the NOE falls off as $1/r^6$, where r is the distance between nuclei, the spins must be close in space. Typically, r must be less than about 4 A. In the *syn* conformation, the distance between a purine H8 proton (or pyrimidine H6) and the H1′ on its own ribose is only about 2.3 A, whereas this distance increases to more than 3.5 A in the *anti* conformation. If the guanine H8 proton is irradiated, an NOE is expected to the H1′ proton if the conformation is *syn,* but little or no NOE is expected if the conformation is *anti.* Patel and co-workers (8) effectively utilized this difference in NOE experiments on high salt poly[d(G-C)], showing that the G residues were indeed *syn* whereas the C residues were *anti,* as expected for Z-form.

Figure 2b shows the NOE difference spectrum of poly[r(G-C)] that is obtained upon irradiating the GH8 proton. A large NOE can be seen to a peak at 5.7 ppm, which is assigned to the GH1′. This is the signature of the *syn* conformation for the guanosine residues, as found in Z structure. As an experimental check, an NOE from the CH6 to the CH5 proton is shown in panel c. This NOE is always expected as these protons are only 2.3 A apart in all conformations. There is no second large NOE to an H1′, indicating that the cytidine residues are all *anti*, again consistent with Z-form. Similar experiments with the polynucleotide in A-form conditions gives only the CH6 to CH5 NOE; there are no aromatic proton to H1′ NOE's, indicating that all nucleotides are *anti*, i.e., A-form (1).

Corresponding to these changes observable by NMR are changes in the CD and UV absorption spectra. The UV absorption spectrum of poly[r(G-C)] in 6M $NaClO_4$ at 45 C shows a lowered extinction coefficient at 260 nm and a shoulder at 295 nm in comparison to the spectrum at 22 C (1). These are both characteristics of the B to Z transition in DNA. The 22 C spectra are identical to those of the polymer in 3M perchlorate at 45 C, and are similar to spectra taken with no added perchlorate; these are all considered to be of A-form.

The solid line in figure 3 shows the CD spectrum of poly[r(G-C)] in 6 M $NaClO_4$ at 22 C. Again, this spectrum is identical to that in 3M perchlorate at 45 C, and similar to the low salt spectrum. Increasing the temperature to 45 degrees results in a cooperative transition to a structure characterized by the dotted line. This is the Z-form spectrum. The change in appearance of these spectra accompanying the transition can not be considered an inversion; the 45 degree spectrum remains positive at all wavelengths above 215 nm. This implies that care must be observed when attempting to determine helix handedness from near UV CD spectra alone. In fact, the A-form spectrum of the RNA polymer is reminiscent of the Z-DNA spectra of poly[d(G-C)], even though the molecules are of opposite handedness. The spectral inversion at long wavelength accompanying the B to Z transition with DNA is probably just fortuitous.

CD measurements on both the DNA and RNA alternating G-C polymers have recently been extended into the vacuum UV, and in this wavelength region the spectra are indicative of handedness (9). The A and B forms both have a single intense positive peak around 186 nm (10) as shown in figure 4a. The Z-form CD of poly[d(G-C)] below 220 nm is very similar to that of poly[r(G-C)] in 6M perchlorate

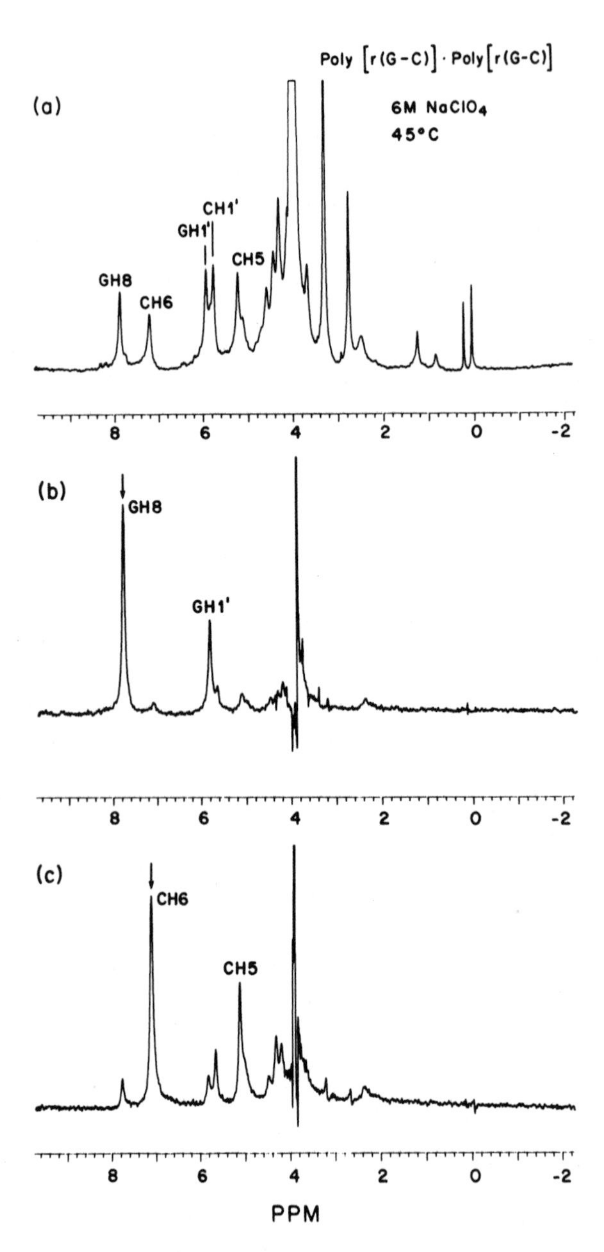

Figure 2. Proton NMR spectra of poly[r(G-C)] (approximately 30 mM in nucleotides) at 45 C in 6 M $NaClO_4$, 10 mM phosphate buffer, pH 7.0, and 0.1 mM EDTA. Spectra were obtained at 500 MHz on the JEOL GX-500 instrument at the Stanford University NMR Center, and are referenced to internal standard trimethylsilyl propionate (TSP). Panel (a) shows the spectrum in D_2O. The transient NOE difference spectrum obtained after irradiating the GH8 proton (indicated by the arrow) with a 50 ms saturation pulse is shown in panel (b). The difference spectrum is produced by subtracting the spectrum with the on-resonance presaturation pulse from a control spectrum with the presaturation pulse 10 kHz off resonance. A large NOE to the GH1′ proton can be observed, indicative of the *syn* conformation. Panel (c) shows the NOE difference spectrum obtained after irradiating the CH6 proton as described for (b). A large NOE can be seen to the CH5 proton. Reproduced, with permission, from reference (1).

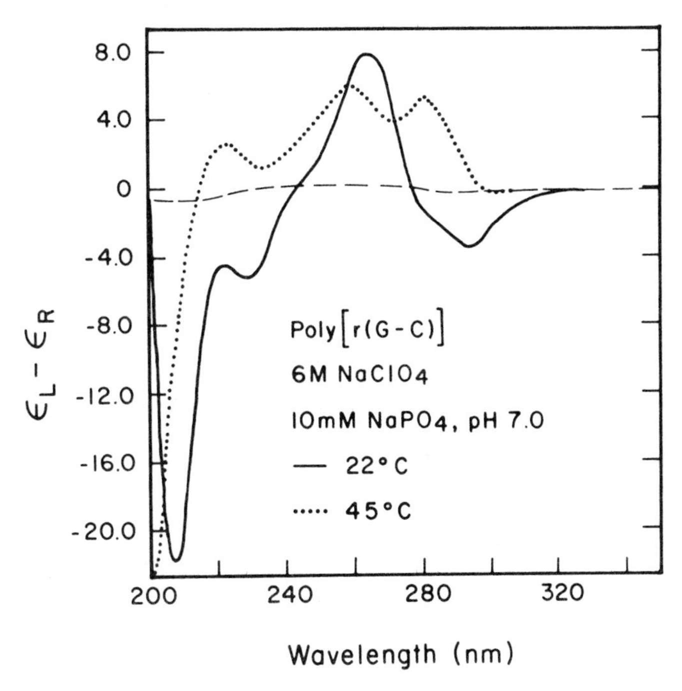

Figure 3. Circular dichroism spectra of poly[r(G-C)] in 6 M $NaClO_4$, and 10 mM phosphate buffer, pH 7.0. Spectra were obtained as described in (1). The dashed line is the base line. The 22 C spectrum is shown by the solid line, and is identical to that for the polymer in 3 M $NaClO_4$. Increasing the temperature to 45 C results in a cooperative transition to the dotted line (Z-form) spectrum.

at 46 C. Figure 4b demonstrates that these spectra are characterized by a large negative peak at 190-195 nm, and a large positive peak below 180 nm. This similarity is further evidence that poly[r(G-C)] is in the Z-form in 6M $NaClO_4$ at 45 C.

Although the longer wavelength CD spectra may not be directly sensitive to handedness, the Z-RNA spectrum is quite distinctive, and can be easily used in surveying other solvent conditions for their ability to induce Z-form. Such experiments have shown that while temperatures above 35 C are necessary to induce the A-Z transition in 6M perchlorate, higher temperatures can compensate for lower perchlorate concentrations (1). Addition of ethanol decreases the transition temperature, so that the RNA is in Z-form at room temperature in 4.8 M $NaClO_4$/20% ethanol. This temperature dependence of the transition corresponds to that of poly[d(G-C)] in LiCL or ethanol/water solutions (8,11,12).

Figure 5 shows the near UV CD spectrum of poly[r(G-C)] in 7M NaBr at room temperature. Comparison with figure 3 will show that this is a Z-form spectrum. Both proton and phosphorus NMR spectra of the polymer in these conditions also indicate that it is in the Z-form. Phosphorus NMR shows the 1.4 ppm peak separation characteristic of Z-form, and proton NOE's demonstrate that guanosines are *syn* and cyti-

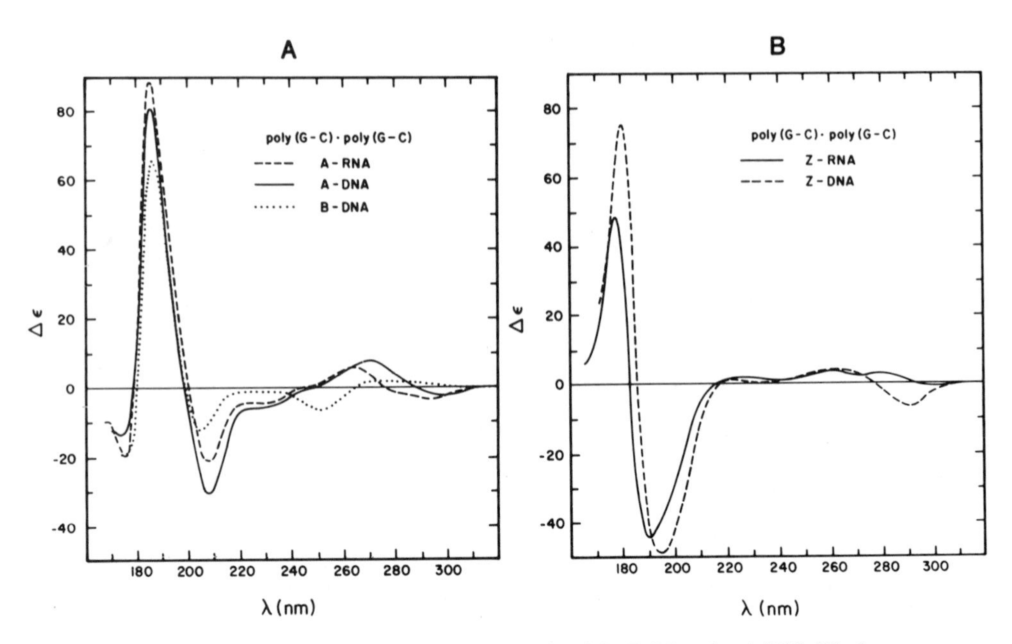

Figure 4. Vacuum UV CD spectra of various forms of poly[r(G-C)] and poly[d(G-C)]. Spectra were obtained as described in (10). Panel (A) shows spectra of the right-handed forms of both the RNA and DNA polymers. The dashed line spectrum (-------) is of poly[r(G-C)] as the A-form in 6M sodium perchlorate, 10 mM phosphate buffer, pH 7.0, 0.1 mM EDTA, at 22 C. The solid line (——) is of poly[d(G-C)] as the A-form in 80% 1,1,1-trifluoroethanol, 0.67 mM phosphate buffer, pH 7, at 22 C. The dotted line (·······) is of poly[r(G-C)] as the B-form in 10 mM phosphate, pH 7.0, at 22 C. These spectra are all very similar below 240 nm. Panel (B) shows the spectra of the left-handed forms of both polymers. The solid line spectrum (——) is of poly[r(G-C)] in conditions as in (A) except at 46 C. This is a Z-RNA spectrum. The dashed line spectrum (-------) is of poly[d(G-C)] as the Z-form in 2 M sodium perchlorate, 10 mM phosphate buffer, pH 7.0, at 22 C. The two Z-form spectra are very similar below 240 nm, and they are very different from the right-handed-form spectra.

dines are *anti*. This finding is important in two ways: first, it shows that sodium perchlorate is not unique in the ability to induce Z-form, and second, provides a solvent system that makes Z-RNA amenable to study by Raman spectroscopy.

Raman spectroscopy has played a vital role in the short history of Z-DNA, as this technique provided the link between crystallographic and solution studies (4). Several studies have shown that certain Raman bands undergo large changes in position and/or intensity during the B to Z transition in DNA (4,13-15). The most characteristic of these changes are the near disappearance of the phosphodiester antisymmetric stretch in the 810-830 cm^{-1} region, the increase of the base-stacking-sensitive G ring mode around 1320 cm^{-1}, and the shift in the G ring breathing from 682 to 625 cm^{-1} accompanying the B to Z transition. Figure 6 shows the Raman spectra of poly[r(G-C)] in 5.5 M NaBr at 22 and 40 C (16). These conditions correspond to the A and Z conformations respectively, as determined by CD. It is apparent that the antisymmetric stretch at 813 cm^{-1} in the A-form spectrum nearly disappears at 40

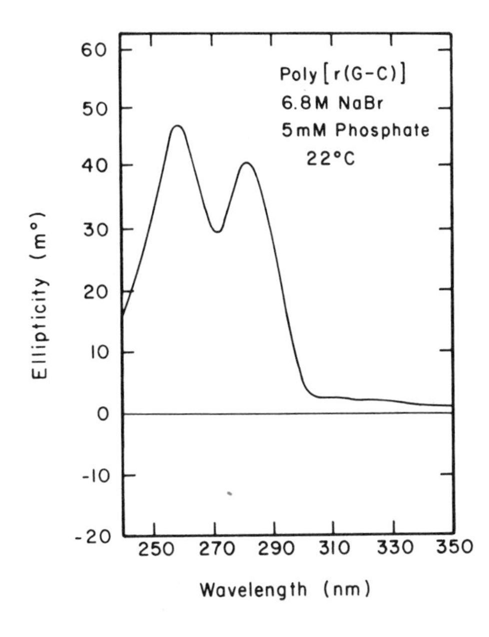

Figure 5. Near UV CD spectrum of poly[r(G-C)] at room temperature in 7 M NaBr, and 5 mM phosphate buffer, pH 7.0. The spectrum was obtained with a JASCO J500C spectropolarimeter. Ellipticities are reported in millidegrees since the polymer concentration could not be determined accurately due to the high absorbance of Br^-. Comparison with figure 3 indicates that the polymer is in Z-form.

C. Also, there is a relatively intense peak at 1322 cm^{-1} in the spectrum at 40 C relative to the A-form spectrum. Thus, the Raman spectra substantiate that poly[r(G-C)] is in the Z-form in 5.5M NaBr at 40 C. The characteristic shift of the G ring breathing mode to lower frequency accompanying the right-to-left transition also occurs, although the Z-form frequency, 641 cm^{-1}, is intermediate between those of the A-form and Z-DNA. This band is sensitive to glycosidic torsion angle, and may indicate a somewhat different conformation for Z-RNA than Z-DNA (16).

Other Conditions That Induce Z-RNA

The preceding discussion has demonstrated how the similarity of Z-form between RNA and DNA has been utilized in identifying and observing Z-RNA. Having available the tools to search for Z-RNA, it is important to discover what conditions favor the conversion to the left-handed form. Clearly, if Z-RNA is to be important biologically, it must exist in conditions other than 6M $NaClO_4$ or 7M NaBr. We have surveyed many of the conditions that are known to facilitate the conversion of DNA to the Z-form, in order to ascertain their effect with the corresponding RNA. A wide variety of solvents and ions which are known to favor Z formation with DNA do not induce a Z structure with RNA, as assayed by CD. Some of these are listed in Table II. The most commonly used salt for Z-DNA formation, NaCl, does not result in Z-RNA with poly[r(G-C)] at any concentration or temperature tested. This is consistent with the work of Westerink et al. (17) and Uesugi et al. (18), who found no transition to Z-form in NaCl for alternating C-G ribooligonucleotides. Apparently, more stringent conditions are required to produce Z-RNA than to produce Z-DNA. Also, the $Co(NH_3)_6^{3+}$ ion, which induces Z-form in DNA at

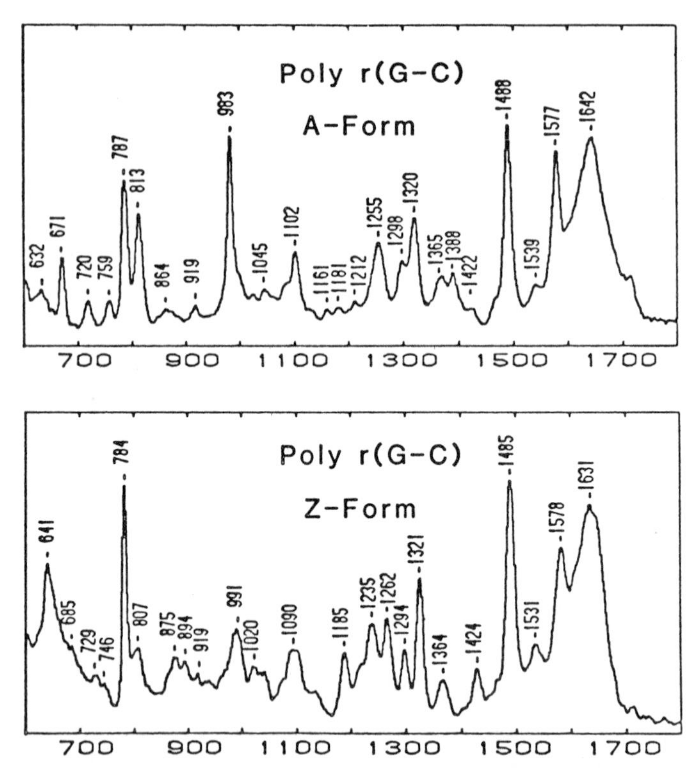

Figure 6. Raman spectra of poly[r(G-C)] in 5.5 M NaBr, 5 mM phosphate buffer, pH 7.0. The concentration of the polymer was approximately 30 mM in nucleotides, with a sample volume of approximately 50 microliters in a glass capillary. The spectra were obtained at a power output of 50 mW from the 514.5 nm line of an argon ion laser, and detected using a Vidicon multichannel detector. The top spectrum was obtained at 22 C; under these conditions the polymer is in the A-form, as indicated by CD and NMR. The bottom spectrum was obtained at 40 C. At this temperature the polymer is in the Z-form, as determined by CD and NMR.

micromolar concentrations (19) by means of specific binding interactions with the nucleic acid, does not effect the corresponding transition in poly[r(G-C)] even at millimolar concentrations. This also implies that Z-RNA may have a somewhat different structure than Z-DNA.

There are a few salts that CD experiments have shown cause an A to Z transition with poly[r(G-C)], and these are listed in table III. The question mark with NaSCN indicates that solvent absorbance precludes taking CD measurements below 280 nm, but the spectra appear Z-like at longer wavelengths. These salt solutions have the common property that they are all chaotropic solvents (20), i.e., salts that destabilize nucleic acid double strands with increasing ionic strength, although chaotropic strength (21) does not seem to correlate with the salt concentration required to induce Z-RNA. Chaotropic salts are a known Z promoting factor with DNA (22). Recent quadrupolar ion relaxation NMR studies (23) of the polynucleotide

Table II
Non Z-RNA Forming Solvent Conditions

Salt[1]	Concentration (Molar)	Temperature (C)	Comment
NaCl	5.0	25-80	A
$Co(NH_3)_6Cl_3$	0.0001	25-60	A
	0.0005	25-50	A
$NaClO_4$	4.0	70	A
	3.0/0.4M $MgCl_2$	25	A
$LiClO_4$	1.8	25-80	A
$KClO_4$	0.05	25-80	A
$NaCF_3COO$	1.0	22-70	A
	2.0	22-70	A
	3.0	22-70	A
$NaCCl_3COO$	2.5	22-70	A
	2.5/5%EtOH	22-70	A
NaSCN	4.0	65	A
	4.0/10%EtOH	22	Aggregate
LiCl	9.0	25-80	A
CsCl	Sat'd	25-80	A
	Sat'd/5mM $MgCl_2$	25	A
CsF	4.0	25-80	A
NH_4Br	5.2	22-70	A
KBr	4.4	22-70	A
$MgBr_2$	1.0	22-70	A
LiBr	1.0	22	Pos. 280 nm band No 295nm band
Urea	8.0	70	A
Guanidinium HCl	7.0	80	A
$NiCl_2$	0.1	70	A
Spermidine	0.002-0.032	25-80	A
$MnCl_2$	0.0005	25-37	A

[1]All solutions at pH 7.

in NaBr solutions demonstrate that the bromide ion is associated with the Na^+-RNA complex, suggesting that the anionic species of these salts may alter the ionic or hydration spheres of poly[r(G-C)].

There is also another type of poly[r(G-C)] CD spectrum evident in some salts, other than A-RNA type or Z-RNA type. An example of this is shown in figure 7, which is the CD spectrum of the polynucleotide in 4M $MgCl_2$. This spectrum is reminiscent of A-form spectra, although the band at 280 nm disappears while the band at 230 nm becomes more positive. The resulting spectrum is very nearly like that of Z-DNA; as stated earlier, Z-DNA and A-RNA spectra are similar. Experiments are underway to determine the nature of the polynucleotide in the form giving rise to these spectra.

Certain covalent modifications are also known to give rise to Z-structure, and Uesugi et al. have shown that the tetranucleotides r(C-m^8G-C-m^8G) and r(C-br^8G-C-br^8G)

Table III
Summary of Conditions Which May Facilitate an A ⇒ Z Transition in RNA

Salt	Temp. (C)	Comment
6.0M $NaClO_4$	35	all Z
4.8M $NaClO_4$	45	all Z
4.8M $NaClO_4$/20% EtOH	25	all Z
4.8M $NaClO_4$/10% EtOH	40	all Z
6.8M NaBr	R.T.[1]	2 hrs. for complete rxn.
5.8M NaBr	45	some Z at R.T.
5.0M NaBr	65	
5.0M NaSCN/5% EtOH	R.T.	Z (?)
4.0M NaSCN/5% EtOH	60	Z (?)
6.0M NaI	R.T.	Z (?)
2.9M $MgBr_2$	R.T.	all Z, pH 6.5
2.0M $MgBr_2$	65	some sample hydrolysis
4.0M $MgCl_2$, pH 4	R.T.	Z-DNA like CD spectra
2.9M $Mg(ClO_4)_2$	R.T.	Z-DNA like CD spectra

[1]Room temperature.

exist in a Z structure (24). Another modification known to facilitate Z-DNA formation is bromination. This reaction can be accomplished conveniently by dissolving the polymer in bromine water (25); this leads to bromination at the 8 position of guanine and, to a lesser extent, at the 5 position of cytosine. Poly[r(G-C)] incubated

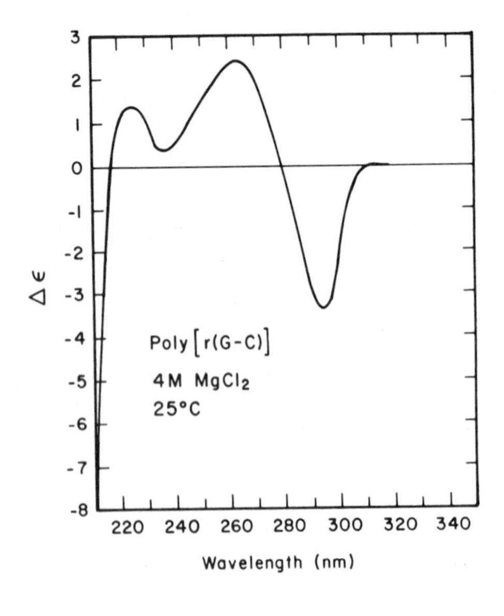

Figure 7. CD spectrum of poly[r(G-C)] at room temperature in 4M $MgCl_2$, pH 4, obtained on a JASCO J500C spectropolarimeter. This spectrum is very similar to that of DNA in the Z-form, although it is not very different from an A-RNA spectrum.

A

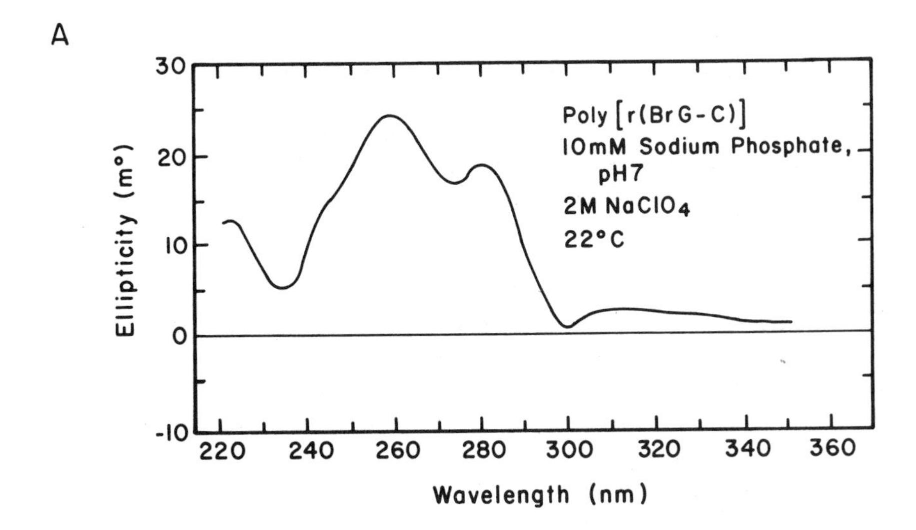

B

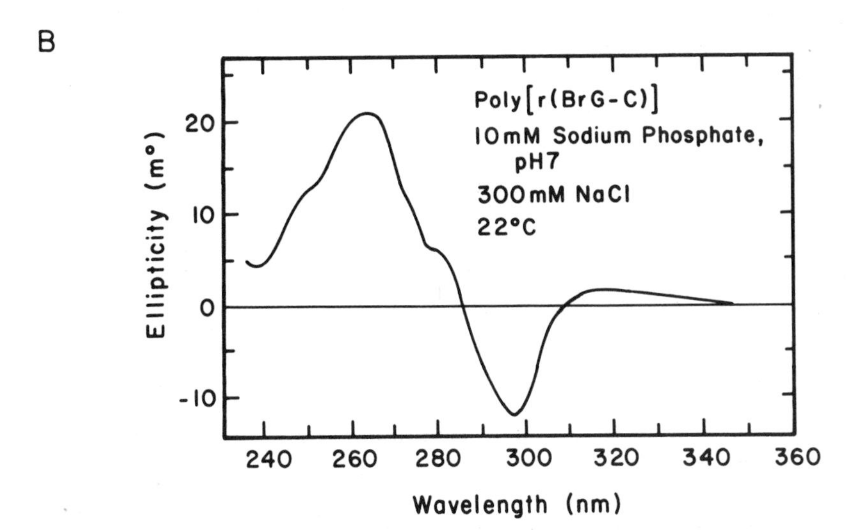

Figure 8. (A) CD spectrum of chemically brominated poly[r(G-C)] in 2 M $NaClO_4$, 10 mM phosphate buffer, pH 7.0, at room temperature. The polymer was brominated by incubating under Z-forming conditions for 10 minutes in bromine water at a relative concentration of 3.5 bromines per nucleotide. Comparison with figure 3 shows that the polymer is mostly in the Z-form, although the perchlorate concentration and temperature have been reduced from values required for Z-form with the non-brominated polymer. Ellipticities are not given as molar ellipticities pending determination of the extinction coefficient of the brominated polymer. (B) CD spectrum of brominated poly[r(G-C)] in 300 mM NaCl, 10 mM phosphate buffer, pH 7.0, at room temperature. The polymer was prepared as described in (A). The positive shoulder at 280 nm indicates that about 5-10% of the polymer is in the Z-form under these near physiological conditions.

under Z conditions in bromine water of 3.5 bromines/nucleotide gives a room temperature CD spectrum in 2M $NaClO_4$ that is mostly (approximately 80%) Z as shown in figure 8a. At high levels of modification, the brominated polymer has a Z-DNA-like

CD spectrum in millimolar NaBr solutions; the appearance of this spectrum is much like that of poly[r(G-C)] in 4M $MgCl_2$. Perhaps most importantly, when this brominated polynucleotide is placed in 300 mM NaCl, the CD spectrum indicates that approximately 5-10% is in the Z-form as demonstrated in figure 8b. If, like its DNA relative, Z-RNA is an immunogen, then Z-RNA existing at near physiological salt concentration in this modified polynucleotide may elicit antibody production.

Imino Proton Exchange Rates

NMR studies have been done to examine the base pair opening dynamics of Z-RNA. When placed in H_2O solutions, the base pairing imino protons of nucleic acids become observable by proton NMR, and the rates of exchange with solvent protons can be determined by NMR spin-lattice relaxation measurements. In an NMR relaxation experiment the resonance of interest is selectively spin labeled, and then observed at some later time. The commonly accepted mechanism for the exchange observed with NMR (26,27) is base pair opening, followed by base catalyzed exchange of the labeled proton for an unlabeled solvent proton, then base closing. Application of the steady-state approximation to this mechanism leads to the following rate equation:

$$k_{ob} = \frac{K_{op}\, k_{ex}[cat]}{k_{cl} + k_{ex}[cat]}$$

where k_{ob}, k_{op}, k_{cl}, and k_{ex} are the observed, base pair opening, base pair closing, and base catalyzed exchange rate constants, respectively. In the limiting case of $k_{cl} \ll k_{ex}[cat]$, this expression reduces to:

$$k_{ob} = k_{op}$$

If these conditions hold, the reaction is said to be open limited, and the observed exchange rate can be identified with the opening rate.

In NMR T1 relaxation experiments, the observed relaxation rates are a sum of magnetic spin-lattice relaxation and chemical exchange. These two contributions can be distinguished by measuring the relaxation as a function of temperature. At low temperature, where high solvent viscosity results in slow molecular tumbling, spin-lattice relaxation rates are high. Similarly, the chemical exchange rates are slow at low temperature, and spin-lattice effects dominate the observed relaxation. At high temperature, where chemical exchange is faster and increased molecular tumbling leads to lower spin-lattice relaxation rates, chemical exchange may dominate the relaxation behavior. This can be seen in figure 9, which compares the relaxation rates of A and Z-RNA measured in $NaClO_4$ (28) to those determined by Mirau and Kearns for B and Z-DNA in NaCl (29). There is a striking difference between Z-RNA and Z-DNA. The latter shows almost no temperature dependence of relaxation rates, implying that imino proton exchange is negligible up to 80 C. In contrast, Z-RNA shows significant proton exchange at high temperature, although

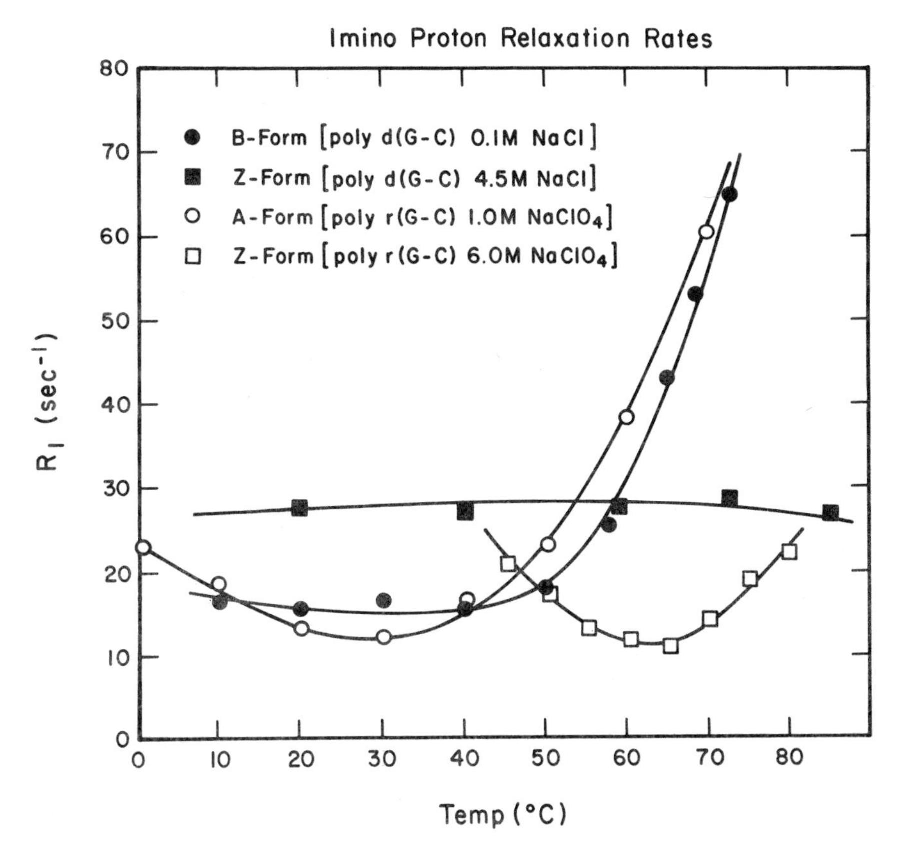

Figure 9. Imino proton NMR spin-lattice relaxation rates as a function of temperature for poly[r(G-C)] in the A and Z forms, and poly[d(G-C)] in the B and Z forms, all at pH 7. The RNA data were obtained at 500 MHz on the JEOL GX-500 at the Stanford University NMR Center using the saturation recovery technique (30). Open circles represent data for A-RNA, obtained in 1.0 M $NaClO_4$, 10 mM phosphate buffer, 0.1 mM EDTA. Open squares represent data for Z-RNA (6.0 M $NaClO_4$, 10 mM phosphate buffer, 0.1 mM EDTA). The DNA data were obtained by Mirau and Kearns, as described in (29). Closed circles represent data for B-DNA, obtained in 0.1 M NaCl. Closed squares represent data for Z-DNA, obtained in 4.5 M NaCl. The displayed rates are the sum of contributions from chemical exchange and magnetic T1 relaxation.

a lower rate than the two right-handed forms. To the extent that the exchange reaction is open limited, Z-RNA appears to have a much higher rate of base opening than Z-DNA. Part of this difference may be related to the different solvent systems used with the DNA and RNA polymers. It has recently been shown that imino proton exchange rates are faster in $NaClO_4$ than in NaCl (31). It is possible that the same properties of perchlorate that are important in increasing the exchange rate, are related to its ability to induce Z structure.

It is interesting to note that the RNA structures show the expected low temperature relaxation behavior, i.e., a decrease in relaxation rate with increasing temperature that follows the solvent viscosity, while the DNA conformations show much less temperature dependence. This is probably due to some sort of pre-melting behavior with DNA. The spin-lattice contribution to relaxation can be subtracted by extrapolating the low temperature profiles to the temperature range where chemical exchange is dominant. These corrected exchange rates have been used in an Arrhenius plot to determine the exchange activation energies. The values are 13 kcal/mol for A-RNA and 11 kcal/mol for Z-RNA (28). Both of these are less than the 20 kcal/mol for B-DNA (29) or the 22 kcal/mol for Z-DNA (31).

The Tetranucleotide r(CpGpCpG) Also Assumes Z-form

If Z-RNA is relevant biologically, it must occur in sequences other than polymeric alternating C-G. Short runs of alternating purine-pyrimidine are relatively common in natural RNA's. Also, the higher resolution NMR spectra possible with a short oligonucleotide can potentially reveal information unobtainable with a polynucleotide. For these reasons it is desirable to observe Z-form in a short RNA. Figure 10 shows the CD spectrum of the tetranucleotide r(CpGpCpG) in 1M $NaClO_4$ and 6M $NaClO_4$, both at 0 C (32). These spectra are similar to those of poly[r(G-C)] in the A- and Z-forms respectively. The ^{31}P NMR spectra of this tetranucleotide at 0 C in 6M $NaClO_4$ (32) is shown in figure 11. Since the tetranucleotide has a melting temperature of less than 20 C, the 0 C temperature was chosen to maximize the amount of double strands present. UV absorption melting curves indicate that under these conditions there is a significant amount of single stranded form present. There are three distinct phosphates in this molecule, and the spectrum can most easily be rationalized in terms of two forms in slow exchange, giving rise to six peaks. Five of the peaks

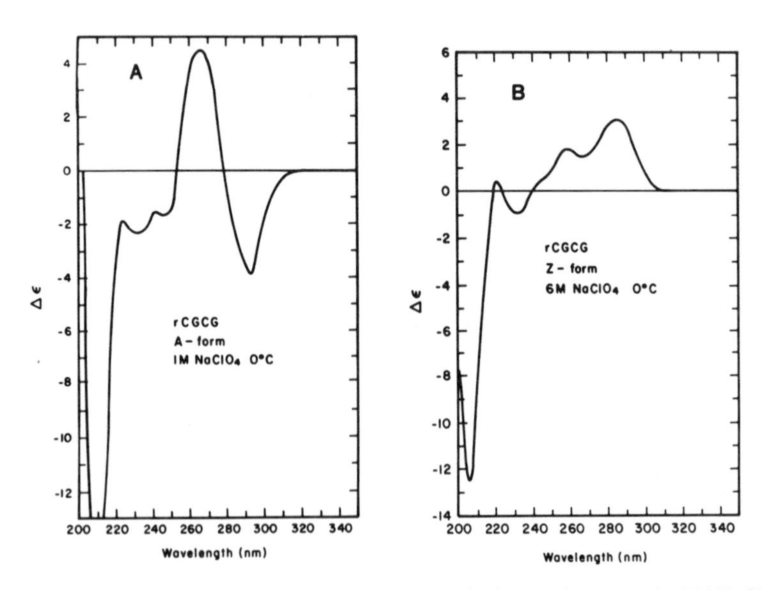

Figure 10. CD spectra at 0 C of the tetranucleotide r(CpGpCpG). (A) Spectrum in 1 M $NaClO_4$, 1 mM EDTA, pH 7. (B) Spectrum in 6 M $NaClO_4$, 1 mM EDTA, pH 7. The spectra were obtained on a JASCO J500C spectropolarimeter, equipped with a thermoelectric cell block for temperature control.

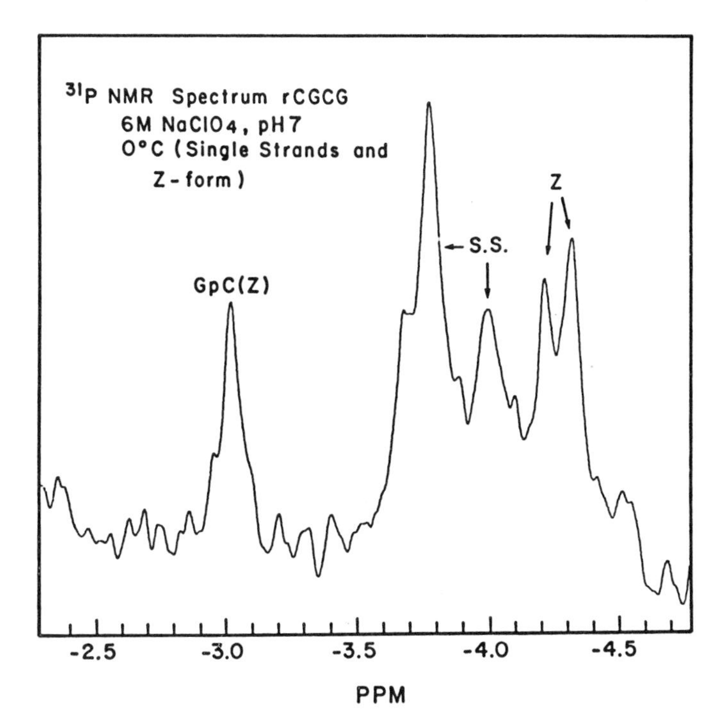

Figure 11. Phosphorus NMR spectrum of the tetranucleotide r(CpGpCpG) at 0 C in 6 M $NaClO_4$, and 10 mM phosphate buffer, pH 7. The tetranucleotide was approximately 0.5 mM in strands. The spectrum was obtained at 121 MHz on the BVX-300 instrument at Berkeley, using proton broad-band decoupling. Chemical shifts are referenced to external trimethyl phosphate (TMP). The spectrum is shown with tentative assignments, where S.S. indicates single strands, and Z indicates Z-form.

(two overlapped) are grouped between −3.7 and −4.4 ppm, while the sixth peak is shifted downfield about 1 ppm from the rest. This downfield shifted resonance is indicative of Z structure. Since only one peak is shifted downfield, and there is only one GpC phosphate in this molecule, it is tempting to assign the Z-RNA downfield ^{31}P resonance to GpC residues. Although this is not a definitive assignment, this does correspond to the Z-DNA case where the downfield resonance has been unambiguously assigned to GpC by means of phosphorothioate substitutions (22).

The proton NMR spectrum of the tetranucleotide under similar conditions (32) also shows evidence of multiple species present as can be seen in figure 12. Comparison of the 0 degree spectrum with that taken at 20 C, where single strands predominate, allows assignment of specific resonances to either single strand or Z-form. GH8 resonances can be distinguished from CH6 resonances since the latter have a 7.5 Hz scalar coupling to the CH5 proton. These criteria are used to derive the partial assignments shown in the figure. The Z-form CH6 protons show an upfield shift, similar to their polymeric counterparts (compare, for instance, with figure 2).

In contrast to these spectra, both proton and phosphorus NMR spectra of the tetranucleotide in 1M $NaClO_4$ show intermediate exchange at temperatures where

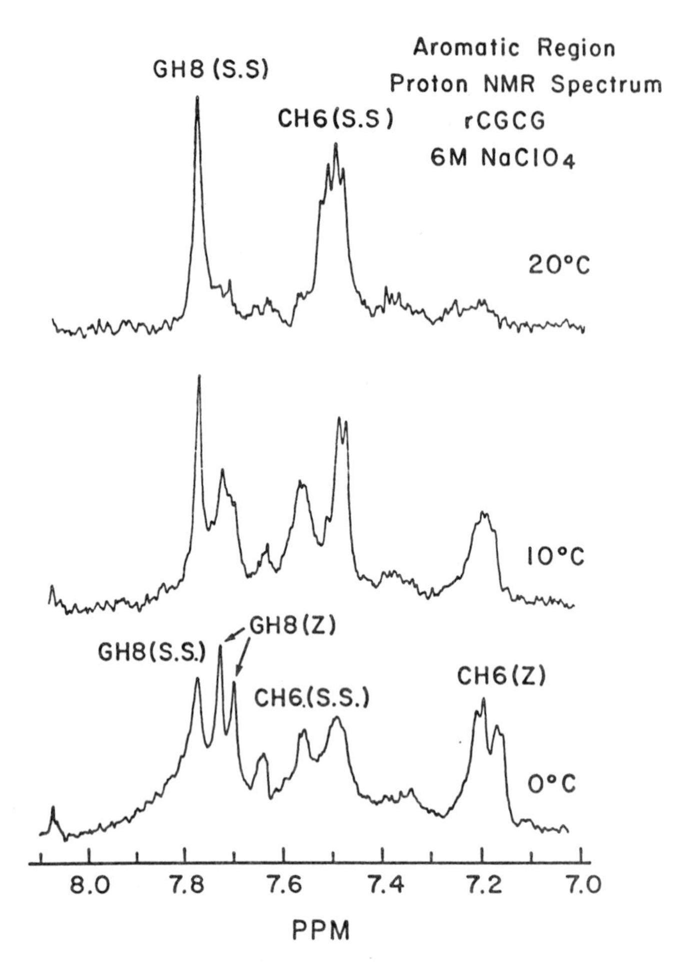

Figure 12. The aromatic region of the proton NMR spectrum of r(CpGpCpG), in 6 M $NaClO_4$, 10 mM phosphate buffer, pH 7. (Top) at 20 C where only single strands (S.S.) are present, (middle) at 10 C, and (bottom) at 0 C where both single strands (S.S.) and Z-form (Z) are present. Some tentative assignments are indicated. Spectra were obtained at 500 MHz on the Bruker AM-500 at UC-Berkeley, and referenced to trimethylsilyl propionate (TSP). As the temperature is lowered a new set of peaks appears, indicating slow exchange between the two forms.

both single and double strands are present. This indicates that the Z to coil transition is much slower than the A to coil transition; one possibility is that Z-form melting proceeds through some intermediate form present in small amounts, e.g., A-form.

Transient NOE measurements have been obtained on some of the protons of r(CpGpCpG) in 6M perchlorate at 0 C (32). Irradiation of a Z-form GH8 proton results in an NOE to a GH1′ proton as shown in the top of figure 13, indicating a *syn* conformation for the guanosine residues in the tetranucleotide. This again is evidence for the tetranucleotide existing in the Z-form. However, NOE peaks can arise through two mechanisms: through-space magnetization transfer (true NOE), and

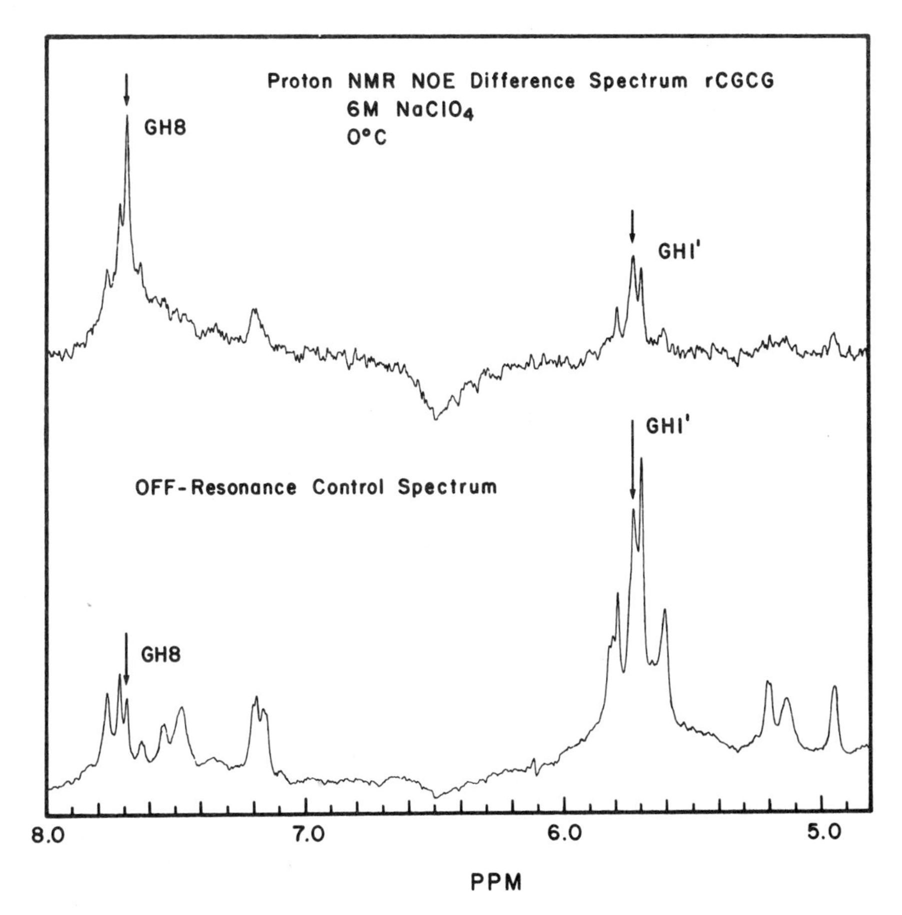

Figure 13. (Top) Proton NMR NOE difference spectrum obtained upon irradiating a Z-form GH8 proton (indicated by the left arrow) of r(CpGpCpG) at 0 C. The length of the presaturation pulse was 1.0 sec. All other conditions are as in the legend to figure 12. The difference spectrum was obtained by subtracting the spectrum in which the GH8 proton was saturated from a control spectrum in which the irradiation was off-resonance (at 6.5 ppm). At the bottom is shown the off-resonance control spectrum. Spectra were acquired in the interleaving mode. A large NOE can be observed to an H1′ proton, indicating that the guanosine is in the *syn* conformation.

magnetization transfer through chemical exchange. Figure 14 shows an example of the latter case—an apparent NOE from a Z-form CH6 to a single-strand CH6. By varying the irradiation time in the NOE experiment, the kinetics of the exchange between the two forms can be determined. These experiments are presently being performed in an effort to determine the Z to single strand exchange rate.

Conclusions

A wide variety of physical techniques have been used to show that RNA, like DNA, can adopt a Z-form under certain conditions. Z-RNA is similar in many ways to Z-DNA. In fact, this had to be so since the known structure of Z-DNA was used as a

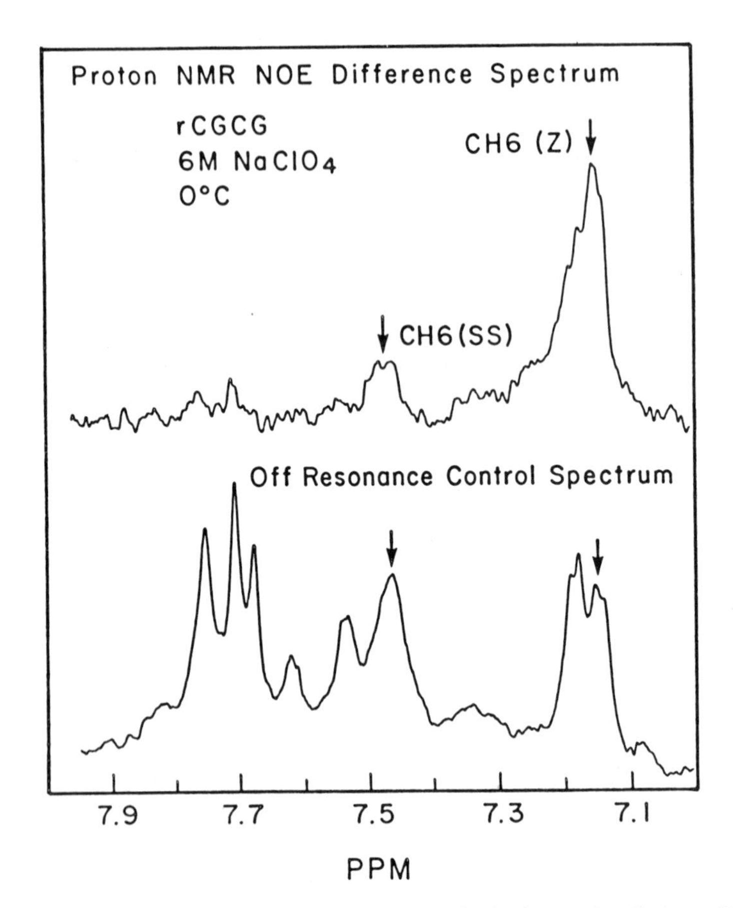

Figure 14. (Top) Proton NMR NOE difference spectrum obtained upon irradiating a Z-form CH6 proton (indicated by the right arrow) of r(CpGpCpG) at 0 C. All other conditions are as in the legend to figure 12. The difference spectrum was obtained as described in the legend to figure 13. At the bottom is shown the off-resonance control spectrum. An apparant NOE is observed to a single strand CH6 proton, indicating magnetization transfer through chemical exchange.

guide in the search for Z-RNA. These similarities include the alternation of the phosphate conformation, the alternation of the glycosidic torsion angle, the solvent accessible location of the GH8 proton, and handedness (at least as indicated by vacuum UV CD). However, there are also some differences between Z-RNA and Z-DNA, most notably a somewhat different guanosine glycosidic torsion angle and/or sugar pucker (as shown by the Raman data), different base pair opening rates, differences related to the disparate appearance of the near UV CD spectra between Z-DNA and Z-RNA, and the greater resistance of RNA to conversion to the Z-form.

Z-RNA is the first new form of double-stranded RNA since the original observation of A-RNA. But Z-RNA is also left handed, and intriguing questions regarding its potential biological role can be asked. Since a short oligonucleotide can exist in

Z-form, and short alternating purine-pyrimidine stretches of double-stranded RNA do occur naturally, it may be possible that a few bases changing from a right-handed to a left-handed form can provide torsional motion to drive cellular functions, e.g., translocation during protein synthesis.

Brominated poly[r(G-C)] is partially Z-form at fairly low salt conditions, so it may be possible to obtain anti-Z-RNA antibodies. This would allow a search for biological roles for Z-RNA along the same lines as is presently being done with Z-DNA. An additional approach with RNA is to use the difference in exchange rates for GH8 protons, since the rate difference is much greater between Z and A-forms than between Z and B-forms. This search is now getting under way.

Acknowledgements

The work at Berkeley was supported in part by NIH grant GM10840 and the US Department of Energy, Office of Energy Research, under contract 03-82ER60090.000. Some of the proton NMR experiments were performed at the Stanford University NMR Center, which is supported by NSF grant GP23633 and NIH grant RR00711. PC was supported by NIEHS grant T32ES07075-06.

References and Footnotes

1. Hall, K., Cruz, P., Tinoco, I., Jr., Jovin, T.M., and van de Sande, J.H., *Nature 311,* 584 (1984).
2. Wang, A.H.-J., Quigley, G.J., Kolpak, F.J., Crawford, J.L., van Boom, J.H., van der Marel, G., and Rich, A., *Nature 282,* 680 (1979).
3. Pohl, F.M., and Jovin, T.M., *J. Mol. Biol. 67,* 375 (1972).
4. Thamann, T.J., Lord, R.C., Wang, A.H.-J., and Rich, A., *Nucl. Acids Res. 9,* 5443 (1981).
5. Rich, A., Nordheim, A., and Wang, A.H.-J., *Ann. Rev. Biochem. 53,* 791 (1984).
6. Benevides, J.M., and Thomas, G.J., Jr., *Biopolymers 24,* 667 (1985).
7. Hall, K., Cruz, P., and Chamberlin, M.J., *Archs. Biochem. Biophys. 236,* 47 (1985).
8. Patel, D.J., Kozlowski, S.A., Nordheim, A., and Rich, A., *Proc. Natn. Acad. Sci. U.S.A. 79,* 1413 (1982).
9. Sutherland, J.C., Griffin, K.P., Keck, P.C., and Takacs, P.Z., *Proc. Natl. Acad. Sci. U.S.A. 78,* 4801 (1981).
10. Riazance, J.H., Baase, W.A., Johnson, W.C., Jr., Hall, K., Cruz, P., and Tinoco, I., Jr., *Nucl. Acids Res., 13,* 4983 (1985).
11. van de Sande, J.H., and Jovin, T.M., *EMBO J. 1,* 115 (1982).
12. Hall, K.B., and Maestre, M.F., *Biopolymers 23,* 2127 (1984).
13. Pohl, F.M., Ranade, A., and Stockburger, M., *Biochim. Biophys. Acta 335,* 85 (1973).
14. Thomas, G.A., and Peticolas, W.L., *Biochemistry 23,* 3202 (1984).
15. Wartell, R.M., Harrell, J.T., Zacharias, W., and Wells, R.D., *J. Biomol. Str. Dyns. 1,* 83 (1983).
16. Trulson, M.O., Cruz, P., Puglisi, J., Tinoco, I., Jr., and Mathies, R. A., in preparation.
17. Westerink, H.P., van der Marel, G.A., van Boom, J.H., and Haasnoot, C. A.G., *Nucl. Acids Res. 12,* 4323 (1984).
18. Uesugi, S., Ohkubo, M., Ohtsuka, E., Ikehara, M., Kobayashi, Y., Kyogoku, Y., Westerink, H.P., van der Marel, G.A., van Boom, J.H., and Haasnoot, C.A.G., *J. Biol. Chem. 259,* 1390 (1984).
19. Behe, M., and Felsenfeld, G., *Proc. Natl. Acad. Sci. USA 78,* 1619 (1981).
20. Hamaguchi, K., and Geiduschek, E.P., *J. Am. Chem. Soc. 84,* 1329 (1962).
21. von Hippel, P.H., and Schleich, T., *Accts. Chem. Res. 2,* 257 (1969).
22. Jovin, T.M., McIntosh, L.P., Arndt-Jovin, D.J., Zarling, D.A., Robert-Nicoud, M., van de Sande, J.H., Jorgenson, K.F., and Eckstein, F., *J. Biomol. Str. Dyns. 1,* 21 (1983).

23. Hardin, C.C., Wolk, S., Puglisi, J., and Tinoco, I., Jr., unpublished.
24. Uesugi, S., Ohkubo, M., Urata , H., Ikehara, M., Kobayashi, Y., and Kyogoku, Y., *J. Am. Chem. Soc. 106,* 3675 (1984).
25. Moller, A., Nordheim, A., Kozlowski, S.A., Patel, D.J., and Rich, A., *Biochemistry 23,* 54 (1984).
26. Teitelbaum, H., and Englander, S.W., *J. Mol. Biol. 92,* 55 (1975).
27. Crothers, D.M., Cole, P.E., Hilbers, C.W., and Shulman, R.G., *J. Mol. Biol. 87,* 63 (1974).
28. Cruz, P., and Tinoco, I., Jr., in preparation.
29. Mirau, P.A., and Kearns, D.R., *J. Mol. Biol. 177,* 207 (1984).
30. Johnston, P.D., and Redfield, A.G., *Nucl. Acids Res. 4,* 3599 (1977).
31. Mirau, P.A., and Kearns, D.R., *Proc. Natl. Acad. Sci. USA 82,* 1594 (1985).
32. Davis, P., Hall, K., Cruz, P., and Tinoco, I., Jr., *Nucl. Acids Res.,* in press (1986).

Biomolecular Stereodynamics IV, Proceedings of the Fourth Conversation in the Discipline Biomolecular Stereodynamics, State University of New York, Albany, NY, June 04-09, 1985, Eds., Ramaswamy H. Sarma & Mukti H. Sarma, ISBN 0-940030-18-7, Adenine Press, ©Adenine Press 1986.

Anomalous Protonation of polypurine:polypyrimidine DNA sequences: A new form of DNA helix

D.E. Pulleyblank and D.B. Haniford
Department of Biochemistry,
University of Toronto,
Toronto, Ontario, Canada M5S-1A8

Abstract

The sensitivity of cloned simple repeating polypurine/polypyrimidine DNA sequences to a variety of single strand specific nucleases is related to anomalous protonation of residues within these sequences in mildly acidic conditions. The protonated form of a plasmid insert of d(TC)n•d(GA)n has been studied in detail. This form is detectable in equilibrium with the normal Watson-Crick base paired form at pH's up to 7. The observed protection of the N-7 atoms of dG residues of the d(TC)n•d(GA)n insert against alkylation at low pH suggests that the protonated conformation is one in which dG residues are either Hoogsteen or reverse Hoogsteen base paired to protonated dC residues of the polypyrimidine strand. A structure in which dA:dT Watson-Crick base pairs alternate with Hoogsteen (syn) dG:dC pairs appears to be the most stereochemically acceptable model that is consistent with the chemical properties of the protonated form of this sequence.

Introduction

Although the Z form of DNA (1,2) was the first well characterized example of a non-Watson-Crick conformation for anti-parallel double stranded DNA, there have been indications that it may not be unique. Another type of structural transition in double stranded DNA was first reported by Johnson and Morgan (3). They noted that a series of simple repeating polypurine/polypyrimidine DNA polymers undergo a large shift in the buoyant density of the polymer when the pH was lowered from neutrality to pH 5.0. During this transition the polymers are protonated at an anomalously high pH. Among the polymers that are susceptible to this type of transition the simple repeating dinucleotide d(TC)n•d(GA)n has been observed with surprising frequency as a non-coding component of eukaryotic genes. Among eukaryotic genomes the distribution of this polymer resembles that of d(TG)n•d(CA)n. The latter polymer can be induced to undergo a transition to a left handed (Z) conformation by superhelical torsion (4-6). Several examples of genomic sequences have been found where the two polymers are juxtaposed (7-9).

Hentschel (10) and Htun *et al.* (11) recently reported that cloned tracts of d(TC)n•d(GA)n are anomalously sensitive to the action of S1 nuclease. Other S1 sensitive oligopurine/oligopyrimidine sequences have been located in 5′ non-coding regions of several eukaryotic genes (12-16). S1 nuclease is considered to be highly specific for single stranded nucleic acids. A model has been proposed to explain the anomalous sensitivity of the d(TC)n•d(GA)n tract in which it was suggested that the two strands of the polymer may slip past each other to produce displaced single stranded loops on opposite strands (10). In that this model assumes strand slippage, it resembles that of Johnson and Morgan (3) who proposed that in the linear polymer the strands disproportionate to form partially protonated multistranded complexes. In the case of the polypurine/polypyrimidine tracts embedded in other sequences the two strands of the polymer would be prevented from completely dissociating by the neighboring sequences.

Unpairing of bases in DNA at room temperature requires the input of approximately 2-6 kcal per base pair (Reviewed in 17). Any model suggesting the formation of single stranded loops should also offer a source for the energy required to break the base pairs that ultimately remain single stranded. Coupling of strand slippage to protonation of one or both strands of the S1 sensitive sequence might provide such an energy source. It does not however provide a source of energy to drive the anomalous protonation. Although several groups have reported polypurine/polypyrimidine tracts in cloned genes that are unusually sensitive to S1 nuclease, (10-16) there has previously been no clear demonstration that this sensitivity is related to anomalous protonation of these sequences.

Our initial objective in studying the S1 sensitivity of cloned tracts of d(TC)n•d(GA)n was to determine whether anomalous protonation of the type observed by Johnson and Morgan was responsible for the S1 hypersensitivity. The results reported below demonstrate that anomalous protonation is in fact associated with the acquisition of S1 hypersensitivity by these tracts. The results have yielded sufficient new information to permit of a model for the protonated form of d(TC)n•d(GA)n to be derived. The nature of the S1 nuclease sensitive conformation of phosphodiester bonds is also discussed.

Materials and Methods

S1 nuclease reactions

Relaxation of plasmid DNA by chicken erythrocyte topoisomerase 1 (18) in the presence of ethidium bromide was used to obtain differentially supercoiled DNA samples. Nuclease nicking reactions were monitored by diluting 50 uL aliquots of a reaction mixture containing 0.5 ug of DNA in the buffer indicated in the figure legends into 1 mL 45 mM LiOH, 1 mM EDTA, 1 ug/mL ethidium bromide. In this strongly alkaline solution nicked and linear plasmid molecules denature spontaneously at 23°C while closed circular molecules remain double stranded. Fluorescence of ethidium bound to the remaining closed circular molecules was measured fluorimetrically (using 520 nm excitation and 600 nm emission filters). Reactions were terminated by the addition of EDTA when <80% of the molecules had received a single stranded break. In some cases the nicked circular molecules were separated from residual supercoiled closed circular molecules by preparative electrophoresis on 1.2% agarose gels. Gel fragments containing DNA were dissolved in saturated lithium bromide containing 10 mM Tris•HCl pH 8.0, 1 mM EDTA. The DNA was adsorbed to glass powder. The adsorbed DNA was washed first with saturated lithium bromide and then twice with 0.1 M ammonium acetate in 50% ethanol. DNA was eluted from the glass with distilled water and was then concentrated by extraction with n-butanol. 70-90% of total counts were recovered.

Two dimensional gel electrophoresis

Two dimensional gel electrophoretic separations of plasmid topoisomers were performed as described previously (6,19) except that the buffer used in the first dimension was 20 mM Tris•acetate 1 mM magnesium acetate adjusted to the desired pH with acetic acid.

Construction of vectors and plasmids

p913, was constructed from the polylinker vectors pDPL13 (20) and pUC9 (21) by fusing a restriction fragment containing the amino terminal portion of the β-lactamase gene and part of the 13 site polylinker from pDPL13 with a restriction fragment containing the carboxy terminal portion of the β-lactamase gene, the origin of replication and the amino terminal portion of the lac Z′ gene from pUC9. p913 contains a polylinker with a BstEII site (5′GGTCACC3′) at the point of fusion. After cleavage by BstEII the site can be labelled on either one or other strand by selective backfilling with reverse transcriptase (Life Sciences Inc.) in the presence of dGTP, dTTP and either ^{32}PdCTP or ^{32}PdATP. Simple repeating polymer inserts which had originally been characterized as inserts in pDPL13 or pDPL6 (6,19) were transferred into the polylinker of p913 by standard gene splicing techniques. pTC45 contains a 45 bp insert of d(TC)n•d(GA)n at the SmaI site of p913.

Localization of S1 nuclease sensitive bonds

S1 nuclease sensitive bonds in the d(TC)n strand of pTC45 were localized after nicking by opening the plasmid at the unique BstEII site and selectively backfilling the end nearest the insert with reverse transcriptase in the presence of dTTP, dGTP and ^{32}PdATP. Electrophoretic analysis of the end labelled DNA was performed on denaturing 8% polyacrylamide gels. The products of DNA sequencing reactions on DNA which had been end labelled at the BstEII site served as standards. Since chemical cleavage results in the destruction of the 5′ nucleoside residue, the enzymatically cleaved fragments are 1 residue longer than the corresponding species in the chemically cleaved channel. Sites of S1 nuclease cleavage on the d(GA)n strand of the nicked pTC45 insert were identified after end labelling the plasmid at the Cla 1 site.

Alkylation of pTC45 by Dimethylsulphate

pTC45 was incubated in 0.5 mL of the buffer indicated, (legend to figure 7) for 1 hr. at 23°C then 1 ul dimethylsulphate was added. After 3 minutes the reaction was terminated by the addition of 5 uL mercaptoethanol. Samples were end labelled at the Cla 1 site. After recutting with Bgl 1 the labelled end was purified as described above. Samples were cleaved at N-7 methylated sites by treatment with piperidine for 30 min at 90°C.

Results and Discussion

As illustrated in Figure 1 a pH dependent structural transition in was detected in 2D electrophoretic gels of topoisomers of pTC45. Since no corresponding transition was observed in the vector, the transition must be associated with the insert. This experiment leads to three conclusions:

1) The apparent pKa for the transition is ~6. Even at pH 7.0 there is some evidence of the protonated state when the gels were run in the presence of 1 mM magnesium acetate. At pH 5 and pH 6 the presence of this ion is necessary to prevent smearing of the bands. Previous work by Johnson has indicated that divalent metal ions stabilize the protonated forms of linear polypurine/polypyrimidine polymers (22). In the absence of magnesium ion no transition was detected in pTC45 at pH 7.0.

2) In the transition state the polymer is underwound and has about 1 right handed helical turn per 20 base pairs. The incomplete unwinding argues against models in which the strands of the polymer become unpaired. The result also argues against models in which the polypurine/polypyrimidine tract forms a left handed helix (23). Although partial unwinding could result from strand slippage of the type suggested by Hentschel (10), the amount of torsional free energy released by the transition is not consistent with this possibility. This free energy was calculated from the mobility shift accompanying the transition as previously described (9,16). At pH 5.0 only about 8 kcals is released during the transition. At higher pH's the release is smaller, indicating that the transition is incomplete under these conditions. The energy absorbed during the transition is sufficient to break only 2-3 base pairs.

3) Since the plasmids remain monomeric under conditions in which the insert has undergone the transition, the insert cannot adopt a multistranded structure involving three or four strands (3,24).

pTC45 shows enhanced sensitivity to S1 nuclease relative to the vector without insert as shown by the experiments illustrated in figures 2 and 3. As shown in figure 2 when the sensitivity was measured as a function of superhelix density at pH 5.0 a strong stimulation of cleavage by negative superhelical coiling was noted. This observation implies that the S1 nuclease sensitive form of the plasmid insert is underwound with respect to B-DNA. It therefore directly associates the anomalous S1 sensitivity of this sequence with the low pH induced conformational transition

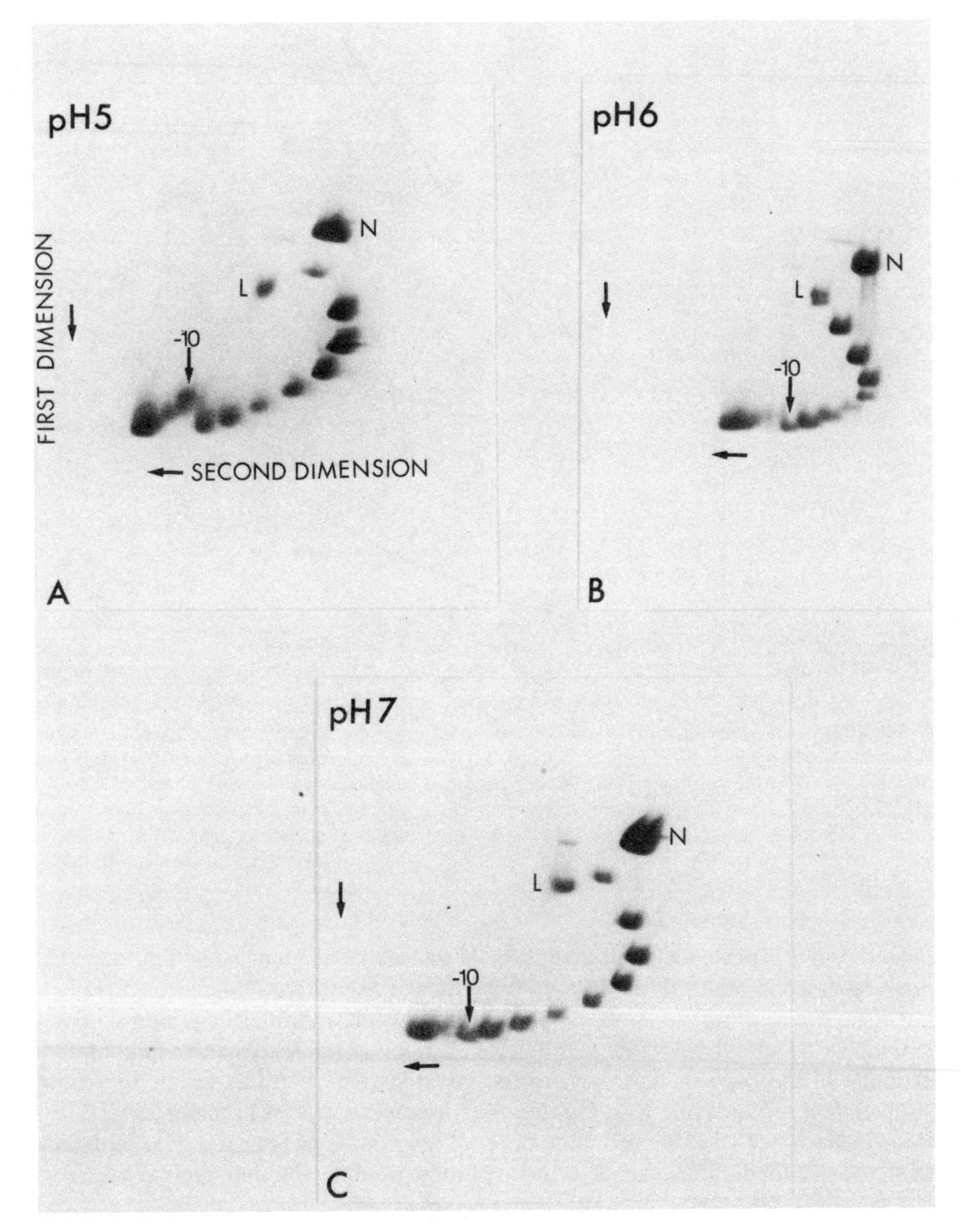

Figure 1. *pH dependence of the torsionally induced structural transition of pTC45.* Two dimensional gel electrophoresis on 1.5% agarose gels of topoisomer series of pTC45. Buffers: first dimension: 20 mM Tris•acetate containing 1 mM Magnesium acetate adjusted to the required pH with acetic acid. Second dimension: 40 mM Tris•acetate pH 8.0, 1 mM EDTA, 1ug/mL chloroquine. N=nicked DNA, L=linear DNA.

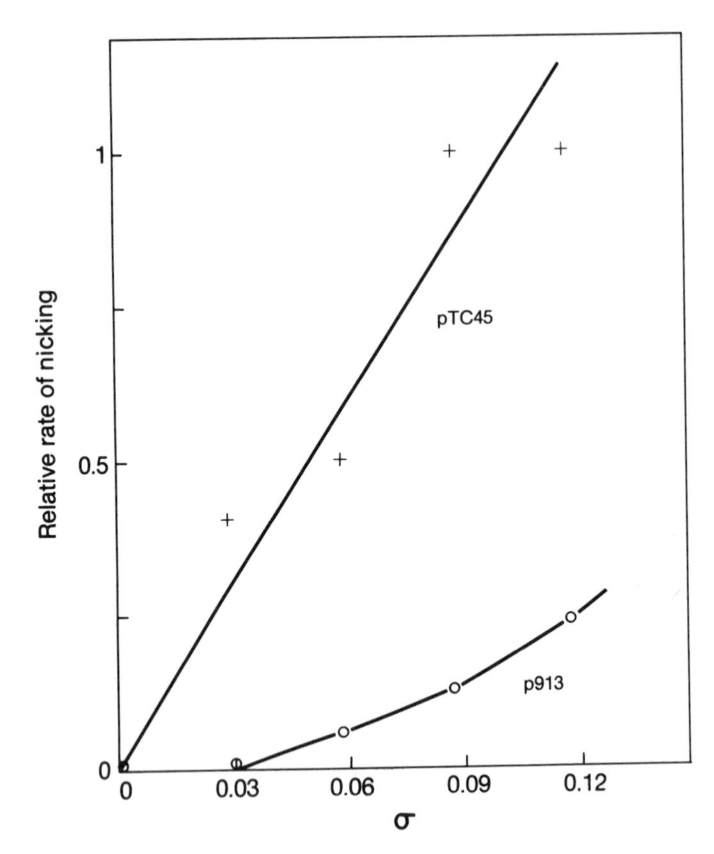

Figure 2. *Effect of unwinding torsion on the rates of S1 nicking of pTC45 and of p913.* Differentially supercoiled DNA samples of pTC45 (+) and p913 (O) were prepared as described in "Materials and Methods". 10 ug of the purified plasmid was nicked with 1-50 units S1 nuclease in 0.5 mL 0.05M NaAc pH 5.0, 0.2 mM $ZnCl_2$ at 23°C. Relative rates of the nicking reactions were calculated from the slopes of semi-log plots of the amount of supercoiled substrate remaining as a function of time.

detected in this plasmid. Similar superhelix density dependent behaviour has also been reported in other plasmids containing S1 hypersensitive sequences (13,15,23). By assuming (see below, figures 4 and 5) that the phosphodiester bonds showing anomalous S1 sensitivity reside within the insert, it is possible to make a quantitative estimate of the ratio of sensitivity of the insert sequence to the sensitivity of the surrounding vector sequences. On this assumption this ratio is in the order of 5,000 fold. This ratio falls within the same order of magnitude as the ratio of sensitivities of single stranded nucleic acids to those of most double stranded nucleic acids.

Figure 3 shows how the sensitivity of the vector and pTC45 change with pH. The −2 slope of the vector sensitivity vs. pH line indicates that the S1 nuclease cleavage of sequences in the vector depends upon two titrable protons with pKa's below 4.5 (25). If the cleavage reaction is itself acid catalysed, one of these protons must

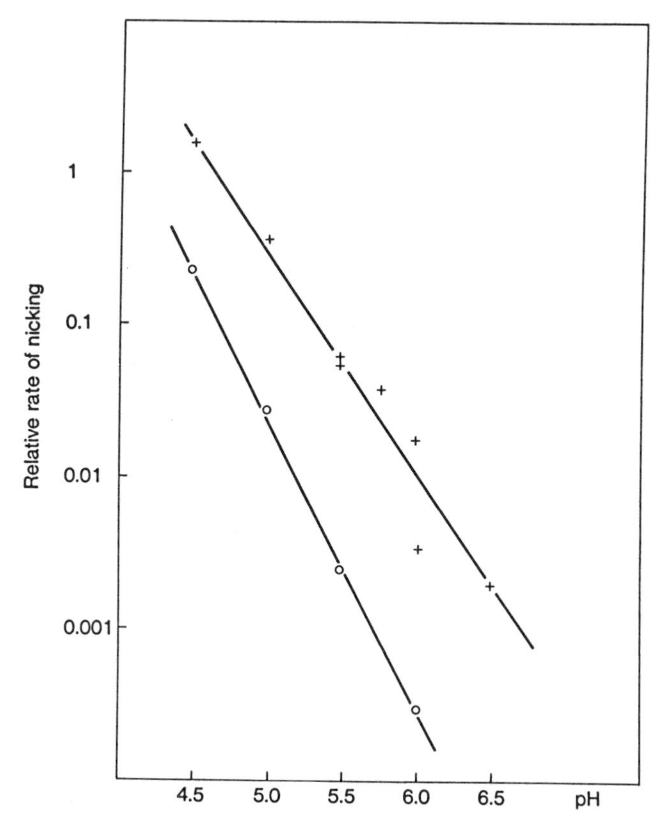

Figure 3. *Effect of pH on the rates of nicking of pTC45 and p913 by S1 nuclease.* The rate of nicking pTC45 and p913 superhelix density −0.07 by S1 nuclease were determined as a function of pH in 0.1 M (sodium ion concentration), NaAc or NaCac buffer containing 0.2 mM $ZnCl_2$. First order rate constants were calculated as described above (Figure 2).

participate directly in the cleavage reaction. An electrostatic contribution to substrate binding is indicated by strongly inhibitory effects of high ionic strength on the activity of S1 nuclease. The second titrable proton may modulate this electrostatic contribution. In the case of pTC45 the slope of the S1 sensitivity vs. pH line is only −1.2. Therefore, (most of) the second titrable proton involved in cleavage of phosphodiester bonds in vector sequences is not required for cleavage of phosphodiester bonds within the d(TC)n•d(GA)n insert. Since the pKa for protonation of the substrate falls within this range of conditions, the effect of the proton bound to substrate would mask an electrostatic effect of the titration of the enzyme (25).

Detailed analysis of the sites of S1 nuclease illustrated in figures 4 and 5 revealed further unexpected aspects of the cleavage reaction. Only the polypyrimidine strand of the d(TC)n•d(GA)n insert of pTC45 was sensitive to first strand cleavage by S1. Furthermore the sites of cleavage within this strand were located exclusively to the 5′ side of dT residues. There was essentially no cleavage 5′ to dC residues.

Two conclusions can be derived from these results.

1) The sensitive conformation of d(TC)n•d(GA)n is not single stranded. S1 nuclease cleaves all phosphodiester bonds in single stranded nucleic acids with approximately equal efficiency. If the polypyrimidine strand were forming a single stranded loop there would be a zone in which every residue would be cleaved equally. Furthermore there would be a corresponding zone of cleavage on the polypurine strand.

2) The conformation of adjacent residues in the S1 sensitive polypyrimidine strand alternates so that only every other residue is in a sensitive configuration.

The experiment shown in Figure 5 shows that a range of single strand specific endonucleases cleave the same region of pTC45 with comparable cleavage specificity. This result argues that these enzymes discriminate between single stranded and double stranded nucleic acids by a common mode of substrate recognition. In the case of *Neurospora crassa* endonuclease, which can be used over a broad range of pH, the polypyrimidine strand of the insert loses most of its anomalous sensitivity when the pH is raised to 8.0. The result confirms the relationship between the nuclease sensitive conformation and protonation of the insert.

The only site on unpaired DNA bases with a pKa near the range of pH in which protonation of d(TC)n•d(GA)n is observed is N-3 of cytosine. The pKa of this site in solution is 4.5. Since the apparent pKa for the transition in d(TC)n•d(GA)n is considerably higher than 4.5, a proton bound to N-3 of cytosine would have to be stabilized through a hydrogen bond to some other nucleophilic site.

Protonation of dC residues in DNA requires a change in the pattern of hydrogen bonds to corresponding dG residues. The three possible arrangements for a

---

Figure 4. *Strand specific nuclease sensitivity of the d(TC)n•d(GA)n insert in pTC45.* Localization of S1 nuclease sensitive bonds in native superhelix density pTC45. Channel 1: fragments 3′ end labelled at the BstEII site, cleaved at thymine residues by osmium tetroxide/piperidine (Friedman and Brown 1978). Channels 2-4: fragments generated by S1 nuclease cleavage at pH 6.0 (0.1 M NaCacodylate, 0.2 mM $ZnCl_2$, 23°C), pH 5.0 (0.1 M NaAc, 0.2 mM $ZnCl_2$, 23°C) and 4.5 (0.1 M NaAc, 0.2 mM $ZnCl_2$, 23°C), respectively before end labelling after BstEII cleavage. Channel 5 contains fragments generated by chemical cleavage at dG residues. Note that the enzymatically cleaved fragments are 1 residue longer than the corresponding species in the chemically cleaved channel.

Figure 5. *Specificity of single strand specific endonucleases for the polypyrimidine strand of the d(TC)n•d(GA)n insert in pTC45.* Channels 1-7: labelled 3′ to the d(TC)n strand, channels 8-11: labelled 3′ to the d(GA)n strand. Channels 1-3: pTC45 superhelix density −0.058, −0.039, −0.019 nicked with S1 nuclease in 0.05 M NaAc pH 5.0, 0.2 mM $ZnCl_2$ 23°C. Note that cleavage at the GAGA sequence at the 5′ end of the insert was detected only in the most supercoiled sample. Channels 4-11: pTC45 superhelix density −0.039, nicked with Mung bean nuclease (Channels 4 and 8), Nuclease P1 (Channels 5 and 9) and *Neurospora crassa* endonuclease (Channels 6 and 10) in 0.05 M NaAc pH 5.0, 0.2 mM $ZnCl_2$ 23°C. Channels 7 and 11 nicked with *Neurospora crassa* endonuclease in 0.05 M Tris•HCl pH 8.0, 2 mM $MgCl_2$. Approximately equal numbers of counts (total) were loaded on each of the channels. Note the almost total absence of first strand cleavage in the d(GA)n strand of the insert by the enzymes tested.

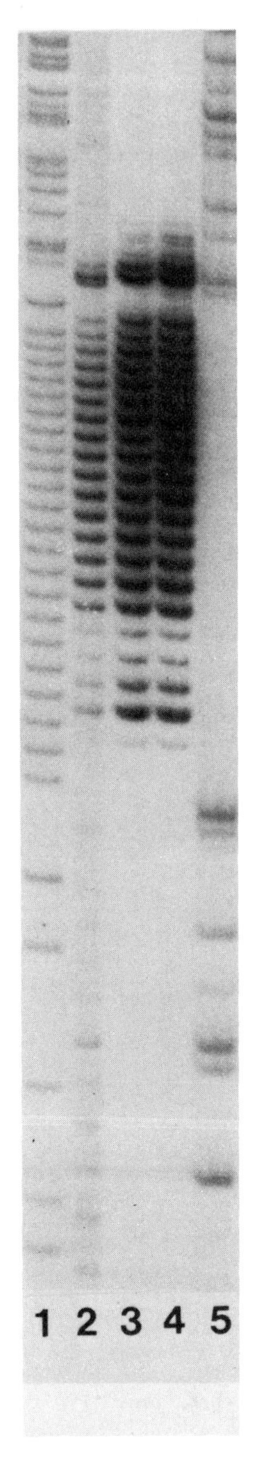

Figure 4

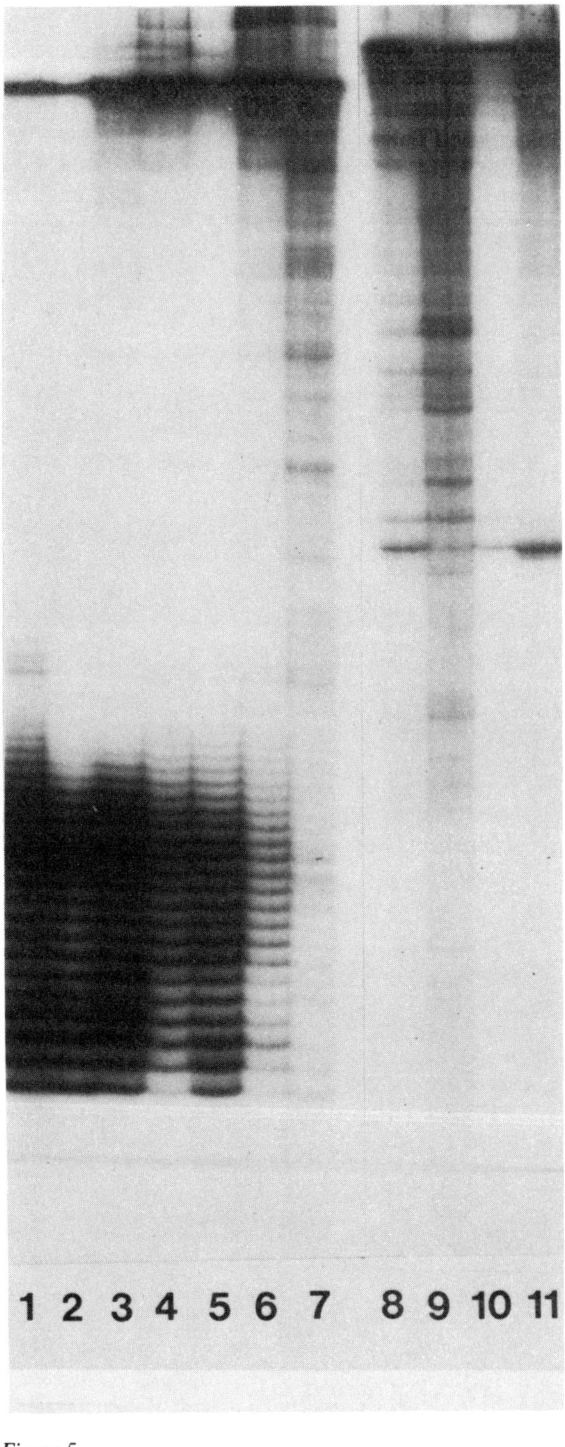

1 2 3 4 5 6 7 8 9 10 11

Figure 5

protonated dG:dCH$^+$ pair are illustrated in figure 6 A-C. Although none of these base pairing arrangements have been found in co-crystals of model compounds, few such crystal structures have been solved. Those that have been solved contain unprotonated Watson-Crick G:C base pairs (26). This paucity of structural data may be attributable to the usual selection of neutral conditions for co-crystallization of these compounds. Guanine crystallizes in a self paired arrangement analogous to Hoogsteen dG:dCH$^+$ pairs (27). Hoogsteen base pairs involving protonated cytosine are a probable element of triple stranded structures (28).

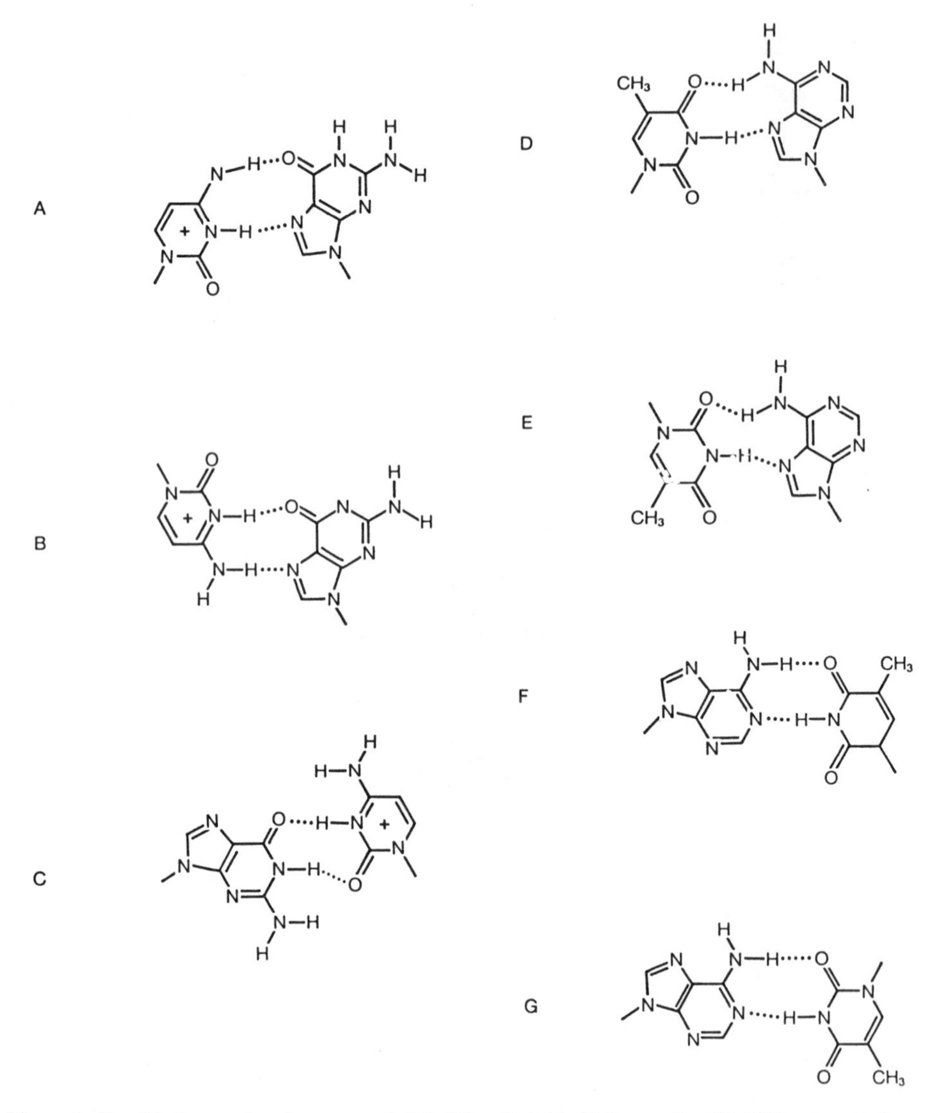

Figure 6. *Possible base pairs of protonated dC$^+$:dG and of dA:dT base pairs.* dG:dCH pairs: A) Hoogsteen, B) Reversed Hoogsteen, C) Slipped Watson-Crick. dA:dT pairs: D) Hoogsteen, E) Reversed Hoogsteen, F) Watson-Crick, G) Reversed Watson-Crick. It is assumed that at *least* two hydrogen bonds are required to define a base pair.

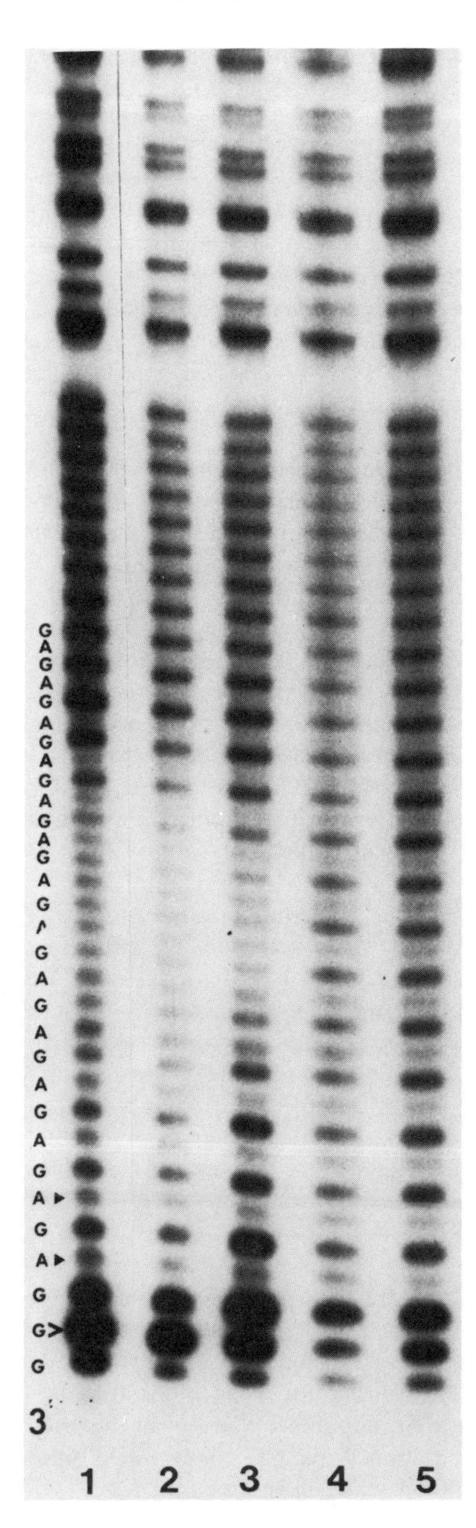

Figure 7. *Protection of dG residues within the d(GA)n• strand of pTC against alkylation by Dimethylsulphate.* pTC45 samples were alkylated with dimethylsulphate, end labelled and piperidine cleaved as described in the experimental section. Channel 1: 0.5 ml 0.1 M Sodium acetate (NaAc) pH 4.5, 2 mM $MgCl_2$. Channel 2: 0.1 m NaAc pH 5.0, 2 mM $MgCl_2$. Channel 3: 0.1 M Sodium cacodylate (NaCac) pH 6.0, 2 mM $MgCl_2$. Channel 4: 0.1 M NaCac pH 7.0, 2 mM $MgCl_2$. Channel 5: 0.1 M NaCac pH 7.0. The central dG residue of the 3′ dG_3 sequence discussed in the text is indicated (>). dA residues which are anomalously sensitive to piperidine cleavage after dimethylsulphate treatment are also indicated (>) (see text).

If altered dG:dC pairs are present within the protonated insert structure altered dA:dT pairs must also be considered. The possible arrangements for dA:dT base pairs are illustrated in figure 6 D-G (26). Examples of each of these base paired arrangements are known from single crystal studies (27,29-32).

Model building studies indicated that the slipped Watson-Crick dG:dCH$^+$ base pair illustrated in figure 6A can be accommodated in a model DNA helix that is similar to the standard B conformation. Although the reverse Hoogsteen arrangements of dG:dC and dA:dT pairs illustrated in Figures 6 A and B can be accommodated in anti-parallel models of double stranded helices these models are severely strained. The Hoogsteen base pair illustrated in figure 3C can be accommodated in antiparallel DNA models if the dG residue is in a syn- configuration.

In either Hoogsteen or Reverse-Hoogsteen base pairs the N-7 atoms of the purine bases should be protected against alkylation by dimethylsulphate. Since N-7 alkylation of purine bases renders the polynucleotide strand sensitive to cleavage by piperidine the exposure of this site can easily be tested. The experiment shown in Figure 7 demonstrates that in pTC45 at native superhelix density the dG residues of the central region of the insert become progressively more protected against alkylation of N-7 as the pH is lowered from 7.0 to 4.5. In keeping with the results of two dimensional gel electrophoresis, slight protection was observed *even* at pH 7.0 when magnesium was present in the reaction mixture. The pH dependence of the protection of dG residues against alkylation is independent evidence for a pH dependent transition in the d(TC)n•d(GA)n insert of pTC45. This result argues for a model for the protonated form of d(TC)n•d(GA)n in which the dCH$^+$:dG pairs are either in a Hoogsteen or in a reverse Hoogsteen arrangement.

The chemical reactivities of residues within and immediately adjacent to the protonated d(TC)n•d(GA)n insert suggest that the dA:dT base pairs are either in a Watson-Crick or in a reversed Watson-Crick arrangement. Within the region of the d(GA)n tract in which dG residues are protected against alkylation, the dA residues are anomalously sensitive to cleavage by piperidine after treatment with dimethyl-sulphate. In non-base paired dA N-1 is most readily alkylated while N-7 and N-3 are less reactive (22-33). Alkylation of DNA at N-1 or N-3 of dA residues without a subsequent depurination reaction does not promote strand cleavage by piperidine. Alkylation of dA residues at N-7 promotes cleavage. The N-7 atoms of dA residues in the protonated form of d(TC)n•d(GA)n therefore appear to be more exposed than the N-7 atoms of dA residues in B-DNA.

Among the four models that the above results suggest are possible alternatives for the protonated form of d(TC)n•d(GA)n, the most stereochemically reasonable is one in which the dCH$^+$:dG pairs are of a Hoogsteen type and dA:dT pairs are of a Watson-Crick Type. This model is illustrated in Figure 8. In order to construct this model with anti-parallel strands it was necessary to flip dG residues from the usual anti-orientation found in B-DNA to a syn-orientation. Among the deoxynucleosides dG is unique in preferring a syn- conformation when in monomeric form (34). Although the polypurine strand of the model resembles Z-DNA in that alternate

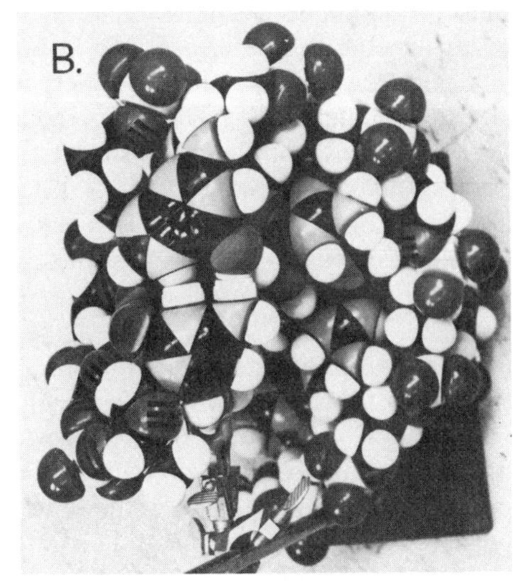

Figure 8. *Alternating Hoogsteen/Watson-Crick CPK Model for the protonated form of d(TC)n•d(GA)n.* In panel A an 8 bp model representing ½ of a full helical turn of alternating Hoogsteen/Watson Crick helix is illustrated from the side of the minor groove. Note the alternation in the orientation of the phosphates with respect to the axis of the helix.

Panel B illustrates the model as viewed from above with a dG:dCH$^+$ base pair uppermost. Note the alternation in base stacking which is good 5′ to dT and dG residues and poor 5′ to dC and dA residues. Note also the close proximity of the 2-amino groups of the dG residues to the corresponding 5′ phosphates which may permit the formation of a third hydrogen bond.

residues are in a syn-conformation, the conformation of the $^{(syn)}$purine deoxyribose residues are not inverted as in the Z conformation. For this reason the model does not adopt a left handed helix but instead forms a right handed helix with a pitch of ~1 turn per 16 residues, a value which is close to the experimentally determined value of 1 right handed turn per 20 residues.

An aspect of all models involving protonated dG:dCH+ pairs is that there are only two hydrogen bonds in any of the protonated dG:dC pairs whereas there are 3 in an ordinary Watson-Crick dG:dC pair. There should be a significant energy cost resulting from this loss of a hydrogen bond. The loss may be minimized if a new hydrogen bond forms between the exocyclic 2-NH_2 group of the dG residue and the adjacent (5′) phosphate, a possibility indicated by the close proximity of these sites on the model. Because of the underwinding in the model the base stacking averages slightly better than in B-DNA. This stacking alternates between almost perfect on the 5′ sides of dG and dT, and rather poor on the 5′ sides of dC and dA. Stacking of purine bases in a similar alternating arrangement with comparably large overlaps, has previously been reported from single crystal studies of guanosine and inosine (27). The selectivity of transitions involving anomalous protonation for polypurine/polypyrimidine sequences may result from the much larger stacking energies found between adjacent purine bases than between either adjacent purines and pyrimidines or between adjacent pyrimidines and pyrimidines (17).

The correlation between the unpairing of bases and sensitivity of DNA to the single strand specific endonucleases is only a weak one. In another study we have shown that poly d(AT)n•d(AT)n can undergo a transition to a cruciform conformation under the influence of negative superhelical torsion (35). The region of this cruciform that is most susceptible to the single strand specific endonucleases is displaced on each strand by ~3 residues to the 3′ side of the center of the palindrome. Since only the center of the palindrome is sensitive to modification by bromoacetaldehyde, any unpaired bases must reside at this point. The S1 sensitive phosphodiester bonds are therefore found in regions that are base paired.

Since all phosphodiester bonds in both single stranded RNA and DNA are susceptible to the action of the single strand specific endonucleases it is unlikely that either the base or the 2′ substituent on the sugar determines the ability of these nucleases to discriminate between single and double stranded nucleic acids. Furthermore the pucker of the sugar ring is unlikely to have large direct effects on the sensitivity because the predominant form in dsDNA is C2′ endo while the predominant form in dsRNA is C3′ endo. The remaining conformational characteristic that these enzymes may utilize to discriminate between single and double stranded nucleic acids is the conformation of the C3′-C4′-C5′-O-P-O-3′C series of bonds in the backbone of the molecule. Several of the possible conformations of this series of bonds can be eliminated because they appear in A, B or Z helices, none of which are very susceptible to the action of these nucleases. The extreme S1 hypersensitivity of the protonated form of d(TC)n•d(GA)n suggests that the conformation of the sensitive phosphodiester bonds conform closely to the one that is sensitive to the action of these nucleases. Although we have not established which aspects of the double stranded protonated structure of d(TC)n•d(GA)n make it susceptible to the action of the single strand specific endonucleases the model does suggest a possible sensitive conformation for the C3′-C4′-C5′-O-P-O-3′C group of bonds.

Steric crowding resulting from the extensive base overlap to the 5′ of dT residues may cause the C4′-C5 bond to flip from the usual (gauche,gauche) conformation found in B-DNA to a (gauche,trans) conformation. Although extensive overlap also occurs to the 5′ side of dG residues, the steric crowding of the dG C5′ methylene group is less severe because of the smaller angle subtended by the C1′ substituent. It is noteworthy that a secondary transition in the polylinker sequence adjoining the 5′ end of the d(TC)n tract results in S1 hypersensitivity of ApG linkages (see figures 3 and 4) which may reflect the combined effects of protonation and proximity of the hypersensitive d(TC) in tract.

A (gauche,trans) of the C4′-C5′ bond conformation should be freely accessible to single stranded nucleic acids, but would arise only in unusual circumstances in double stranded nucleic acids. Examples of these circumstances may be the displaced sensitivity of the d(AT)n•d(AT)n hairpin and junctional sensitivity found in the vicinity of Z-DNA tracts (36). Although a (gauche,trans) conformation is found in Z-DNA, the inversion of the deoxyribose moiety results in a different orientation of the phosphate oxygen atoms from that predicted for the S1 sensitive bonds of protonated d(TC)n•d(GA)n. Z-DNA is only slightly more sensitive than ordinary B-DNA to the action of the single strand specific endonucleases.

There have been indications of a possible association between S1 nuclease hypersensitive sites and both eukaryotic and prokaryotic promotor sequences (12-16,37). Although these regions do not have the simple sequence d(TC)n•d(GA)n they are commonly enriched in polypurine/polypyrimidine tracts. We note that the general model proposed here for the protonated form of d(TC)n•d(GA)n can be adapted to other polypurine/polypyrimidine sequences either—by allowing Hoogsteen base pairs between dA and dT residues or—by relaxing the requirement that alternate residues be Hoogsteen base paired. The previous studies by Johnson and Morgan have demonstrated acid induced structural transitions in a series of repeating polypurine/polypyrimidine polymers that range in composition from d(TTTC)n•d(GAAA)n to d(G)n•d(C)n. In work that will be reported elsewhere we have demonstrated that cloned tracts of each of these polymers are hypersensitive to S1 nuclease.

Kowhi-Shigematsu *et al.* (14) have reported that the region of the β-globin gene which exhibits anomalous sensitivity to S1 nuclease is also sensitive to modification by bromoacetaldehyde under mildly acidic conditions. Either Hoogsteen or reversed Hoogsteen base pairing of the dA:dT base pairs would be expected to expose the dA residues of the pTC45 insert to modification by this reagent. We have examined the d(AG)n strand of the pTC45 insert for sensitivity to bromoacetaldehyde using direct chemical methods to detect sites of modification. Although the dA residues of this strand of the d(TC)n•d(GA)n sequence did not exhibit any anomalous sensitivity to this reagent, slightly enhanced reactivity of dC residues on the opposite strand was observed.

Some circumstantial evidence suggests that Hoogsteen base pairs may also occur in response to supercoiling within prokaryotic promotor sequences. S1 nuclease sensitive sites have been found in the *E. coli* lac promotor (15) and in the tyrT promotor (37). Furthermore Hale *et al* (38,39) have reported that a variety of *E. coli* promotors exhibit an anomalous sensitivity to modification by a water soluble carbodiimide, a reagent which reacts specifically with N-1 of dG or N-3 of dT residues (40). N-1 of dG is exposed in a $dG{:}dCH^{+}$ Hoogsteen base pair.

Acknowledgements

We wish to thank the Medical Research Council of Canada for an operating grant. David Haniford is recipient of an Ontario Graduate Scholarship.

References and Footnotes

1. Wang, A.H.J., Quigley, G.J., Kolpack, F.J., Crawford, J.L., van Boom, J.H., van der Marel, G. and Rich, A., *Nature 282,* 680-686 (1979).
2. Pohl, F.M. and Jovin, T.M., *J. Mol. Biol. 57,* 375-397 (1972).

3. Johnson, D. and Morgan, A.R., *Proc. Nat. Acad. Sci. USA. 75,* 1637-1641 (1978).
4. Peck, L.J., Nordheim, A., Rich, A. and Wang, J.C., *Proc. Nat. Acad. Sci. U.S.A. 79,* 4560-4564 (1982).
5. Klysik, J., Stirdivant, S.M., Larson, J.E., Hart, P.A. and Wells, R.D., *Nature 290,* 627-677 (1982).
6. Haniford, D.B. and Pulleyblank, D.E., *Nature 290,* 627-577 (1983).
7. Richards, J.E., Gilliam, A.C., Shen, A., Tucker, P.W., and Blattner, F.R., *Nature, 306,* 483-487 (1983).
8. Cheng, H.L., Blattner, F.R., Fitzmaurice, L., Mushinski, J.F. and Tucker, P.W., *Nature 296,* 410-415 (1982).
9. Nordstrom, J.L., Hall, S.L. and Kessler, M.M.,*Proc. Nat. Acad. Sci. U.S.A. 82,* 1094-1098 (1985).
10. Hentschel, C.C., *Nature 295,* 714-716 (1982).
11. Htun, H., Lund, E. and Dahlberg, J.E., *Proc. Nat. Acad. Sci. U.S.A. 81,* 7288-7292 (1984).
12. Weintraub, H.,*Cell 32,* 1191-1203 (1983).
13. Nickol, J.M. and Felsenfeld, G., *Cell 35,* 467-47728 Cell 35, 467-477 (1983).
14. Kohwi-Shigematsu, T., Gelinas, R. and Weintraub, H. *Proc. Nat. Acad. Sci. USA. 80,* 4389-4393 (1983).
15. Schon, E., Evans, T., Welsch, J. and Efstradiatis, A., *Cell 35,* 837-848 (1983).
16. Ruiz-Carillo, A., *Nucleic Acids Res. 12,* 6473-6492 (1984).
17. Saenger, W., in *Principles of Nucleic Acid Chemistry,* Springer-Verlag, N.Y., Berlin, Heidelberg, Tokyo Series Ed. C.R. Cantor.
18. Pulleyblank, D.E. and Ellison, M.J., *Biochemistry 21,* 1155-1161 (1982).
19. Haniford, D.B. and Pulleyblank, D.E., *J. of Biomol. Struc. and Dynam. 1,* 593-609 (1983).
20. Gendel S., Straus, N., Pulleyblank, D. and Williams, J., *FEMS Microbiol. Lett. 19,* 291-294 (1983).
21. Vieira, J. and Messing, J., *Gene 19,* 259-268 (1982).
22. Johnson, D.A. Ph.D. thesis University of Alberta. (1976).
23. Cantor, C.R. and Efstradiatis, A.,*Nucleic Acids Res. 12,* 8059-8072 (1984).
24. Lee, J.S., Johnson, D.A. and Morgan, A.R., *Nucleic Acids Res. 8,* 4309-4320 (1980).
25. Dixon, M. and Webb, E.C., *Enzymes* 3'rd Edition, Academic Press N.Y., San Francisco. pp 155-156 (1979).
26. Voet, D. and Rich, A., *Prog. in Nucleic Acid Res. and Mol. Biol. 10,* 183-260 (1969).
27. Bugg, C.E., Thewalt, U.T. and Marsh, R.E., *Biochim. Biophys. Res. Commun. 33,* 436-440 (1968).
28. Morgan, A.R. and Wells, R.D., *J. Mol. Biol. 37,* 63-80 (1968).
29. Hoogsteen, K., *Acta Cryst. 16,* 907-916 (1963).
30. Haschenmeyer, A.E.V. and Sobell, H.M., *Acta Cryst. 18,* 525-540 (1965).
31. Sobell, H.M., *J. Mol. Biol. 18,* 1-7 (1966).
32. Wing, R., Drew, H., Takano, T., Broka, C., Tanaka, S., Ikatura, K. and Dickerson, R.E., *Nature 287,* 755-758 (1980).
33. Singer B., *Prog. Nucleic Acids Res. and Mol. Biol. 15,* 219-284 (1975).
34. Guschlbauer, W., *Jerusalem Symp. Quant, Chem. Biochem. 4,* 297-309 (1972).
35. Haniford, D.B. and Pulleyblank, D.E., *Nucleic Acids Res.* in press. (1985).
36. Kilpatrick, M.W., Klysik, J., Singleton, C.K., Zarling, D.A., Jovin, T., Hanau, L.H., Erlanger, B.F. and Wells, R.D., *J. Biol. Chem. 295,* 7268-7273 (1984).
37. Drew, H.R., Weeks, J.R., and Travers, A.A., *EMBO J. 4,* 1025-1032 (1985).
38. Hale, P., Woodward, R.S. and Lebowitz, J., *Nature 284,* 640-644 (1980).
39. Hale, P., Woodward, R.S. and Lebowitz, J., *J. Biol. Chem. 258,* 7828-7839 (1982).
40. Metz, D.H. and Brown, G.L., *Biochemistry 8,* 2312-2328 (1969).

Biomolecular Stereodynamics IV, Proceedings of the Fourth Conversation in the Discipline Biomolecular Stereodynamics, State University of New York, Albany, NY, June 04-09, 1985, Eds., Ramaswamy H. Sarma & Mukti H. Sarma, ISBN 0-940030-18-7, Adenine Press, ©Adenine Press 1986.

Application of the Precession Method to Fibre Diffraction: Structural Variations in lithium B-DNA

M. Sundaramoorthy, P. Parrack and V. Sasisekharan*
Molecular Biophysics Unit
Indian Institute of Science
Bangalore 560012, India

Abstract

The precession method of x-ray diffraction has been applied to the study of the B form of lithium DNA. The advantages of using the precession method generally for fibres are pointed out. The results presented here show that the diffraction data are incompatible with a uniform helical structure. Additionally, accurate measurements of the structural parameters reveal variations in the values of these parameters, dictated by salt content and relative humidity. These are indicative of the large flexibility inherent in the DNA structure giving rise to a polymorphism within the B-form.

Introduction

The precession method due to Buerger (1) to record x-ray diffraction patterns has been extensively used in the study of single crystals. For the study of diffraction of x-rays by fibres, stationary film methods are usually employed using the conventional flat-plate camera. However, the precession technique applied to fibres has certain advantages. Here we show the applicability of this method to fibres by comparison with "rotation-precession" photographs taken for a single crystal. As an example, precession patterns of lithium B-DNA are discussed. Incidentally, this is the first time that precession patterns for B-DNA fibres are obtained and analysed. Interestingly, our results show that there is a considerable amount of unexplained data, necessitating a re-examination of the presently accepted tenfold regular helical model for B-DNA.

Materials and Methods

The fibres were prepared from calf thymus DNA as described earlier (2). Precession patterns (D = 10 cm) were recorded after perfectly aligning the fibres normal to the beam. This was found necessary in order to obtain sharp, clear diffraction patterns. Again for these purposes, a zero-level screen was used throughout. The crystal used for comparison studies was glycyl-DL-phenylalanine, $C_{11}H_4N_2O_3$, which crystallises

*for correspondence and requests for reprints

in the orthorhombic space group Pbca with $a = 9.241(2)$A, $b = 28.41(2)$A and $c = 8.602(2)$A as determined by Marsh *et al* (3). It was mounted on the *b* axis and continuously rotated to simulate the orientation of crystallites about the cylindrical axis in fibres. Fig. 1a shows the familiar rotation photograph for glycyl-DL-phenylalanine using a flat plate camera. The precession pattern using a zero-level screen and recorded for the crystal under continuous rotation about the *b* axis is shown in Fig. 1b, this we shall designate as the rotation-precession pattern for the crystal. Throughout, a Ni-filtered Cu-K_a beam was used. Humidity of the fibres was controlled as described in ref. 2.

The rotation-precession of a single crystal

The advantages of using the precession method rather than the stationary film one are well-known in the case of single crystals. This method has earlier been applied to fibre diffraction studies (4,5). However, we would like to reemphasize the usefulness of the precession method by bringing out certain points of comparison between flat plate and precession methods, for a better appreciation of the application to DNA fibres, described in this paper.

The most important feature of a precession pattern is that the reciprocal lattice is obtained on the film in an undistorted manner with a one-to-one correspondence between each reflection recorded on the film and a reciprocal lattice point. Such a recording of a precession pattern however, is not possible in the case of a fibre owing to an inherent cylindrical symmetry about the fibre axis. Thus a one to one correspondence between reciprocal lattice points and reflections on the film cannot be obtained for a fibre. In spite of this, the precession method can be made use of as one obtains a projection of the reciprocal lattice on the film with linear correlation and without geometrical distortion. The theory of the application of the precession method to objects with cylindrical symmetry and having periodicity only in one dimension is being worked out in our laboratory. Here a comparison of the rotation (Fig. 1a) and rotation-precession (Fig. 1b) patterns for the crystal is made and the following essential differences are brought forth: (i) The rotation-precession photograph has a parallel array of reflections while the corresponding rotation photograph consists of reflections at intersections of curved (hyperbolic) lines. (ii) Layer lines in the rotation-precession photograph of Fig. 1b are equally spaced while those in the rotation photograph are not. (iii) The rotation pattern (Fig. 1a) does not contain reflections corresponding to the axis of rotation except for a few lower order ones. However, the geometry of the precession motion allows the intersection of the Ewald sphere with all reciprocal lattice points. As a consequence, the rotation-precession pattern (Fig. 1b) contains recordings of all the reflections.

Indexing of the reflections in Fig. 1a needs the use of a Bernal chart (Fig. 2a). However, in Fig. 1b the indexing of reflections is simpler and easier since they are distributed in parallel arrays of points. For a crystal with the a^* axis oriented parallel to the x-ray beam, only the 0kl reflections would be recorded in the precession pattern taken with a zero-level screen. In the case of rotation-precession, owing to

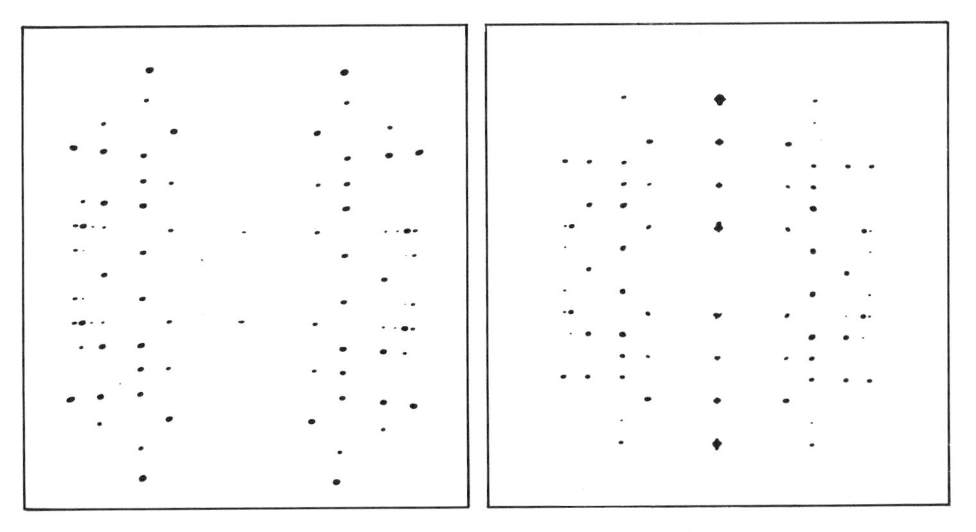

Figure 1a. Rotation photograph of glycyl-DL-phenylalanine. The crystal was rotated about the *b*-axis.

Figure 1b. Rotation-precession (see text) pattern for the same crystal.

rotation (about the *b* axis), both the a^* and c^* axes become parallel to the beam, as well as directions intermediate between them. Thus, 0kl, hk0 and hkl reflections—all of them are recorded (Fig. 2b). The other important feature is that the correlation between reflections and reciprocal lattice points is linear. Although reciprocal lattice points mapped this way are not the same as for the usual zero-level precession for a crystal, the distribution of reflections (Fig. 2b) does contain some noteworthy features, which would be useful for DNA fibre patterns. As already stated, the rows are equally spaced and each row has reflections with constant k index. So also, each column has reflections having constant h,l values with different k, and there are sets of equally spaced columns with multiples of h,l or hl reflections respectively.

The precession method for fibres

It is easy to visualise that the crystal rotation and rotation-precession patterns are similar to the flat plate and precession patterns respectively for fibres. For, fibres in general have cylindrical symmetry about the fibre axis and thus could be likened to crystals rotated about an axis of symmetry. This similarity can be readily seen by a comparison of Fig. 3a with Figs. 1a and 2a, and of Fig. 3b with Figs. 1b and 2b.

It is well known that the structural parameters n (the number of residues per turn) and h (unit rise) are obtained from measurements on a flat plate pattern for a DNA fibre. Because of the disposition of reflections in parallel arrays, n and h are calculated more easily and more accurately from precession patterns. A direct measurement on the film suffices to calculate n from the ratio of the spacings, without referring to the geometry of the camera arrangement. Meridional reflections,

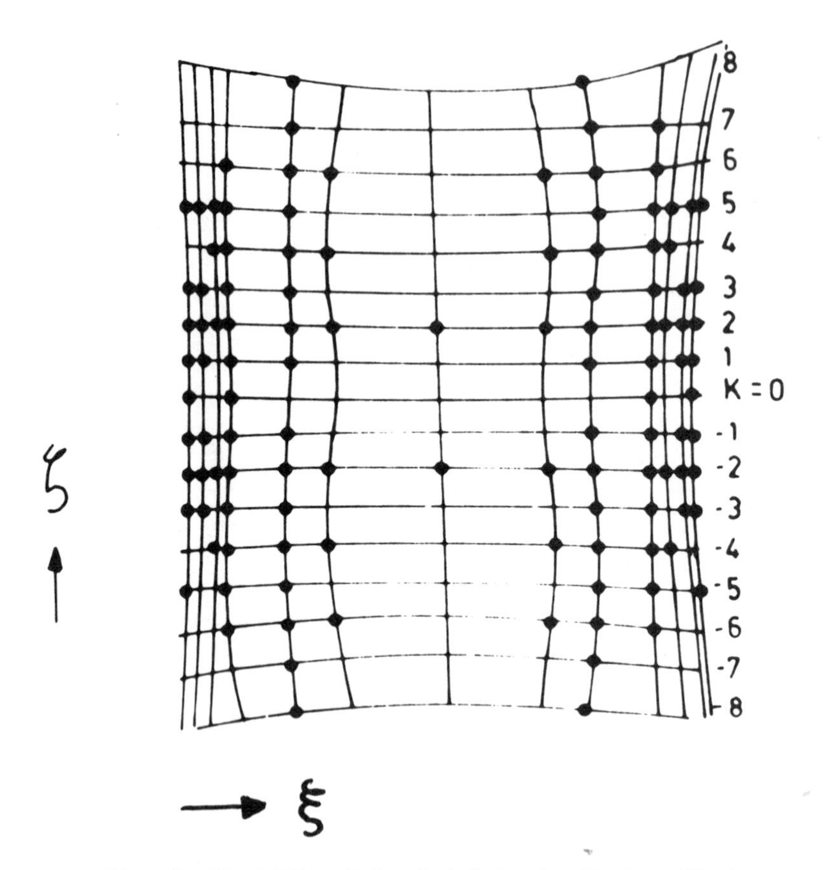

Figure 2a. The ξ,ζ (Bernal) chart for indexing the reflections of fig. 1a.

which are crucial to structure determination, are recorded straightaway on the film. Meridional and near-meridional reflections are separated and therefore, easily distinguished. Contemporary methods (using flat plate) require the fibre to be tilted through a particular angle (with the normal to the beam) to catch the meridian on any particular layer line. This angle of tilt has to be calculated from a knowledge of the structural parameters. For precession however, the geometry of recording causes the necessary tilt for every layer line. This is achieved without an a priori knowledge of any structural parameter whatsoever.

The calculation of h and c (helix pitch) are more accurate for the precession method than for the flat plate. For the latter, $d = \lambda(D^2 + \rho^2)^{1/2}/\zeta$ while for precession, $d = \lambda D/\zeta$. Here, λ = wavelength of x-rays, D = specimen to film distance, and $\rho^2 = \xi^2 + \zeta^2$, ξ and ζ denoting the film coordinates. For a tenfold helix, $h = d_{10}$, the d-value for the tenth layer line ($l = 10$), and $c = d_1$. Calculations show that the relative error in measurement of d (i.e. $\Delta d/d$) is smaller for the precession method.

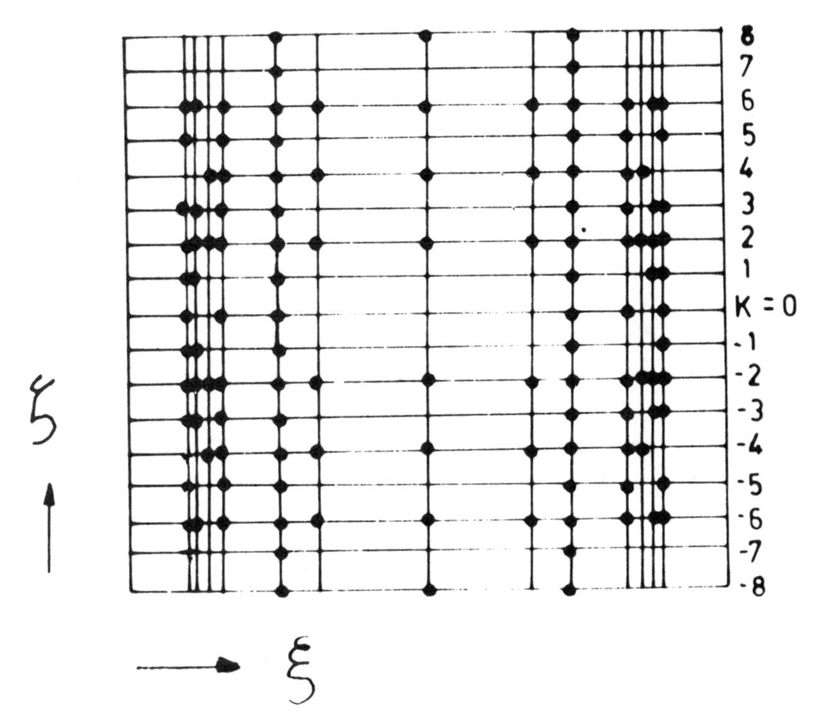

Figure 2b. The "rotation-precession" chart for fig. 1b. This is analogous to the Bernal chart for flat-plate. Note that the array of reflections are parallel (orthogonal in this case, the crystal being orthorhombic) with equally spaced layer lines.

Application to B-DNA

The precession pattern for the B-form (see Fig. 3b) shows some interesting features. Meridional reflections occur at the 4th and the 6th layer lines (the implications of this have been discussed earlier, see refs. 2,6). The most interesting result is that the layer lines are not equally spaced. Measurements show the spacings to deviate significantly from the expected values for the presently accepted tenfold helical model (see table I). Repeated measurements made on several such patterns show this deviation which cannot be explained on the basis of experimental errors. This result for B-DNA is indeed surprising, in view of the fact that the crystal rotation-precession as well as the Na A-DNA precession patterns consist of reflections arranged on equally spaced layer lines. The unequal spacings suggest the possibility of the helix parameter *n* being non-integral. It is interesting to note that non-integral helices (with $n < 10$) like 29_3, 39_4 etc. would also have ten observable layer lines, unequally spaced. However, occurrence of meridional reflections on the 4th and the 6th layer lines rules out this possibility. This is because such reflections are not compatible with ideal helical structures, whether integral or non-integral. Further, we had found that the intensities of reflections 004, 006 and 0010 showed a variation with salt concentration. This was interpreted as arising from salt-induced variations

Figure 3a. Flat plate diffraction pattern for calf thymus DNA at $\Delta = 1.6$ (Δ denotes the ratio of Li^+ ion concentration to the DNA concentration in the fibre[2]), and 66% r.h. The pattern is crystalline with reflections distributed on hyperbolic lines, similar to that in fig. 1a.

in the structure (2,6,7). Measurements on precession patterns obtained for different salt amounts and humidities show that there is indeed a small but significant variation in the values of the structural parameters, with the salt ratio and relative humidity. In view of the structure being different from a helical one, the terms "helical repeat" and "number of residues per turn" lose their significance. The only parameter that one can speak of is the fibre axis repeat, c. As in ref. 2, we designate the number of non-equivalent residues in a c-axis repeat by n' and the average height per residue by h'. An approximation to a helical structure is assumed for the sole purpose of determining $n(= n')$ and $h(= h')$ from the precession patterns. Some typical values obtained from our measurements are shown in table II. The values for h are accurate up to ± 0.005 A for 66% and up to ± 0.01 A for 98% r.h. The most interesting observation from table II is that the presently accepted values $n = 10$ and $h = 3.4$ A are not satisfied together. Though $n = 10$ is obtained for high humidity patterns, h is considerably less. On the other hand, patterns for lower humidity give h values close to 3.4 A, but in these cases, $n < 10$. Although a range of values for these parameters have been found earlier for the C-form of lithium DNA (8), such a variation has not so far been reported for the B-form. Here we find that even for the B-form, we can speak of a range of values : $n = 9.5$ to 10, $h = 3.35$A to 3.41A, and $c = 32.5$A to 33.5A.

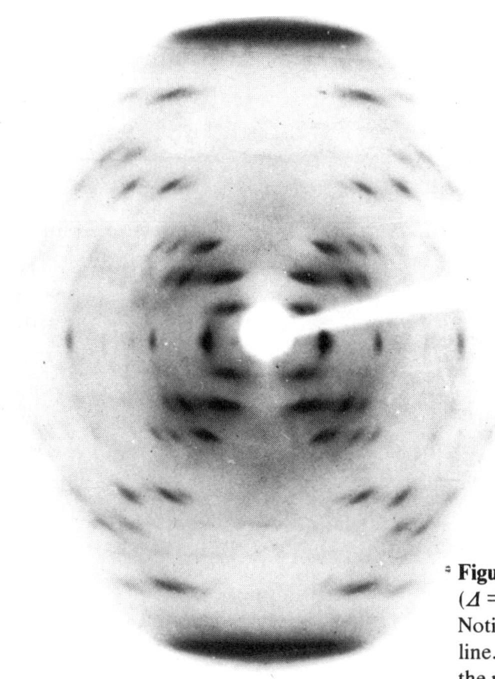

Figure 3b. Precession pattern for the same salt ($\Delta = 1.6$) at 66% r.h., taken with a zero-level screen. Notice the meridional reflection on the 6th layer line. The layer lines are straight and parallel, and the reflections on the higher order layers are much more sharp and well-defined than in fig. 3a.

Table I
Spacings in mm for different layer lines

l	Expected value for a 10_1	Measured value
1	9.10	9.4
2	18.20	18.7
3	27.36	27.6
5	45.60	45.7
6	54.72	54.7
8	72.96	72.9
10	—	91.0

Table II
Some typical values for the structural parameters

Δ	r.h.	h(A)	c(A)	$n=c/h$
1.4	66%	3.41	32.72	9.6
1.6	66%	3.39	32.81	9.7
1.6	98%	3.35	33.50	10.0

Note: h is calculated for the strongest layer line, assumed to contain a J_0 reflection (this is generally taken as the 10th layer). c is calculated for the first layer line observed. If the strongest layer line does not contain J_0, the h value, and consequently the value of n, will change (see ref. 2).

Conclusions

We have been able to apply the precession method to fibre diffraction of DNA with certain advantages. Coupled with a better control of salt in fibres, the method gives rise to the following interesting consequences. The diffraction data are not compatible with a perfect helical structure for B-DNA. Approximating the structure to a regular helical one, we find that our results do not show $n = 10$ and $h = 3.4$A simultaneously, though both the values are separately obtained, depending upon the conditions of the experiment. Finally, there is a variability in the structure, characterised by a series of values for n, h and c. It may be of interest to know, why these effects in the B-DNA structure escaped the notice of earlier workers. In this regard, we must point out the smallness of the effects that we are observing, with the better accuracy provided by the precession method. However, if this variability is inherent in the B-DNA structure, one would expect these results from conventional flat plate method as well. In principle, this should be so, but we are unable at present to comment on whether in practice they would be observed—i.e., whether these variations are so small that they fall within the errors of measurements in the flat plate method. We are carrying out these investigations with flat plate patterns of salt-controlled fibres at 100 mm distance, results of which would be published elsewhere.

The significance of $n \neq 10$ in the fibre (i.e., in the solid state) is not clear to us, since it raises questions in terms of packing of the DNA molecule in the unit cell (9). Theoretical model-building studies also become increasingly difficult with irregular structures. Work is going on in our laboratory to build a satisfactory model which would explain our observations. Lithium gives the C pattern at low salt, and the B-form at higher salt. The C form is known to show a variety of n,h values. The current observation that the B-form also is multivalued for n and h indicate the existence of fine variations in the structure, dictated by salt and humidity: for a fixed r.h.(room humidity, ~55%) when the salt is varied from $\Delta = 1.0$ to $\Delta = 1.9$, a variation of $n = 8.7$ to 9.8 is observed. h and c also show a progressive variation, and the structure undergoes a C⇒B transition. With fixed salt, variation in humidity induces such transitions: for $\Delta = 1.0$, when r.h. is varied from 50% to 98%, a C to B transition is observed, with none of the patterns showing good crystallinity. For any Δ between 1.2 and 1.9, a variation of r.h. results in a transition from a crystalline to a paracrystalline B form. Here, n is always found to have a value less than 10 (see also footnote with table II) at room humidity, which increases to 10 at 98% r.h. These results point towards the flexibility inherent in DNA proposed from theoretical studies (10). This flexibility provides for small but significant changes in the structure when under the influence of different environments. Binding of several peptides to DNA have been shown to cause fine structural changes, detectable by fibre diffraction (11). The influence of bound water molecules, the effect of which remains ill-understood, may also be capable of causing such changes. Different solution studies also have given different values for n. It is possible that n assumes a range of values in a single molecule. On the other hand, several molecules having slightly different n values might coexist. Measurements would then give an average of these values.

It may be noted that lithium ions, used as the cations for our DNA preparation, have a large hydration shell. As a result, they could serve as carriers for water molecules. Probably, the variations in the structure are brought about by bound water. Change in the salt concentration changes the hydration, and consequently we observe structural changes in DNA. This correlates well with the fact that increase of salt with fixed r.h. and increase of r.h. at fixed salt concentration have similar effects on the structural variations.

The range of n values obtained for lithium extends from $n = 8.7$ (C-pattern) to $n = 10$ (high-humidity B-form). Several interesting questions now emerge. Does n change continuously in this range, depending upon Δ and r.h.? Again, how is the crystallinity of the pattern related to the values of the structural parameters: is there any such correlation? Our method described here which makes use of the precise control of salt and the accurate parameters obtained by the application of the precession method to fibres, could be used as a powerful tool to answer such questions. Work in our laboratory is in progress in this direction.

Acknowledgement

This work was partially supported by the Department of Science and Technology, Govt. of India, under Thrust Area Program.

References and Footnotes

1. Buerger, M.J., *The Precession Method in X-ray Crystallography* John Wiley & Sons, New York, London, Sydney (1964).
2. Parrack, P., Dutta, S. and Sasisekharan, V., *J. Biomol. Struc. Dynam. 2,* 149 (1984).
3. Marsh, R.E., Ramakumar, S. and Venkatesan, K. *Acta Cryst. B 32,* 66 (1976).
4. King, M.V. *Acta Cryst. 21,* 629 (1966).
5. Sasisekharan, V., Zimmerman, S.B. and Davies, D.R. *J. Mol. Biol. 92,* 171 (1975).
6. Dutta, S., Parrack, P. and Sasisekharan, V. *FEBS Lett. 176,* 110 (1984).
7. Parrack, P., Sundaramoorthy, M. and Sasisekharan, V. *Proc. Int. Symp. Biomol. Struct. Interactions, Suppl. J. Biosci.,* Vol. 8, in press (1985).
8. Marvin, D.A., Spencer, M., Wilkins, M.H.F. and Hamilton, L.D. *J. Mol. Biol. 3,* 547 (1961).
9. Dover, S.D. *J. Mol. Biol. 110,* 699 (1977).
10. Gupta, G., Bansal, M. and Sasisekharan, V. *Proc. Natl. Acad. Sci. USA 77,* 6486 (1980).
11. Portugal, J., Aymami, J., Fornells, M., Subirana, J.A., Pons, M. and Giralt, E., in *Jerusalem Symp. Quant. Chem. Biochem.* (Eds. B. Pullman and J. Jortner), D. Reidel Publishing Co., Dodrecht, Boston, Lancaster, Vol. 16, p. 317 (1983).

Biomolecular Stereodynamics IV, Proceedings of the Fourth Conversation in the Discipline Biomolecular Stereodynamics, State University of New York, Albany, NY, June 04-09, 1985, Eds., Ramaswamy H. Sarma & Mukti H. Sarma, ISBN 0-940030-18-7, Adenine Press, ©Adenine Press 1986.

DNA and RNA Structures in Crystals, Fibers and Solutions by Raman Spectroscopy with Applications to Nucleoproteins*

George J. Thomas, Jr.†, James M. Benevides and Betty Prescott
Department of Chemistry
Southeastern Massachusetts University
North Dartmouth, Massachusetts 02740

Abstract

The results of DNA structure determination from single crystal x-ray diffraction analysis at atomic or near atomic resolution have been combined with the data of laser Raman spectroscopy to assign conformation-sensitive Raman marker bands of the DNA backbone and of different nucleosides in A, B and Z-DNA. The identification of distinct conformation markers is aided by the use of a recently developed Fourier deconvolution method for improvement of resolution in Raman spectra of condensed phases. The transferability of single crystal Raman data among samples of different morphology, including fibers, gels and dilute solutions, is exploited to determine the preferred conformers of DNA for experimental conditions which vary significantly from those in the crystal and which may in certain cases more closely approximate those *in vivo*. This analysis is extended to RNA models, as well as to RNA/DNA hybrid structures. Single crystals containing the following nucleic acid sequences have been investigated: d(CGCGCG), d(CGCATGCG), d(^{m}CGTA^{m}CG), r(GCG)d(CGC) and d(CCCCGGGG). In applications to DNA fibers several new or unorthodox structures are revealed, including A-helical structures for poly(dA-dT)•poly(dA-dT) and poly(rA)•poly(dT). In aqueous solutions of RNA/DNA hybrids, oligonucleotides as well as polynucleotides can form "mixed" structures in which both A and B type conformers are present.

In applications to biological structures: (i) non-uniform secondary structure is indicated for both high and low molecular weight DNA in aqueous solution, including calf thymus DNA and the 17-base pair operator sequence, O_L1, which provides the tightest binding site for the phage lambda cI repressor; (ii) B-DNA is confirmed as the secondary structure for the packaged genome of the double stranded DNA phage P22; and (iii) C3′endo nucleoside conformations are detected for encapsidated genomes in certain viruses containing single-stranded DNA (bacteriophage Pf3) or single-stranded RNA. In native TMV, however, at least two-thirds of the nucleosides differ from the usual C3′endo/anti conformation which is observed for protein-free TMV RNA.

*This is part 31 in the series Raman Spectral Studies of Nucleic Acids. Paper 30 in this series is reference 8 (Benevides & al., 1985).
†To whom correspondence may be addressed.

Abbreviations: ^{m}C = 5-methylcytidine, d = deoxyribosyl (5′⇒3′) phosphate backbone, r = ribosyl (5′⇒3′) phosphate backbone

Introduction

Single crystal x-ray diffraction analysis has provided new insight into the molecular structures of DNA. Recent reviews (1-3) survey the stereochemical details revealed at atomic or near atomic resolution for A, B and Z-DNA double helices. Although the x-ray results obtained from crystalline DNA fragments suggest probable structures for DNA *in vivo,* other methods are required to investigate the structures of aqueous DNA directly. Ideally, such methods should be capable of revealing fine details of structure and should be applicable over a wide range of sampling conditions including condensed and protein-associated nucleic acids. Many of these capabilities are shared in varying degrees by spectroscopic methods, including laser Raman spectroscopy (4).

An important advantage of Raman spectroscopy is the low level of interference usually encountered from biological solvents. Liquid water is among the weakest Raman scatterers known. Additionally, Raman spectra may be obtained from samples of virtually any morphology, including single crystals, amorphous solids, fibers, gels and solutions. Ordinarily, the data are directly transferable among samples of different types, so long as the respective molecular structures are identical or nearly so. Therefore, structural conclusions reached from single crystal studies can often be extended to aqueous biomolecules which yield the same Raman scattering frequencies and intensities. This advantage suggests the usefulness of Raman spectroscopy as a means of determining whether structures revealed at atomic resolution in crystals are maintained upon transfer of the molecules to an aqueous medium or to another significantly different molecular environment. Such a strategy has been successfully exploited in confirming the identical crystal and aqueous structures of yeast phenylalanyl tRNA (5), as well as in confirming the identity of crystalline and aqueous structures of Z-DNA (6,7). A similar approach can provide information on the secondary structures of DNA or RNA molecules which are packaged in viral capsids (4). We have extended the combined x-ray and Raman approach to a number of other crystals from which high resolution diffraction data have been obtained and for which the molecular structures have been solved (7,8).

The x-ray/Raman analyses have permitted the identification of several spectral bands or "Raman lines" with frequencies and intensities characteristic of specific phosphodiester backbone geometry, including nucleoside sugar pucker (C3′endo vs. C2′endo) and glycosidic bond orientation (anti vs. syn). In many cases, these spectral lines, referred to as Raman conformation markers, are sufficiently well separated from one another to permit the detection of two or more different conformers in the same molecule. In other cases, different Raman conformation markers are extensively overlapped and cannot be instrumentally resolved from one another. We have employed a recently described Fourier deconvolution method (9,10) to enhance the separation of such bands in order to estimate quantitatively the proportions of different nucleoside or backbone conformers present in DNA. Deconvolution procedures are also particularly useful in applications of the Raman method to nucleoproteins and viruses in order to resolve overlapping bands from protein and nucleic acid constituents. The results of several applications to DNA and RNA viruses are considered in this paper.

The nucleic acid models employed in this work include: (i) The following oligonucleotide duplexes for which molecular structures have been solved by x-ray diffraction: $[d(CGCGCG)]_2$, $[d(CGCATGCG)]_2$, $[d(^{m}CGTA^{m}CG)]_2$ and $[r(GCG)d(TATACGC)]_2$ (11-16); (ii) $[r(GCG)d(CGC)]_2$ and $[d(CCCCGGGG)]_2$, for which the crystal structures have been examined and preliminary molecular structures indicated (8,14,15); (iii) a number of polynucleotide fibers and gels which serve as "secondary" standards by virtue of the close correspondence between their Raman markers and those of the above listed single crystals. In most cases, the polynucleotides have also been characterized by fiber x-ray diffraction methods (17). In this last category are poly(dA-dT)•poly(dA-dT), poly(dG-dC)•poly(dG-dC), poly(dA)•poly(dT), poly(dG)•poly(dC) and corresponding ribopolymers.

In recent work, the Raman spectrum has been employed as a means of monitoring the kinetics of deuterium exchange of the purine 8C-H group in polynucleotides and nucleic acids (18). The kinetics of 8C-H exchange have been shown to be dependent upon both the kind and amount of nucleic acid secondary structure. The equilibrium Raman results discussed here are shown to be consistent with the results of dynamic Raman studies of the same polynucleotides and nucleic acids.

Materials and Methods

1. Oligonucleotides.

Single crystals of $[d(CGCGCG)]_2$, $[d(CGCATGCG)]_2$, $[d(^{m}CGTA^{m}CG)]_2$, $[r(GCG)d(CGC)]_2$ and $[d(CCCCGGGG)]_2$ were grown by Dr. A. H.-J. Wang, Department of Biology, Massachusetts Institute of Technology. These oligomeric DNAs, as well as the oligomers $[d(^{m}CG^{m}CG^{m}CG)]_2$ and $[r(GCG)d(TATACGC)]_2$ were synthesized and generously provided by Dr. J. H. van Boom and collaborators, Gorlaeus Laboratories, University of Leiden. The crystallization and x-ray structural analyses of most of these DNA models have been described in detail elswhere (13-16).

2. Polynucleotides.

Poly(dA-dT)•poly(dA-dT) (#658-92), poly(rA)•poly(dT) (#675-81), poly(dG)•poly(dC) (#709-57), poly(dG-dC)•poly(dG-dC) (#692-63), poly(dI)•poly(dC) (#730-51), poly(dI-dC)•poly(dI-dC) (#760-107), poly(dG) (#427840), poly(dC) (#663-82), poly(dA) (#658-42) and poly(dT) (#668/105) were purchased from Pharmacia-PL Biochemicals. Poly(dA)•poly(dT) (#111F-05981), poly(rI) (76C-0168), poly(rA) (41F-0403), poly(rG) (44C-1726) and poly(rU) (42F-0275) were obtained from Sigma Chemical Co. Solutions were prepared either by directly dissolving the polynucleotide duplex at the concentrations specified below in 0.1M NaCl at pH 7.5, or by annealing stoichiometric equivalents of single stranded polynucleotides to form appropriate duplexes. In the case of poly(dG)•poly(dC), the duplex was first dissociated at pH

13 then re-annealed through progressive dialysis against media of gradually decreasing pH down to pH7.5. Additionally, poly(dG)•poly(dC) was prepared by annealing stoichiometric equivalents of poly(dG) and poly(dC) in alkaline solution. All methods produced GC complexes of apparently identical secondary structure as judged from the Raman spectra.

Fibers of polynucleotides were drawn as described previously (19) and were sealed in hygrostatic Raman cells maintained at the desired relative humidity (RH).

3. Nucleic Acids and Viruses.

The operator O_L1 was the gift of Dr. Michael Weiss, Department of Biology, Massachusetts Institute of Technology. This seventeen base-pair DNA contains the 5′⇒3′ sequence d(TATCACCGCCAGTGGTA). Bacteriophage P22 and P22-DNA were the gifts of Professor Jonathan King, Department of Biology, M.I.T. TMV

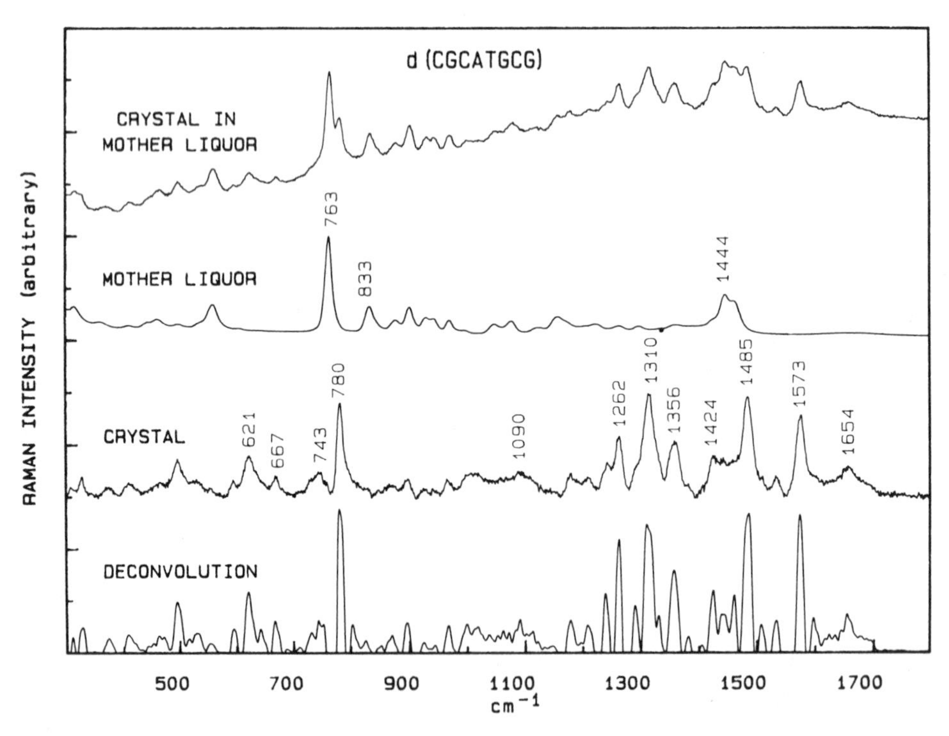

Figure 1. From top to bottom: (a) Raman spectrum of $[d(CGCATGCG)]_2$ as a single crystal (approximate dimensions: 0.1 × 0.5 × 0.5 mm) suspended in mother liquor. (b) Raman spectrum of mother liquor [2-methyl-2,4-pentanediol (40 wt-%) + 25mM sodium cacodylate (pH7) + 2mM spermine +10mM $MgCl_2$]. (c) Corrected spectrum of crystal $[d(CGCATGCG)]_2$ obtained by subtraction of (b) as well as the background of non-zero slope from (a), followed by amplification of the ordinate scale. (d) Fourier deconvolution of (c) using a Gaussian desmearing function of 15 cm^{-1} halfwidth to represent the slit function, with 20 iterations of the constrained deconvolution scheme described in reference 9.

was obtained from Professor Gerald Stubbs, Vanderbilt University and Pf3 from Dr. Loren A. Day, Public Health Research Institute, New York.

4. Raman Spectroscopy.

Raman spectra of solutions, fibers and crystals were obtained by methods previously described in detail (7,19,20) and summarized below.

The crystals of DNA oligomers, authenticated by x-ray diffraction as referenced in the preceding section, were transferred with approximately 10μl of mother liquor (usually 2-methyl-2,4-pentanediol (40%) + 25mM sodium cacodylate (pH7) + 2mM spermine + 10mM $MgCl_2$) to a Raman sample cell (Kimax #34507 glass capillary) which was thermostated at 32°C. The 514.5nm line from a Coherent Model CR-2 argon laser was focussed on the crystal and the Raman scattering at 90° was collected and analyzed by a Spex Ramalog spectrometer under the control of a North Star Horizon II microcomputer. Spectral data were typically collected at increments of 1.0 cm^{-1} with an integration time of 1.5 sec. A slit width of 8.0 cm^{-1} was employed for each scan of the spectrum from 300 to 1800 cm^{-1} unless otherwise indicated below. Spectra of solutions of the oligomers and polymers were obtained similarly.

Each Raman spectrum displayed below is the average of several scans, of 1.5 cm^{-1} or less repeatability, from which the fluorescent background and scattering by the solvent or mother liquor have been removed using computer subtraction techniques as described (7). The spectrum of solvent or mother liquor was always recorded with the same instrument settings employed for solution or crystal. In some of the spectra the noise was smoothed by a least squares fit of third order polynomials to overlapping fifteen point regions. This procedure did not measurably alter the Raman line frequencies or intensities. Raman frequencies are believed accurate to within ± 2 cm^{-1} for sharp lines.

In Fig. 1 we illustrate the experimental approach to obtaining and refining the vibrational Raman spectrum of the crystalline Z-DNA model $[d(CGCATGCG)]_2$.

5. Deconvolution of Raman Spectra.

The use of Fourier deconvolution for improvement of resolution in Raman spectra of nucleic acids has been described in detail elsewhere (9,10). Here, deconvolution was employed to separate or resolve partially overlapped bands which could not be adequately resolved instrumentally. Generally, the spectral envelope in the 600-900 cm^{-1} interval was deconvoluted with a Gaussian function of 15 cm^{-1} halfwidth. This desmearing function corresponds closely to the instrument slit function. The number of cycles of Fourier deconvolution in each case was dictated by the extent of overlap, but was always within the range 10 to 20. The iterative deconvolution was terminated when the bands of interest were sufficiently well separated to permit accurate measurement of their integrated areas, as long as the agreement between the input bandshape and the deconvolution result was 95% or greater.

Results and Discussion

1. Crystals of DNA Oligomers.

The Raman spectra of some recently examined DNA crystals are shown in Fig. 2. The nucleoside conformations determined or inferred from x-ray studies of these and other crystals are summarized in Table I. Each Raman spectrum of Fig. 2 was refined in the same manner indicated in Fig. 1 (c). Reference to earlier model compound studies (21) permits assignment of the prominent Raman lines to vibrational modes of either a nucleic acid base (A, C, G, T) or backbone (bk) residue, as indicated in Fig. 2 for many of the spectral peaks. These and related data provide a basis for identification of Raman markers corresponding to specific conformations of individual nucleosides and/or specific environments of the bases. The actual Raman frequencies and assignments for a number of these markers are given in Table II.

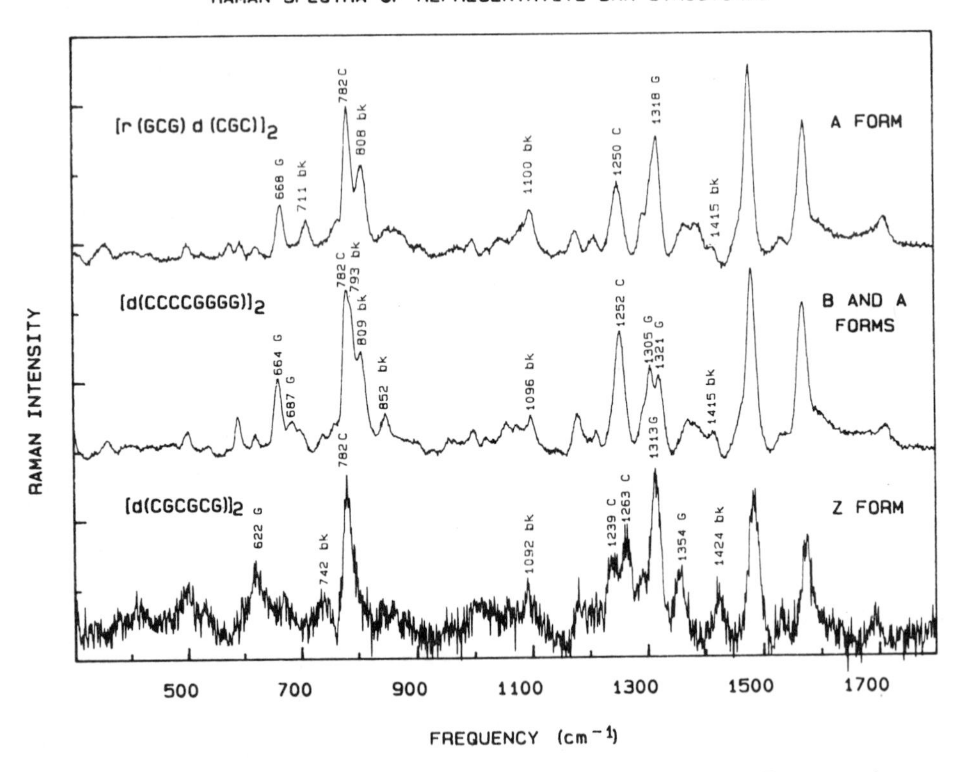

Figure 2. Raman spectra of DNA crystals representing Z, B and A form structures. From top to bottom: $[r(GCG)d(CGC)]_2$ (A conformation), $[d(CCCCGGGG)]_2$ (both B and A conformations), $[d(CGCGCG)]_2$ (Z conformation). Data were corrected for background and mother liquor as in Fig. 1(c), but no smoothing or deconvolution has been employed. Labels indicate Raman frequencies in cm^{-1} units with assignments to base (A, C, G or T) and/or backbone (bk) residues.

Table I
DNA Crystals and Their Nucleoside Conformations[a]

DNA Helix		Nucleoside Conformations				References
[d(CGCGCG)]$_2$	5′	2′endo/a	C-G	2′endo/s	3′	11
		3′endo/s	G-C	2′endo/a		
		2′endo/a	C-G	3′endo/s		
		3′endo/s	G-C	2′endo/a		
		2′endo/a	C-G	3′endo/s		
	3′	2′endo/s	G-C	2′endo/a	5′	
[d(CGCATGCG)]$_2$	5′	2′endo/a	C-G	2′endo/s	3′	12
		3′endo/s	G-C	2′endo/a		
		2′endo/a	C-G	3′endo/s		
		3′endo/s	A-T	2′endo/a		
		2′endo/a	T-A	3′endo/s		
		3′endo/s	G-C	2′endo/a		
		2′endo/a	C-G	3′endo/s		
	3′	2′endo/s	G-C	2′endo/a	5′	
[d(^{m}CGTA^{m}CG)]$_2$	5′	2′endo/a	^{m}C-G	2′endo/s	3′	13
		3′endo/s	G-^{m}C	2′endo/a		
		2′endo/a	T-A	3′endo/s		
		3′endo/s	A-T	2′endo/a		
		2′endo/a	^{m}C-G	3′endo/s		
	3′	2′endo/s	G-^{m}C	2′endo/a	5′	
[r(GCG)d(CGC)]$_2$	5′	3′endo/a	rG-C	3′endo/a	3′	8,14
		3′endo/a	rC-G	3′endo/a		
		3′endo/a	rG-C	3′endo/a		
		3′endo/a	C-rG	3′endo/a		
		3′endo/a	G-rC	3′endo/a		
	3′	3′endo/a	C-rG	3′endo/a	5′	
[d(CCCCGGGG)]$_2$[b]	5′	3′endo/a	C-G	3′endo/a	3′	8,14,15
		3′endo/a	C-G	3′endo/a		
		2′endo/a	C-G	3′endo/a		
		3′endo/a	C-G	2′endo/a		
		2′endo/a	G-C	3′endo/a		
		3′endo/a	G-C	2′endo/a		
		3′endo/a	G-C	3′endo/a		
	3′	3′endo/a	G-C	3′endo/a	5′	

[a]Abbreviations: a=anti, s=syn, ^{m}C=5-methylcytosine, r=ribonucleoside.
[b]Nucleoside conformations assumed identical to [d(GGCCGGCC)]$_2$. See reference 12.

Most of the DNA sequences so far examined are deficient in A and T residues. Therefore, single crystal studies have provided much more information on G and C markers than on A and T markers, as is evident from Table II. Until additional AT crystals become available, we are pursuing two approaches towards improving the base of information on conformation-sensitive Raman lines of A and T. In the first instance, we have applied computer subtraction methods to the available crystal data, i.e. the AT contributions are determined by difference after subtracting the

Table II
Deoxynucleoside Conformation Markers from Raman Spectra of DNA Crystals and Fibers

Frequency (cm^{-1})[a]	Assignment Base	Conformer	Model Compounds[b]
666 ± 2	gua	C3′endo/anti	r(GCG)d(CGC), d(CCCCGGGG), $(dG)_n(dC)_n$
1318 ± 2	gua	C3′endo/anti	r(GCG)d(CGC), aq[r(CGCGCG)]
682 ± 2	gua	C2′endo/anti	aq[d(CCCCGGGG)], aq[(dG-dC)]n
1333 ± 2	gua	C2′endo/anti	d(CCCCGGGG), aq[d(CGCGCG)],
625 ± 2[c]	gua	C3′endo/syn	d(CGCGCG), d(CGCATGCG), d(^{m}CGTA^{m}CG)
1316 ± 2	gua	C3′endo/syn	d(CGCGCG)
670 ± 2	gua	C2′endo/syn	d(CGCGCG), d(CGCATGCG), d(^{m}CGTA^{m}CG)
642 ± 2	thy	C3′endo/anti	$(dA\text{-}dT)_n$
777 ± 2	thy	C3′endo/anti	$(dA\text{-}dT)_n$
1239 ± 2	thy	C3′endo/anti	$(dA\text{-}dT)_n$
665 ± 2	thy	C2′endo/anti	aq[$(dA\text{-}dT)_n$], r(GCG)d(TATACGC), d(CGCATGCG), d(^{m}CGTA^{m}CG)
748 ± 2	thy	C2′endo/anti	aq[$(dA\text{-}dT)_n$], r(GCG)d(TATACGC), d(CGCATGCG), d(^{m}CGTA^{m}CG)
1142 ± 2	thy	C2′endo/anti	$(dA\text{-}dT)_n$
1208 ± 2	thy	C2′endo/anti	$(dA\text{-}dT)_n$
727 ± 2	ade	C3′endo/anti	$(dA\text{-}dT)_n$, r(GCG)d(TATACGC)
1335 ± 2	ade	C3′endo/anti	$(dA\text{-}dT)_n$
727 ± 2	ade	C2′endo/anti	aq[$(dA\text{-}dT)_n$], aq[r(GCG)d(TATACGC)]
1339 ± 2	ade	C2′endo/anti	$(dA\text{-}dT)_n$
624 ± 2	ade	C3′endo/syn	d(CGCATGCG), d(^{m}CGTA^{m}CG)
729 ± 2	ade	C3′endo/syn	d(CGCATGCG), d(^{m}CGTA^{m}CG)
780 ± 2[d]	cyt	C3′endo/anti	r(GCG)d(CGC), aq[r(GCG)d(TATACGC)]
782 ± 2	cyt	C2′endo/anti	d(CGCGCG), d(CGCATGCG), d(^{m}CGTA^{m}CG), aq[d(CGCGCG)], aq[d(CGCATGCG)]
1250 ± 2[e]	cyt	C2′endo/anti	d(CCCCGGGG), d(CGCGCG), r(GCG)d(CGC), aq[r(GCG)d(TATACGC)]

[a]Frequencies are those of non-deuterated structures only. Many of the markers listed are substantially shifted to lower frequency by deuteration of ring amino and imino groups in D_2O media.
[b]See references cited in text. All oligomer sequences refer to the crystal, except where indicated "aq" (aqueous solution).
[c]A weak band from C3′endo/syn A may underly the more intense band of G at this frequency.
[d]Sensitive to substitution of ribose for deoxyribose in the nucleoside.
[e]Exhibits diminished intensity in C3′endo/anti conformations and shifts to ca. 1265 cm^{-1} when paired with syn G in Z-DNA.

known GC contributions from spectra of DNAs which contain both GC and AT. Secondly, we have investigated polynucleotides containing only AT pairs, and have manipulated the experimental conditions in order to drive the polynucleotide into a given conformation which could be identified by comparison of the Raman frequencies of the polynucleotide backbone with the known conformation markers established from the DNA crystal studies. This second approach will be discussed in the next section. Here, we apply the difference method to Z-DNA, by subtracting the spectrum of crystalline $[d(CGCGCG)]_2$ from that of crystalline $[d(CGCATGCG)]_2$ to obtain Raman markers of A and T nucleosides in the C3′endo/syn and C2′endo/anti conformations, respectively.

Fig. 3 shows the result of subtraction of the spectrum of the GC hexamer $[d(CGCGCG)]_2$ from that of the octamer $[d(CGCATGCG)]_2$. The difference spectrum contains no bands from the bases G and C, and bands from the sugar-phosphate backbone exhibit diminished intensity. The prominent positive bands in the difference spectrum should correspond to those of a hypothetical Z-DNA containing only AT pairs in alternating sequence. Several of these are listed in Table II. Confirmation of the AT assignments is found in the spectrum of $[d(^mCGTA^mCG)]_2$ (7). Indeed, comparison of Fig. 3 with the data of Benevides et al. (reference 7, Fig. 5) shows that the Raman intensity profile in the interval 1600-1700 cm^{-1} is very similar for both [d(CGCATGCG)] and $[d(^mCGTA^mCG)]$, due primarily to the contributions (carbonyl group vibrations) of the two thymidine residues in each oligomer. Both Z oligomers also contain a relatively prominent shoulder near 730 cm^{-1} (due to A), an intense line near 1255 cm^{-1} (T) and a trio of intense Raman lines between 1300 and 1375 cm^{-1} common to AT structures. (Note, however, that this triplet is somewhat obscured in $[d(^mCGTA^mCG)]_2$ by interference from comparably intense Raman lines of the methylated cytidine residue at 1300 and 1360 cm^{-1} (7).)

An important result from Fig. 3 is the assignment of the 624 cm^{-1} line to the C3′endo/syn conformer of deoxyadenosine. In previous work (7), this band was not

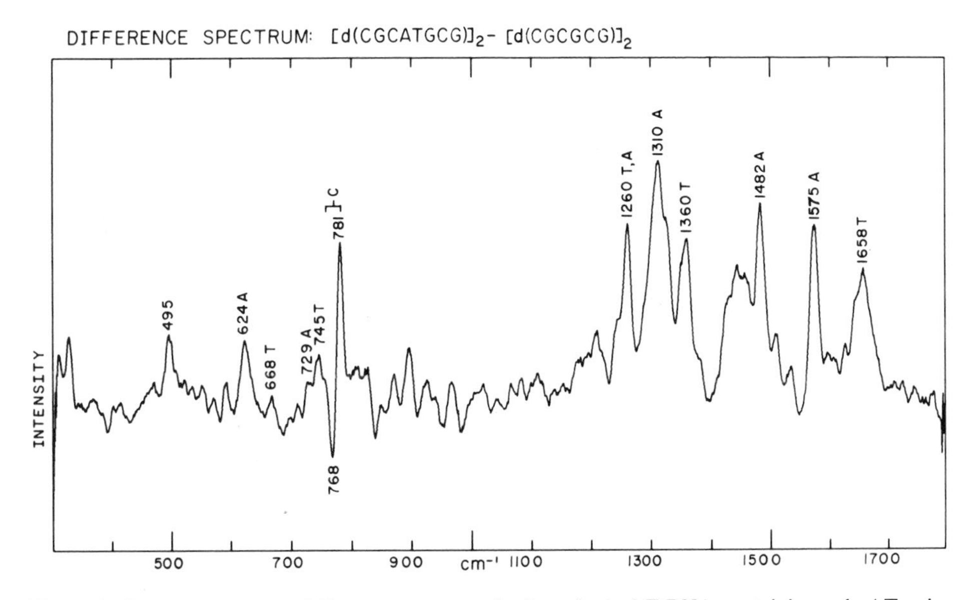

Figure 3. Computer generated Raman spectrum of a hypothetical Z-DNA containing only AT pairs, obtained by subtraction of the spectrum of crystalline $[d(CGCGCG)]_2$ from that of crystalline $[d(CGCATGCG)]_2$. Labels as in Figure 2.

revealed. The present assignment (Table II) is supported by the fact that the Z structures of both $[d(CGCATGCG)]_2$ and $[d(^{m}CGTA^{m}CG)]_2$ generate more intense Raman scattering at 624 cm^{-1} than does the Z structure of $[d(CGCGCG)]_2$ (6). Specifically, the intensity ratio of Raman lines at 624 and 670 cm^{-1}, which in the absence of an adenosine contribution should correspond to the ratio of C3′endo/syn G to C2′endo/syn G, exhibits the anticipated 2:1 ratio in $[d(CGCGCG)]_2$ but has the higher values of 2.6:1 in $[d(CGCATGCG)]_2$ and 1.4:1 in $[d(^{m}CGTA^{m}CG)]_2$ (6,7). The latter higher values are thus consistent with a contribution from syn adenosine at 624 cm^{-1}.

The same strategy can in principle be employed to infer the Raman markers of AT sequences incorporated into an A helix structure. For example, the Raman spectrum of a GC oligomer of the A form (such as crystalline $[r(GCG)d(CGC)]_2$) could be subtracted from that of an A form oligomer containing both AT and GC pairs (crystalline $[r(GCG)d(TATACGC)]_2$) to yield the spectrum of a hypothetical A form DNA containing only AT pairs. Unfortunately, the latter crystal is unavailable at present. However, the aqueous decamer has been investigated by NMR spectroscopy and is reputed also to assume the A conformation (16,18). Fig. 4 shows the difference spectrum obtained with aqueous $[r(GCG)d(TATACGC)]_2$ as minuend and crystalline $[r(GCG)d(CGC)]_2$ as subtrahend. Comparison of this Raman difference spectrum with data recently obtained on poly(dA-dT)•poly(dA-dT) (20) shows clearly that the TATA box of aqueous $[r(GCG)d(TATACGC)]_2$ does *not* assume an A conformation. All of the prominent peaks in this spectrum point to a B conformation for the four AT pairs of the aqueous decamer. Further discussion of the A and B forms of poly(dA-dT)•poly(dA-dT) is given below (Section 3.).

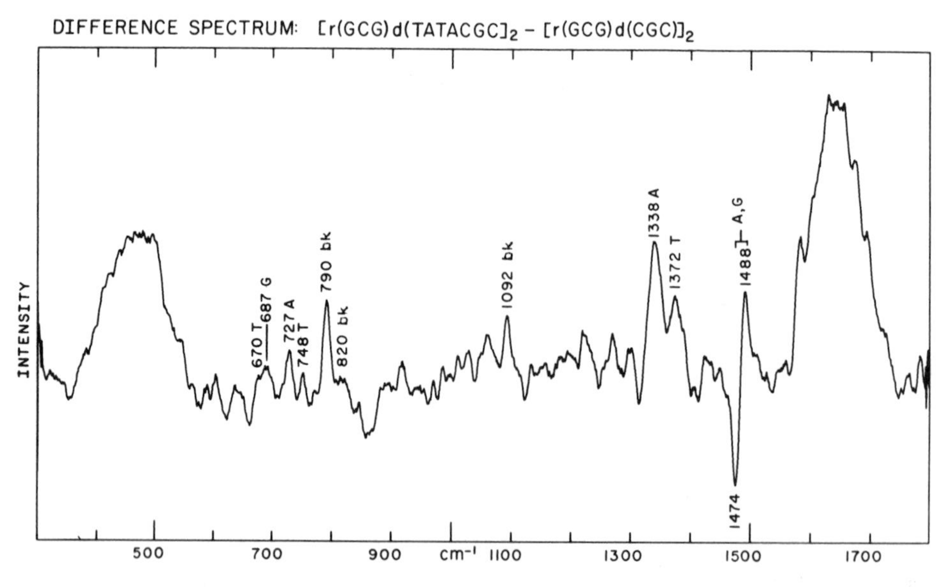

Figure 4. Computer generated Raman spectrum obtained by subtraction of the spectrum of crystalline $[r(GCG)d(CGC)]_2$ from that of aqueous $[r(GCG)d(TATACGC)]_2$. Labels are as in Figure 2.

Comparative analysis of Raman and x-ray diffraction data of crystalline mononucleotides by Tsuboi and co-workers (22,23), and of Raman and NMR data on aqueous nucleotides by Benevides et al. (24) have also been carried out for the purpose of identifying Raman conformation markers. Pertinent results of the nucleotide studies are consistent with the data in Table II.

2. Solutions of DNA Oligomers.

The B helix is generally the preferred secondary structure of high molecular weight DNA in aqueous solution and the same is apparently true of many of the aqueous DNA oligomers investigated in this work, despite the fact that none of the oligomers assumes what could be regarded as a B-DNA structure in the crystal state. Comparison of the Raman spectra of crystalline and aqueous DNA oligomers should provide, therefore, an indication of the conformation dependent Raman lines for both A-to-B and Z-to-B transitions. As an example of the A-to-B transition we show in Fig. 5 the difference curve obtained by subtraction of the spectrum of crystalline

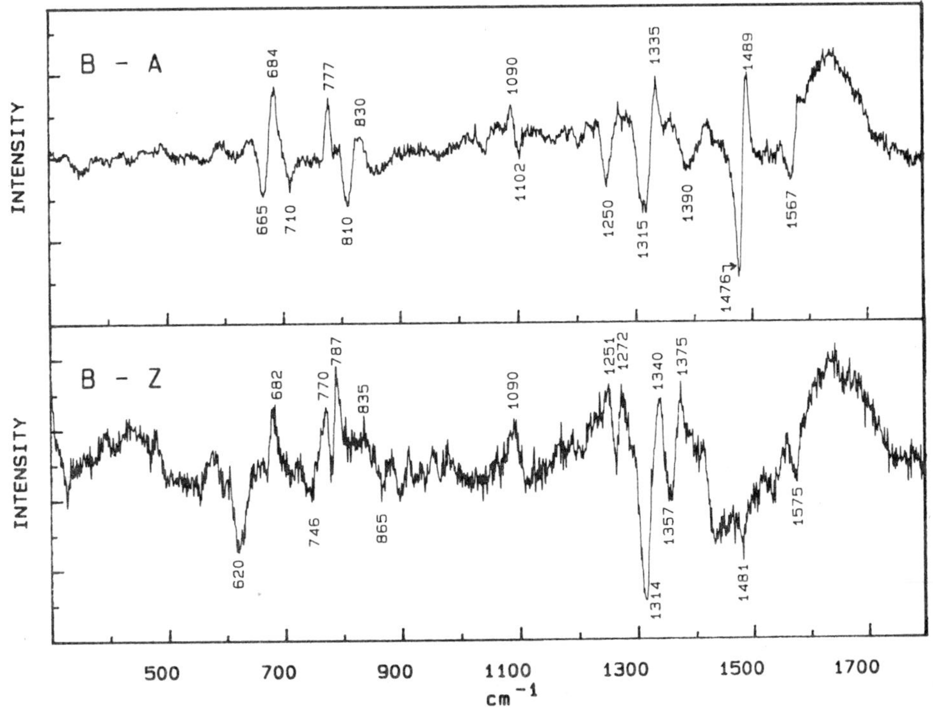

Figure 5. Top: Computer difference spectrum showing the A-to-B structure transition of DNA, obtained by subtraction of the A-DNA spectrum (crystalline $[r(GCG)d(CGC)]_2$) from the B-DNA spectrum (aqueous $[d(CCCCGGGG)]_2$). Bottom: Computer difference spectrum showing the Z-to-B structure transition of DNA, obtained by subtraction of the Z-DNA spectrum (crystalline $[d(CGCATGCG)]_2$) from the B-DNA spectrum (aqueous $[d(CGCATGCG)]_2$). Labels as in Figure 2.

$[r(GCG)d(CGC)]_2$ from that of aqueous $[d(CCCCGGGG)]_2$. Also in Fig. 5, the Z-to-B transition is represented by the difference curve resulting from subtraction of spectra of crystalline and aqueous $[d(CGCATGCG)]_2$. Additional examples have been given elsewhere (6,7,25).

In view of the unavailability of crystalline B-DNA models containing A and T, we have again assumed an indirect approach in order to determine Raman markers which identify the C2′endo/anti nucleoside conformations of A and T. Both $[d(^{m}CGTA^{m}CG)]_2$ and $[d(^{m}CG^{m}CG^{m}CG)]_2$ are expected to adopt the B helix geometry in aqueous solution and their difference spectrum should therefore correspond to that of a hypothetical B-DNA structure of AT pairs in alternating sequence. The difference spectrum is shown in Fig. 6 and the appropriate Raman markers are tabulated in Table II. These results are substantially confirmed (below, Section 3.) in studies of polynucleotides.

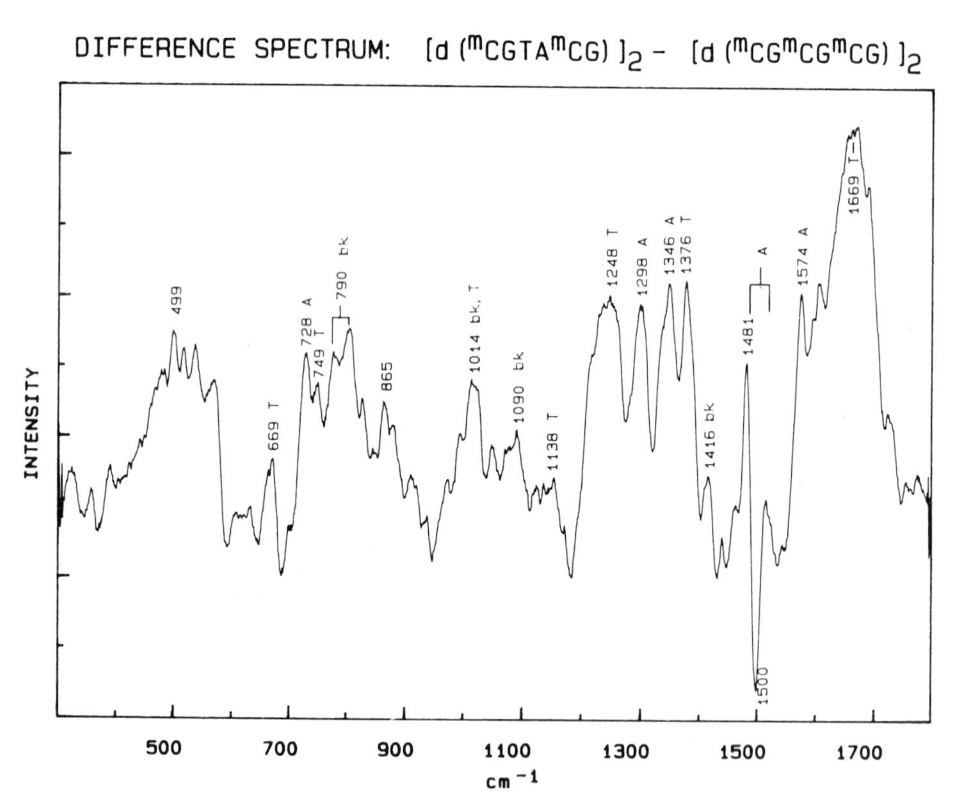

Figure 6. Computer generated Raman spectrum of a hypothetical B-DNA containing only AT pairs, obtained by subtraction of the spectrum of aqueous $[d(^{m}CG^{m}CG^{m}CG)]_2]$ from that of aqueous $[d(^{m}CGTA^{m}CG)]_2$. Labels as in Figure 2.

At this point it is appropriate to consider a limitation of the "spectrum-by-difference" approach applied above to AT sequences. In each example, the AT sequence is boxed between GC neighbors. Since Raman line intensities are known to be sensitive to both the kind and number of stacking interactions (26,27), the spectral intensities for these hypothetical AT structures can be expected to differ from those which would be observed for AT sequences not boxed between GC neighbors, or for non-alternating AT sequences. On the other hand, the frequencies (cm^{-1} values) of the Raman lines are not significantly affected by stacking interactions. Therefore, the data listed in Table II should be generally applicable insofar as the frequencies of the Raman markers are concerned. The only conspicuous exception to this general rule is the line exhibiting an inflection point near 1490 cm^{-1} in Fig. 6. This line corresponds to a ring vibration of guanine and it exhibits a markedly higher frequency than expected (*ca.* 1500 cm^{-1}) in the Z form hexamer when guanine is stacked between T and ^{m}C. Accordingly, we note the sensitivity of this purine ring frequency to DNA base sequence.

3. Polynucleotide Fibers and Solutions.

Polyribonucleotides, polydeoxyribonucleotides and their complexes have been extensively studied by Raman spectroscopy in the past decade and comprehensive reviews have been given recently (4,28). Nevertheless, polynucleotides remain especially interesting structures for Raman investigation and new aspects of DNA and RNA conformation continue to be revealed by refinements in the collection and analysis of their Raman spectra. Particularly useful are the Raman marker bands identified from crystal studies (Table II), since these can permit the confirmation or rejection of specific nucleoside conformers in polynucleotides. Also, the strikingly different characteristic Raman lines obtained for different conformations of the polynucleotide backbone (19,29,30) permit the differentiation of A, B and Z backbone geometries in nucleic acids. The important Raman markers associated with vibrations of the nucleic acid backbone and the sources of the data are summarized in Table III.

Here, we consider two novel structures revealed by Raman spectra of oriented fibers and gels of AT-containing polymers, and consider a probable explanation for the unusually diffuse Raman scattering of the B-DNA backbone when compared with A-DNA or A-RNA.

a. Poly(dA-dT)•poly(dA-dT).

The relative instability of the A helix for DNA deficient in GC has been noted by Dickerson and co-workers (31). We have been able to obtain, however, a metastable structure for poly(dA-dT)•poly(dA-dT) at low RH which is not of the B family and which can be inferred as a structure of the A form on the basis of Raman markers identified above (Tables II and III) (20). The spectrum of this A form of poly(dA-dT)•poly(dA-dT) is shown in Fig. 7. For comparison, we include in Fig. 7 the spectrum of the aqueous or B form of poly(dA-dT)•poly(dA-dT), and the corresponding difference spectrum. Data are included for both non-deuterated (Fig. 7a) and deuterated (Fig. 7b) samples.

Table III
Raman Markers Associated with Vibrations of the Nucleic Acid Backbone

A. Vibrations of the OPO Diester Group.[a,b,c]

Nucleic Acid	Form	Marker (cm^{-1})	Geometry[d] (α,ζ)	References
poly(rA) (ss)	A	811		32
poly(rI) (ss)	A	814		32
poly(rC) (ss)	A	810		47
poly(rG) (qs)	A	815	−81,−85	50,52
poly(rI) (qs)	A	816	−69,−103	46
poly(rC)•poly(rC+)	A	811		47
poly(rA)•poly(rU)	A	813	−67,−71	45,49
poly(rA)•2poly(rU) (ts)	A	813	−76,−69; −59,−75	46,49
poly(rA)•poly(dT)	A	812	−70,−70	32,49
poly(rA-rU)•poly(rA-rU)	A	813	−50,−78	48,49
poly(rG)•poly(rC)	A	813	−50,−78	45,49
poly(rI)•poly(rC)	A	812	−70,−70	32,49
r(CG)	A	812		27
r(GC)	A	812	−76,−69	27,51
RNA	A	813	−67,−71	44,49
poly(dA-dT)•poly(dA-dT)	A	806		20
poly(dG)•poly(dC)	A	808	−50,−78	8,49
DNA	A	806	−90,−45	19,49
r(GCG)d(CGC)	A	808		8
r(GCG)d(TATACGC)	A	808	−69,−75	32
d(CCCCGGGG)	A	809	−67,−77;−110,−91	8
poly(dA)	B	842		32
poly(dT)	B	832		32
poly(dG)	B	830		32
poly(dA)•poly(dT)	B	829	−60,−59;−41,−121	32,49
poly(dA-dT)•poly(dA-dT)	B	839	−45,−145;−72,−153	20,49
poly(dG)•poly(dC)	B	830	−50,−78	8,49
poly(dG-dC)•poly(dG-dC)	B	829	−30,−158;−66,−169	25,49
poly(dI-dC)•poly(dI-dC)	B	828	−51,−154;−76,−141	32,49
poly(rA)•poly(dT)	B	838		20
DNA	B	834	−39,−157	19,49
r(GCG)d(TATACGC)	B	828		32
d(CCCCGGGG)	B	852	−42,−101;−19,−73	15
DNA	C	875	−65,−179	49,53
poly(dG-dC)•poly(dG-dC)	Z	748	52,−65;−140,82	25,49
d(CGCGCG)	Z	742	47,−69;−137,80	3,19
d(CGCATGCG)	Z	745		7
d(mCGTAmCG)	Z	745	72,−60;−150,75	7,13

B. Vibrations of the PO_2^- Dioxy Group.[b,c]

Nucleic Acid	Form	Marker (cm^{-1})	Reference

Table III continued

RNA and ribo models	A	1099±2	25,29
DNA and deoxy models	A	1099±2	19,29
DNA and deoxy models	B	1092±2	19,29
DNA (calf thymus)	C	1090±5	30,53
DNA (alternating GC)	Z	1095±2	25

C. Vibrations of the Methylene Groups.[b,c]

Nucleic Acid	Form	Marker (cm^{-1})	Reference
RNA and ribo models	A	1419±2	32
DNA and deoxy models	A	1418±2	25
DNA and deoxy models	B	1422±2	25
DNA (calf thymus)	C	1417±2	53
DNA (alternating GC)	Z	1425±2	25

[a]Abbreviations: All structures are double stranded, except where indicated as follows: ss=single stranded, ts=triple stranded, qs=quadruple stranded.

[b]Frequencies cited in earlier work may differ slightly from the more accurate values listed here. Deuteration effects are small or negligible for all frequencies listed.

[c]The physical state of each sample (i.e. crystal, fiber or solution) is as discussed in the text and in the references cited.

[d]Values of the backbone dihedral angles (α,ζ) of the purine are given first when values differ for purine and pyrimidine. Average values are given when two or more similar conformations are present in oligomers, copolymers and multistranded complexes.

From Fig. 7 a number of important frequency shifts can be identified with the B to A transition of the alternating AT sequence. These will be discussed in more detail elsewhere (32). We list here only the prominent markers characteristic of the A and B backbones of AT-containing DNA (Table III) and point out the substantial agreement between Raman frequencies of the alternating copolymer (Fig. 7) and corresponding oligonucleotide structures (Figs. 2 and 4). The spectral resolution in the 600-870 cm^{-1} region is improved by deconvolution of the B form spectrum in Fig. 8.

b. Poly(dG-dC)•poly(dG-dC).

The secondary structure of poly(dG-dC)•poly(dG-dC) in aqueous solutions of low ionic strength or in fibers at high RH is generally recognized as that of B-DNA (17,25,33). The Raman spectrum of the B form of aqueous poly(dG-dC)•poly(dG-dC) is included in Fig. 8. The same spectrum is obtained from a fiber of the alternating GC structure at high RH. This spectrum contains essentially the same Raman marker bands as spectra of GC oligomers in the B conformations discussed earlier. It also bears many similarities to Raman spectra of AT structures in the B form, *with one important exception:* The Raman line assigned to the symmetric stretching vibration of backbone phosphodiester groups occurs at 840 cm^{-1} in poly(dA-dT)•poly(dA-dT)

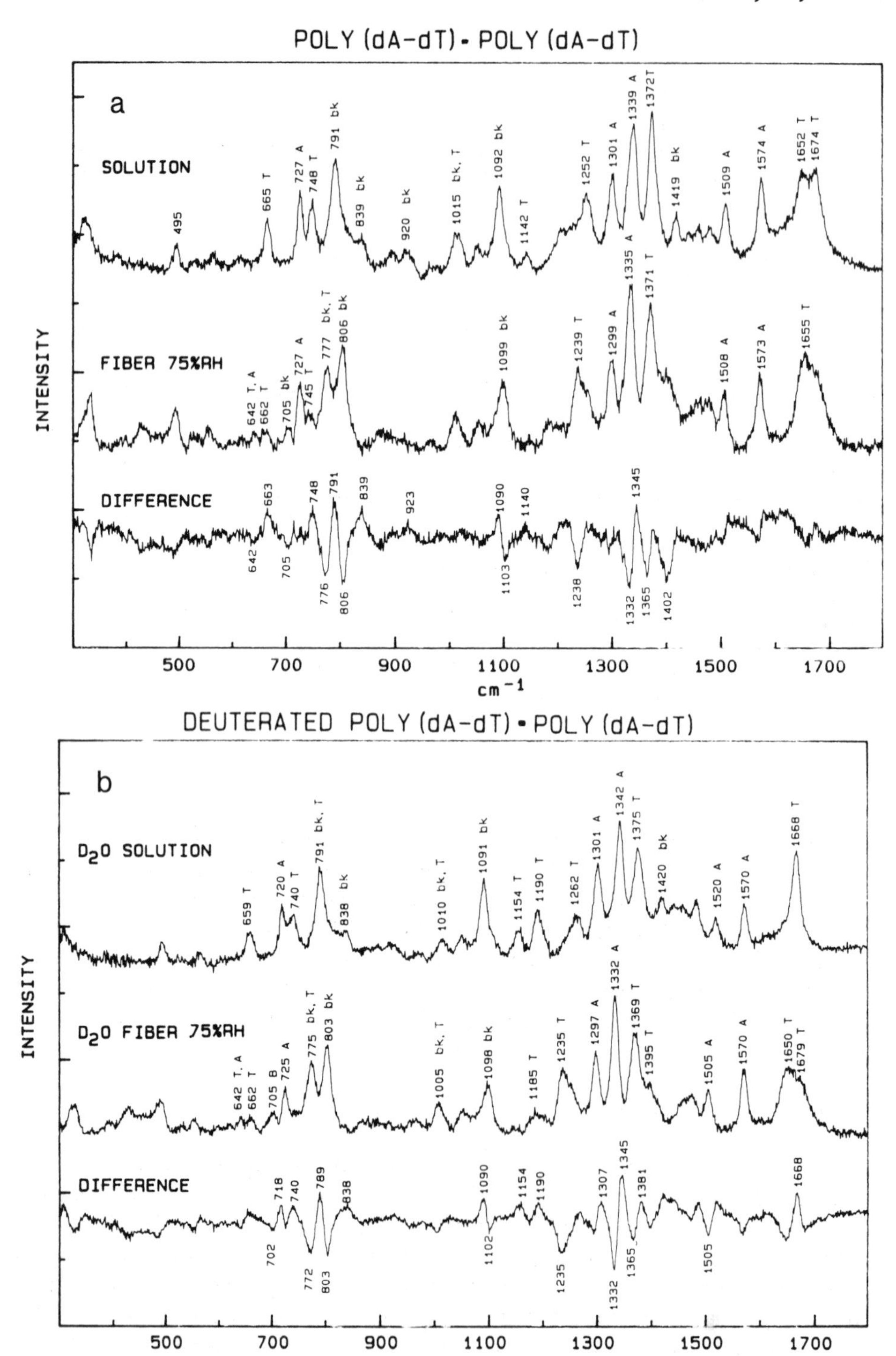

Figure 7. Raman spectra of poly(dA-dT)•poly(dA-dT). (a) Non-deuterated samples in the B form (H_2O solution), A form (fiber at 75% H_2O RH) and their difference spectrum. (b) Deuterated samples in the B form (D_2O solution), A form (fiber at 75% D_2O RH) and their difference spectrum. Labels as in Figure 2.

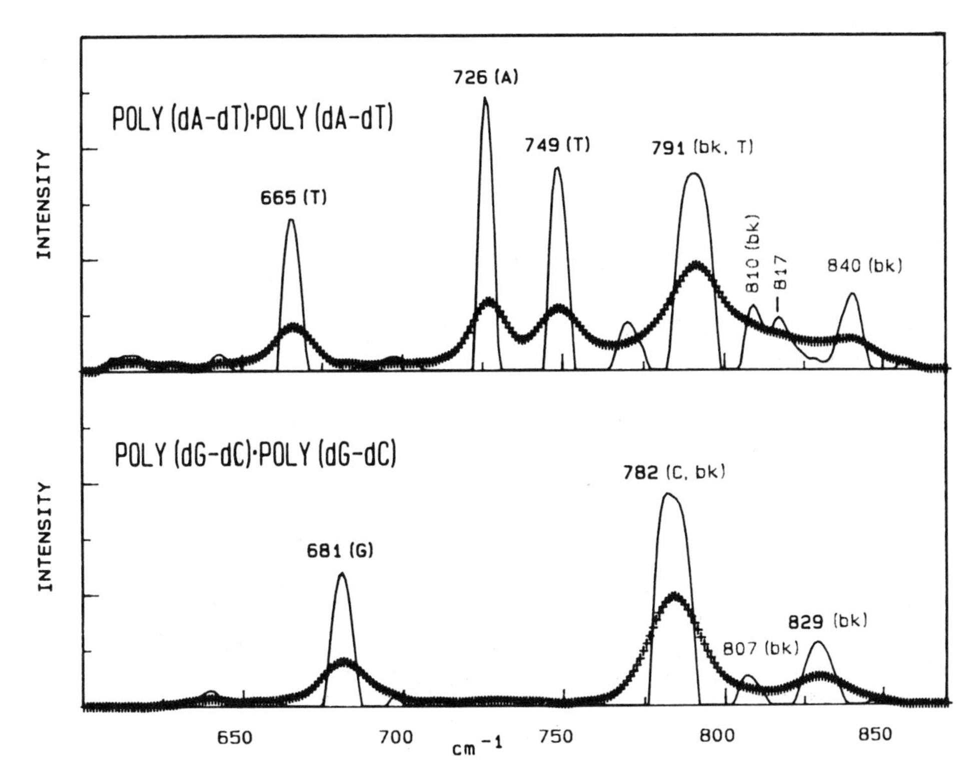

Figure 8. Observed and deconvoluted Raman spectra in the region 600-870 cm^{-1} of the B forms of poly(dA-dT)•poly(dA-dT) (top) and poly(dG-dC)•poly(dG-dC) (bottom). Labels as in Figure 2.

but at 829 cm^{-1} in poly(dG-dC)•poly(dG-dC). The difference is significant and is clearly revealed by both the unrefined data and the deconvolution results shown in Fig. 8.

Recent results on poly(dG)•poly(dC) (8) confirm the non-identical Raman markers of "B forms" of GC and AT containing nucleic acids.

Since the OPO vibration exhibits different frequencies in the AT and GC structures, the detailed molecular geometries of the phosphodiester backbones must be different from one another. We of course refer to both of these structures as DNA of the B form because of the evident similarity of most of the spectral markers to one another and to those of B-DNA (19,30), but it is clear that their respective backbone conformations cannot be identical. It is of interest to compare quantitatively these Raman marker bands of poly(dA-dT)•poly(dA-dT) and poly(dG-dC)•poly(dG-dC) with the corresponding band in B-DNA of irregular base sequence, such as calf thymus DNA.

c. *Comparison of B forms of poly(dA-dT)•poly(dA-dT), poly(dG-dC)•poly(dG-dC) and native DNA.*

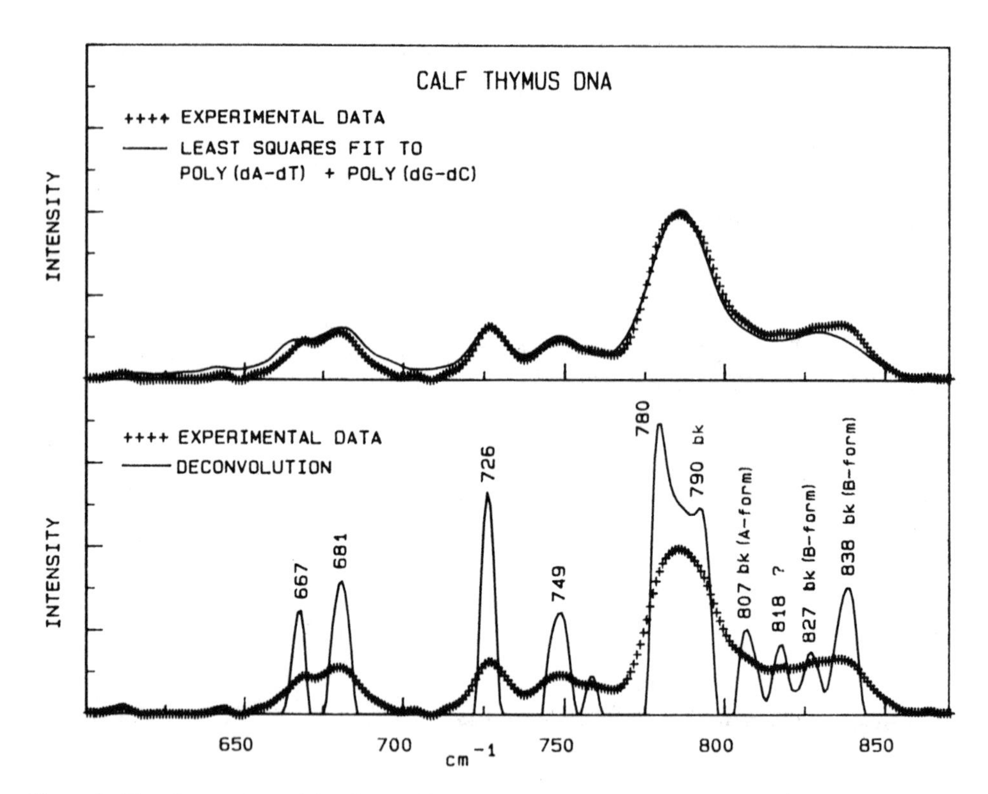

Figure 9. Top: Comparison of the observed Raman spectrum of B form calf thymus DNA with the sum of spectra of poly(dA-dT)•poly(dA-dT) and poly(dG-dC)•poly(dG-dC). The least squares "best" fit represents 48% GC + 52 %AT. (See text.) Bottom: Comparison of the observed spectrum of B form calf thymus DNA with the deconvoluted spectrum. Labels indicate frequencies of the DNA bases and backbone according to the assignments indicated in Figure 8 and in the text.

The Raman spectra of DNA fibers in different structural modifications have been discussed extensively (19,34,35). We note here the pattern of bands in the 600-870 cm^{-1} region of aqueous (B-form) calf thymus DNA, shown in Fig. 9. (The same profile is obtained from fibers at high RH.) The band of B-DNA assigned to the OPO mode occurs at 834 ± 2 cm^{-1}, intermediate between the frequency values observed for AT and GC copolymers. In fact, deconvolution of the DNA spectrum (Fig. 9, bottom) suggests that the 834 cm^{-1} band may result from a superposition of unresolved, overlapping bands of higher (838 cm^{-1}) and lower (827 cm^{-1}) frequencies. The deconvolved peaks of B-DNA are seen from Fig. 9 to agree remarkably well with the actual peak positions in poly(dA-dT)•poly(dA-dT) and poly(dG-dC)•poly(dG-dC).

The deconvolution of Fig. 9 also suggests two additional bands in the B form of calf thymus DNA at 807 and 818 cm^{-1}. The former occurs at the position expected for the A-DNA backbone marker (Table III), and may indicate a minor contribution from such a conformation to aqueous DNA structure. Previous studies have shown that the 807 cm^{-1} A-DNA marker has an intrinsic peak intensity roughly five to six times greater than that of the 834 cm^{-1} B-DNA marker (19). Accordingly, the deconvolution intensity in Fig. 9 suggests 10% or less A form in aqueous DNA at the conditions of this experiment. We tentatively assign the 818 cm^{-1} peak in the deconvolution of Fig. 9 to an adenine ring vibration for reasons discussed below. This line is probably not a reliable conformation marker and may have contributed to the difficulty encountered by others in interpreting Raman spectra of adenine-containing polynucleotides (36).

An alternative approach to quantitative analysis of the B-DNA spectrum is that of curve fitting. Included in Fig. 9 (top) is a least squares fit of the B-DNA spectrum to a sum of the spectra of poly(dA-dT)•poly(dA-dT) and poly(dG-dC)•poly(dG-dC). The optimum fit occurs for mixing coefficients corresponding to 52% AT and 48% GC, in reasonably good quantitative agreement with the base composition of calf thymus DNA and in qualitative agreement with the deconvolution result.

Accordingly, from both deconvolution and curve fitting analyses, the B backbone marker band of DNA *ca.* 834 cm^{-1} is consistent with the presence of two somewhat different backbone conformations. The latter in turn are consistent with those observed in AT and GC copolymers. The present results, of course, do not exclude other less simple interpretations of the 834 cm^{-1} band. Obviously, a larger number of structural variants or "B-family conformers" would also well explain the observed B-DNA band. The important conclusion here is that the B form structures of poly(dA-dT)•poly(dA-dT) and poly(dG-dC)•poly(dG-dC) are not identical to one another and that a mixture of two or more such minor structural variants can account for the data obtained from B-DNA.

d. Comparison with non-alternating polydeoxyribonucleotides.

Raman spectroscopy indicates that the secondary structure of aqueous poly(dA)•poly(dT) is not distinguishable from that of poly(dA-dT)•poly(dA-dT) in solutions of moderate ionic strength at 20°C (32). Others have interpreted the Raman data differently, suggesting a mixture of both A and B-type polynucleotide backbones in poly(dA)•poly(dT), especially at lower temperatures (36). We believe such an interpretation is incorrect for two reasons. First, the difference Raman spectrum between poly(dA)•poly(dT) and poly(dA-dT)•poly(dA-dT) contains no bands which could be interpreted as markers of different nucleoside or backbone conformations. All of the differences between spectra of the two AT structures can be attributed to the different base stacking interactions in the two kinds of sequences (32). This is consistent with NMR and CD results (37). Second, the "A marker" proposed by Peticolas and coworkers (36) for poly(dA)•poly(dT) occurs at a frequency (*ca.* 816 cm^{-1}) which is too high to be attributable to an A form polydeoxyribonucleotide. (See, for example, Section f., below, as well as Fig. 7.) Our interpretation of the weak band ca. 816 cm^{-1} in poly(dA)•poly(dT) is different than that proposed by others.

We note that a Raman band at 816 cm^{-1}, also exhibiting temperature dependent intensity, has been reported previously for single stranded poly(rA) (27,29) and poly(dA) (32). Since the same band occurs in both poly(dA) and poly(rA), it may originate in whole or in part from the adenine residue rather than exclusively from the backbone. The enhanced intensity reported by others at 816 cm^{-1} in the low temperature spectrum of poly(dA)•poly(dT) (36) could be explained by molecular inhomogeneity, i.e. excess poly(dA) in the sample. In support of this interpretation we note (i) the peak near 818 cm^{-1} revealed by deconvolution of the calf-thymus DNA spectrum (Fig. 9), (ii) the peak near 817 cm^{-1} revealed by deconvolution of the poly(dA-dT)•poly(dA-dT) spectrum (Fig. 8) and (iii) the absence of a peak between 810 and 829 cm^{-1} in the deconvolution of poly(dG-dC)•poly(dG-dC) (Fig. 8).

Poly(dG)•poly(dC) assumes either the A or B backbone conformation in solutions and fibers, depending upon the activity of water. At high solute concentration or low RH, the A form is preferred; conversely, in dilute solutions the B form is preferred, as has been demonstrated from Raman and CD studies (8).

e. Uniformity of secondary structure among polyribonucleotides.

As shown in Table III, the backbone conformation marker of polyribonucleotides does not vary significantly among structures of different base composition. An average value of 813 cm^{-1} and deviation of ± 2 cm^{-1} is obtained for AU, GC, IC and heterogeneous RNAs. Evidently, the A conformation occurring in synthetic and native RNA molecules is not significantly affected by base composition or sequence. This conclusion is not new (29), but is extended here to a larger number of representative structures.

f. Comparison of A-RNA and A-DNA conformations.

As in the case of A-RNA, the OPO backbone marker of A-DNA is essentially invariant to base composition and sequence, occurring at 807 ± 2 cm^{-1} in virtually all oligomeric and polymeric A-DNA models and in native DNA fibers. On the other hand, from the criterion of Raman spectroscopy, the A form of RNA (OPO marker at 813 ± 2 cm^{-1}) is not identical to the A form of DNA (OPO marker at 807 ± 2 cm^{-1}). This conclusion is clearly evident from Table III, and has been discussed previously in the interpretation of Raman data from crystals of RNA/DNA hybrids (8).

Summary

The large number of Raman markers contained in Tables II and III may be given the more compact representation of Fig. 10. This illustration shows the conformational dependence of both the position (cm^{-1} value) and intensity (peak height) of the major Raman markers which have so far been identified.

Applications to Complex Structures

The correlations established from model compounds (summarized in Tables II and III) can be applied to DNA or RNA structures in nucleoproteins, viruses and in

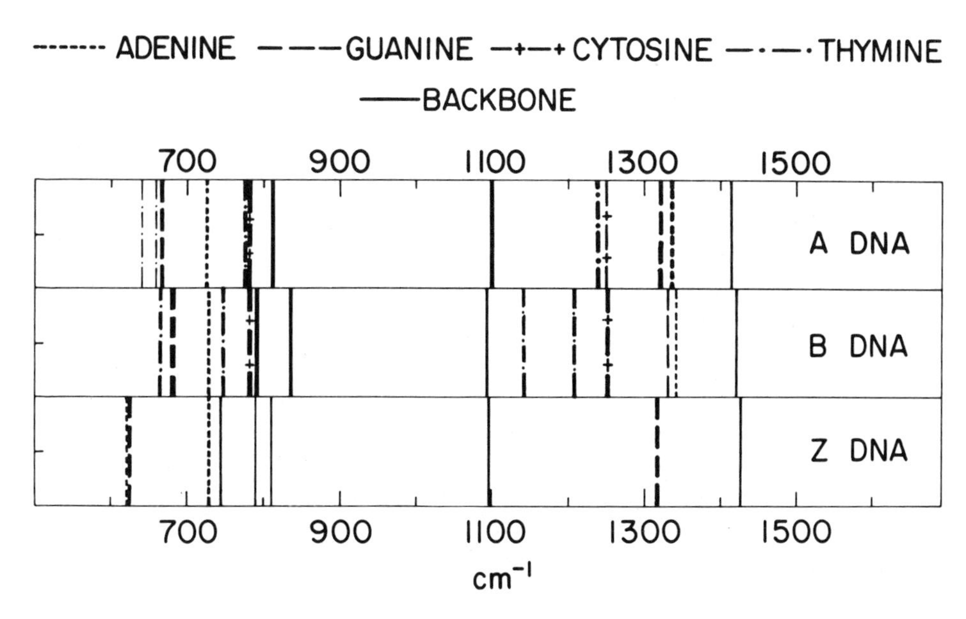

Figure 10. Graphical representation of the important Raman markers for nucleoside and backbone conformations of DNA and RNA. Each line represents a Raman marker, the frequency and/or intensity of which may vary with conformation changes. The relative Raman intensities are indicated by the thicknesses of the lines in each spectrum.

other complexes to deduce nucleoside and backbone conformations of the nucleic acids. Several examples are next considered.

1. RNA/DNA Hybrid Structures.

Poly(rA)•poly(dT) is representative of a hybrid RNA/DNA structure of some controversy (38). The molecular structure of a comparable hybrid oligomer from single crystal x-ray diffraction analysis has not yet been reported. The Raman spectra of the solution and fiber of poly(rA)•poly(dT) are shown in Fig. 11. Reference to Tables II and III shows that the aqueous duplex is a conformational hybrid, in which the RNA strand (poly(rA)) is of the A form with C3′endo/anti adenosine, while the DNA strand (poly(dT)) is of the B form with C2′endo/anti thymidine. The same structure occurs in the fiber at high RH. At low RH, however, the hybrid is converted completely to an A form. In this A form, the OPO marker occurs as a single peak at 812 cm^{-1}, indicating an A structure more similar to that of RNA models than to A-DNA. Hence, the dT strand is drawn into the RNA conformation by the rA strand at these experimental conditions.

2. Bacteriophage Lambda cI Operator, O_L1.

The DNA operator O_L1 is a 17 base pair fragment which is one of the strongest binding sites of the phage lambda cI repressor. The Raman spectrum of this double

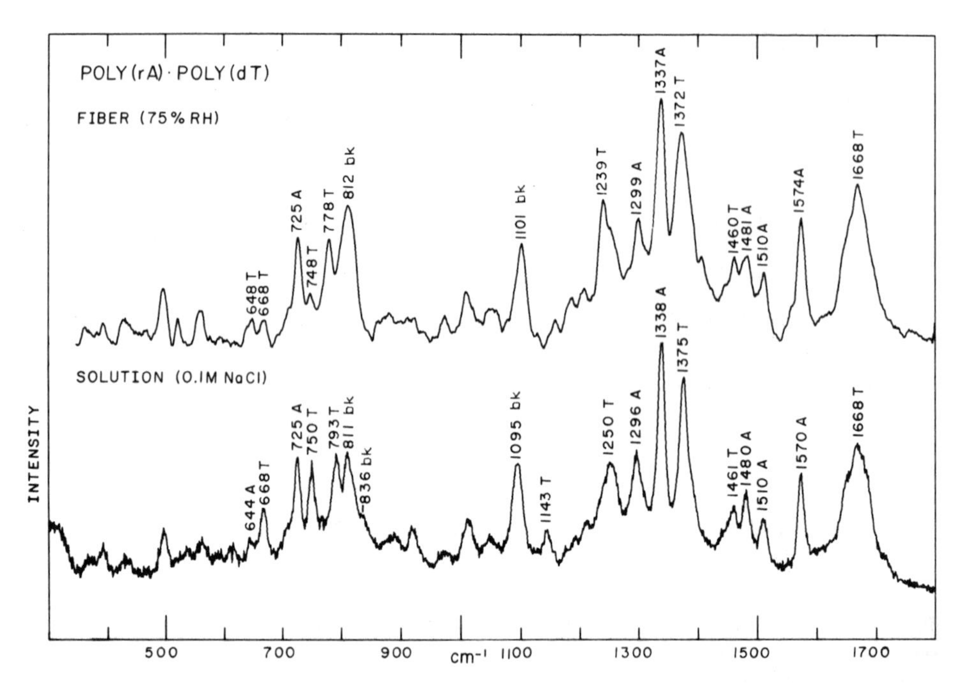

Figure 11. Raman spectra of poly(rA)•poly(dT). Top: Fiber at 75% RH, exhibiting Raman markers characteristic of the A form for both the rA and dT strands. Bottom: Aqueous solution, exhibiting Raman markers chracteristic of the A form for the rA strand and of the B form for the dT strand. Labels indicate frequencies of the bases (A, T) and backbone (bk).

stranded DNA fragment, [d(TATCACCGCCAGTGGTA)]$_2$, is shown in Fig. 12. Three peaks are discernible in the region of OPO marker bands, at 807, 825 and 838 cm^{-1}. The first of these corresponds to an A-DNA marker, the last to a B-DNA marker. The peak at 825 cm^{-1} has no conterpart in the models studied but most likely originates from a B-type conformer, perhaps associated with a specific nucleotide sequence of O_L1 not represented by any of the previouly studied model compounds. The rich pattern of bands in the 650-775 cm^{-1} interval confirms the presence of more than one type of nucleoside conformation in the 17-mer. Deconvolution of the band envelope from 650-700 cm^{-1} (Fig. 12) suggests at least four markers due to G and T conformers (39).

3. RNA of the Tobacco Mosaic Virus.

The TMV RNA molecule, when free of viral coat protein, assumes the same A geometry (C3′endo/anti nucleosides) as other RNA molecules in solution. In the assembled virus, however, the RNA interlaces the capsid subunits and is constrained to adopt a conformation which differs from that of the protein-free molecule (40). This conformation change can be monitored by Raman spectroscopy (41).

The Raman bands of TMV in the 800-900 cm^{-1} region are shown in Fig. 13, together with the result of Fourier deconvolution of the complex band shapes.

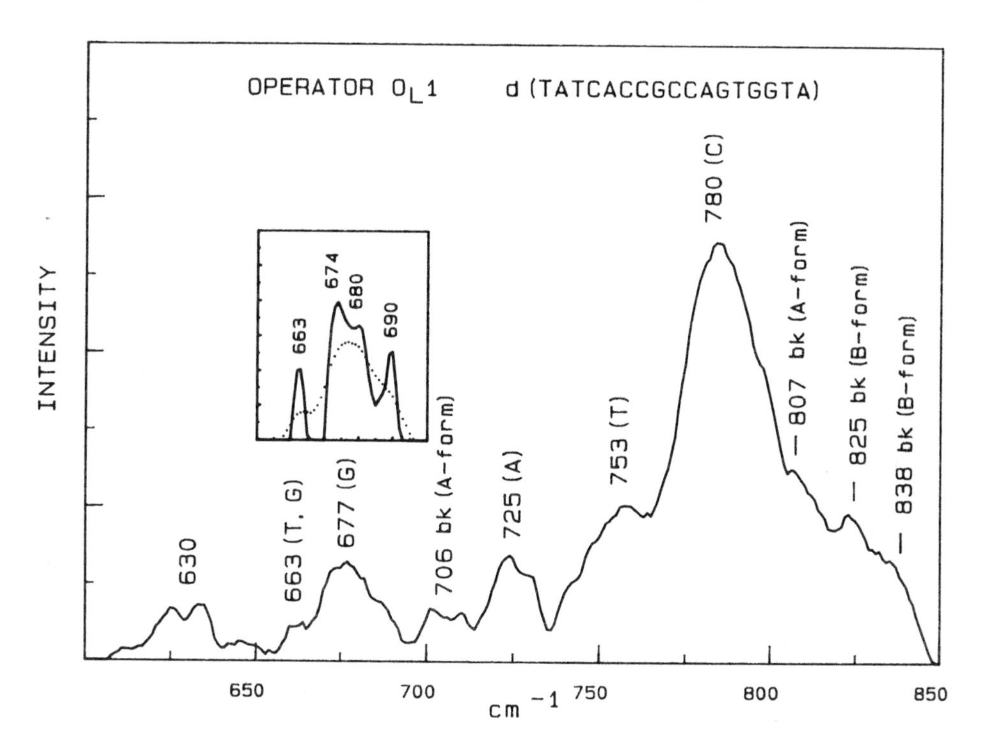

Figure 12. Raman spectrum of the operator sequence O_L1 in the region 600-870 cm^{-1}. The inset shows the partial deconvolution of the bands due primarily to guanine in the 650-700 cm^{-1} interval. Labels as in Figure 2. See also reference 39.

Since TMV is approximately 95% by weight protein, the Raman spectrum is dominated by bands due to protein. Nevertheless, Raman spectra of the capsid and disk of TMV show that the protein has no bands from 800 to 825 cm^{-1}, permitting conformation markers of the RNA backbone to be filtered through this window. Deconvolution resolves peaks at 812 and 820 cm^{-1} due to RNA OPO groups. (The remaining bands at higher frequencies are due to protein aromatics, including the 828 and 851 cm^{-1} peaks which are due to tyrosines.) The shift of two-thirds of the Raman OPO intensity from 812 cm^{-1} in protein free RNA to 820 cm^{-1} in the encapsidated genome indicates that at least two out of three RNA nucleotides do not maintain the gauche (g^-g^-) conformation normally associated with the RNA backbone. The corresponding nucleosides may also depart from the usual C3′endo/anti conformation. (Unfortunately, the nucleoside markers below 800 cm^{-1} cannot be easily examined because of extensive overlap with Raman lines due to viral protein.) The peak remaining at 812 cm^{-1} may arise from residues which retain the normal g^-g^- conformation, or from a different backbone conformer which accidentally exhibits the same OPO vibrational frequency.

The deviation of encapsidated TMV RNA from the usual A geometry is unique among a large number of single stranded RNA plant and bacterial viruses examined by Raman spectroscopy (4).

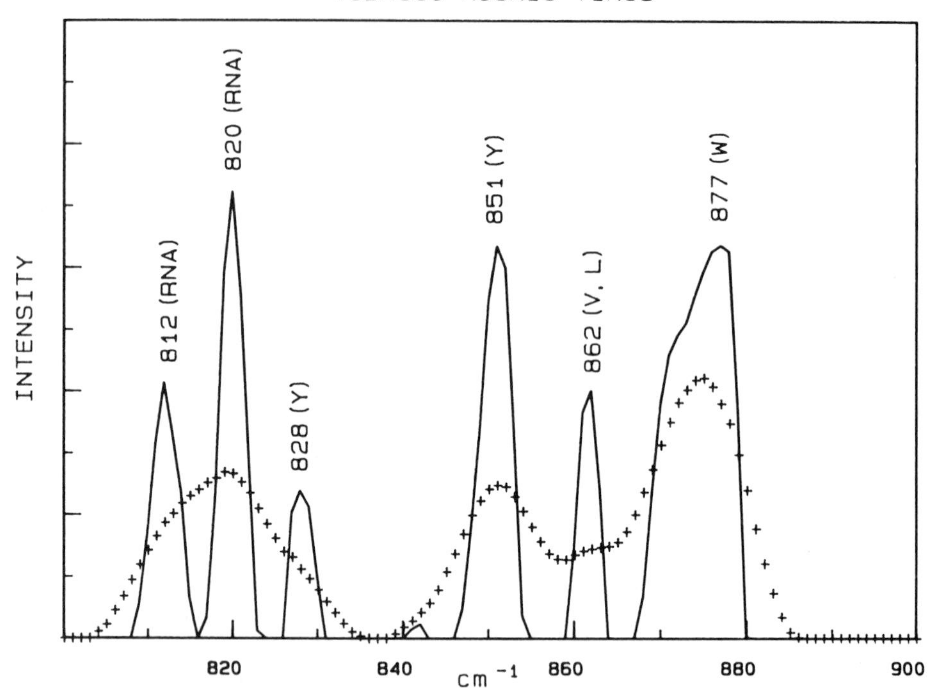

Figure 13. Raman spectrum of the single stranded RNA plant virus, TMV, in the region 800-900 cm^{-1}. The lines at 812 and 820 cm^{-1} are due to the RNA backbone. Other lines are due to various amino acid side chains of the coat protein as indicated. See also references 10 and 40.

4. DNA of the Filamentous Bacterial Virus, Pf3.

The filamentous bacteriophages, like TMV, are composed predominantly of protein. Substantial signal averaging is therefore required to detect their relatively weak DNA Raman lines (42). As shown in Fig. 14, a number of DNA peaks can be discerned in the 600-800 cm^{-1} region of the Raman spectrum of Pf3, and these are better resolved through deconvolution. Fig. 14 indicates that Pf3 does not contain C2′endo/anti conformers of G, and does not contain the characteristic backbone markers of B form DNA. We deduce from the Pf3 data and the tabulations discussed above that the probable nucleoside conformation of Pf3 DNA is C3′endo/anti.

5. DNA of Bacteriophage P22.

P22 exhibits all of the Raman markers of B-DNA (43), as shown in Fig. 14. The same conclusion is reached from the Raman spectra of other double stranded DNA bacteriophages (G. J. Thomas, Jr., unpublished results.)

Conclusions

Raman spectroscopy may be profitably exploited to determine details of nucleic

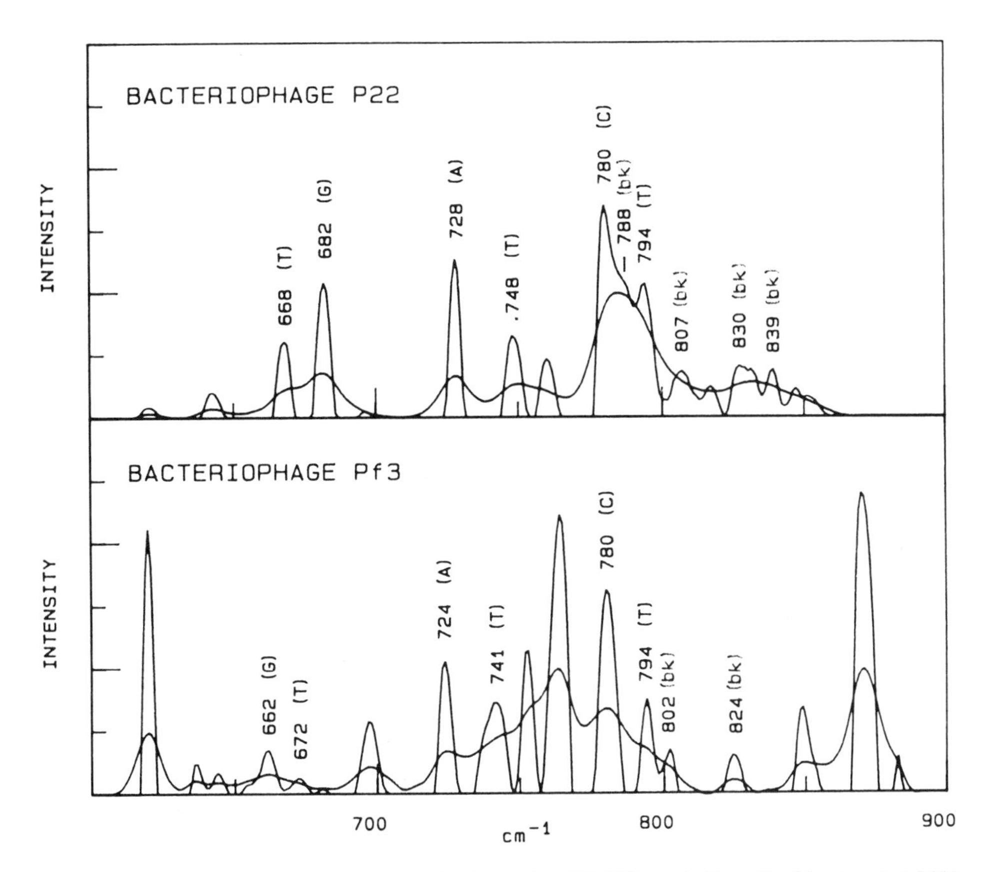

Figure 14. Raman spectra of DNA viruses in the region 600-900 cm^{-1}. Top: Double stranded DNA phage P22. Bottom: Single stranded DNA filamentous phage Pf3. Unlabelled lines are due the viral protein and are discussed in references 41 and 42. Other labels as in Figure 2.

acid secondary structure which cannot be investigated by other physical methods. A number of specific conformation markers have been identified for both the base-sugar and sugar-phosphate moieties.

Applications to crystals, fibers and solutions reveal the stability of preferred nucleoside and backbone conformations for different morphological states of DNA and RNA. The A-RNA structure is the most invariant to changes in base sequence and base composition. The A-DNA structure differs in a small but significant manner from that of A-RNA, probably in the conformation of the C5′-O-P-O-C3′ network, rather than in the pucker of the sugar ring.

In poly(rA)•poly(dT), the A-RNA backbone exerts a dominant role over A-DNA in the following sense. At high relative humidity, the RNA chain maintains its A backbone geometry irrespective of the B conformation of the DNA chain. At low relative humidity, the DNA chain is converted to an A geometry which more closely approximates that of A-RNA than that of A-DNA.

In polydeoxyribonucleotides and oligodeoxyribonucleotides, considerable variability of structure is observed among backbone "B" conformations. This difference is especially evident between alternating copolymer duplexes of AT and GC. The B form of DNA can be represented as a superposition of a family of two or more B variants, including those of the AT and GC alternating copolymers.

Variability of DNA conformation is reflected also in viral DNA genomes, including those of single and double stranded DNA viruses, and DNA fragments such as the operator O_L1 of the bacteriophage lambda cI repressor/operator system. RNA backbone conformation is less polymorphic than that of DNA, but can be perturbed by intimate association with protein as occurs, for example, in the TMV virion.

The present structure classification of aqueous nucleic acids from equilibrium Raman sudies is confirmed by dynamic Raman measurements which monitor the deuterium exchange kinetics of purine 8CH groups (18).

Acknowledgements

The generosity of collaborators in providing many of the samples investigated is gratefully acknowledged. We thank especially A. Rich (M.I.T.), A. H.-J. Wang (M.I.T.), J. H. van Boom (Leiden), M. Weiss (M.I.T.), J. King (M.I.T.), G. Stubbs (Vanderbilt) and L. A. Day (Public Health Research Institute, N.Y.). This work was suppported by Grants from the National Intitutes of Health (AI11855 and AI18758) and the Southeastern Massachusetts University Foundation.

References and Footnotes

1. Shakked, Z. and Kennard, O., *In Biological Macromolecules and Assemblies, Volume 2:* Nucleic Acids and Interactive Proteins, Jurnak, F. A. and MacPherson, A., Eds., Wiley Interscience, New York, pp. 1-36 (1985).
2. Dickerson, R. E., Kopka, M. L. and Pjura, P., In *Biological Macromolecules and Assemblies, Volume 2:* Nucleic Acids and Interactive Proteins, Jurnak, F. A. and MacPherson, A., Eds., Wiley Interscience, New York, pp. 37-126 (1985).
3. Wang, A. H.-J. and Rich, A., In *Biological Macromolecules and Assemblies, Volume 2:* Nucleic Acids and Interactive Proteins, Jurnak, F. A. and MacPherson, A., Eds., Wiley Interscience, New York, pp. 127-170 (1985).
4. Thomas, G. J., Jr., In *Biological Applications of Raman Spectroscopy, Volume 1,* Spiro, T. G., Ed., Wiley, New York, Chapter 5, in press (1986).
5. Chen, M. C., Giege, R., Lord, R. C. and Rich, A., *Biochemistry 17,* 3134-3138 (1978).
6. Thamann, T., Lord, R. C., Wang, A. H. J. and Rich, A., *Nucl. Acids Res. 9,* 5443-5457 (1981).
7. Benevides, J. M., Wang, A. H.-J., van der Marel, G. A., van Boom, J. H., Rich, A. and Thomas, G. J., Jr., *Nucl. Acids Res. 12,* 5913-5925 (1984).
8. Benevides, J. M., Wang, A. H.-J., van Boom, J. H., Kyogoku, Y., Rich, A. and Thomas, G. J., Jr., *Biochemistry,* in press (1986).
9. Thomas, G. J., Jr. and Agard, D. A., *Biophys. J. 46,* 763-768 (1984).
10. Thomas, G. J., Jr., *Spectrochim. Acta 41A,* 217-221 (1985).
11. Wang, A. H. J., Quigley, G. J., Kolpak, F. J., Crawford, J. L., Van Boom, J. H., Van der Marel, G. and Rich, A., *Nature 282,* 680-686 (1979).

12. Fujii, S., Wang, A. H.-J., Quigley, G. J., Westerink, H., van der Marel, G., van Boom, J. H. and Rich, A., *Biopolymers 24,* 243-250 (1985).
13. Wang, A. H.-J., Hakoshima, T., van der Marel, G., van Boom, J. H. and Rich, A., *Cell 37,* 321-331 (1984).
14. Wang, A. H.-J., personal communication (1985).
15. Wang, A. H.-J., Fujii, S., van Boom and Rich, A., *Proc. Natl. Acad. Sci., USA 79,* 3968-3972 (1982).
16. Wang, A. H.-J., Fujii, S., van Boom, J. H., van der Marel, G. A., van Boeckel, S. A. A and Rich, A., *Nature 299,* 601-604 (1982).
17. Leslie, A. G. W., Arnott, S., Chandrasekaran, R. and Ratliff, R. L., *J. Mol. Biol. 143,* 49-72 (1980).
18. Benevides, J. M. and Thomas, G. J., Jr., *Biopolymers 24,* 667-682 (1985).
19. Prescott, B., Steinmetz, W. and Thomas, G. J., Jr., *Biopolymers 23,* 235-256 (1984).
20. Thomas, G. J., Jr. and Benevides, J. M., *Biopolymers 24,* 1101-1104 (1985).
21. Lord, R. C. and Thomas, G. J., Jr., *Spectrochim. Acta 23A,* 2551-2591 (1967).
22. Nishimura, Y., Torigoe, C., Katahira, M. and Tsuboi, M., *Nucl. Acids Res. 12,* 6901-6908 (1984).
23. Nishimura, Y., Tsuboi, M., Nakano, T., Higuchi, S., Sato, T., Shida, T., Uesugi, S., Ohtsuka, E. and Ikehara, M., *Nucl. Acids Res. 11,* 1579-1588 (1983).
24. Benevides, J. M., LeMeur, D. and Thomas, G. J., Jr., *Biopolymers 23,* 1011-1024 (1984).
25. Benevides, J. M. and Thomas, G. J., Jr., *Nucl. Acids Res. 11,* 5747-5761 (1983).
26. Small, E. W. and Peticolas, W. L., *Biopolymers 10,* 1377-1416 (1971).
27. Prescott, B., Gamache, R., Livramento, J. and Thomas, G. J., Jr., *Biopolymers 13,* 1821-1845 (1974).
28. Tsuboi, M. and Peticolas, W. L., In *Biological Applications of Raman Spectroscopy, Volume 1,* Spiro, T. G., Ed., Wiley, New York, Chapter 4, in press (1986).
29. Thomas, G. J., Jr. and Hartman, K. A., *Biochim. Biophys. Acta 312,* 311-322 (1973).
30. Erfurth, S. C., Kiser, E. J. and Peticolas, W. L., *Proc. Natl. Acad. Sci. USA 69,* 938-941 (1972).
31. Dickerson, R. E., Drew, H. R., Conner, B. N., Kopka, M. L. and Pjura, P. J., *Cold Spring Harbor Symp. Quant. Biol. 47,* 13-24 (1982).
32. Benevides, J. M., Prescott, B. and Thomas, G. J. Jr., in preparation.
33. Pohl., F. M., Ranade, A. and Stockburger, M., *Biochim. Biophys. Acta 335,* 85-92 (1973).
34. Erfurth, S. C., Bond, P. J. and Peticolas, W. L., *Biopolymers 14,* 1245-1257 (1975).
35. Martin, J. C. and Wartell, R. M., *Biopolymers 21,* 499-512 (1982).
36. Thomas, G. A. and Peticolas, W. L., *J. Am. Chem. Soc. 105,* 993-996 (1983).
37. Edmondson, S. P. and Johnson, W. C., *Biopolymers 24,* 825-841 (1985).
38. Arnott, S., Chandrasekaran, R., Millane, R. and Park, H. S., *Proc. Natl. Acad. Sci. U.S.A.,* submitted (1985).
39. Prescott, B., Benevides, J.M., Weiss, M.A. and Thomas, G.J. Jr., *Spectrochim. Acta,* in press (1986).
40. Stubbs, G., In *Biological Macromolecules and Assemblies, Volume 1:* Virus Structures, Jurnak, F. A. and McPherson, A., Editors., Wiley, New York, pp.149-202 (1984).
41. Fish, S. R., Hartman, K. A., Stubbs, G. J. and Thomas, G. J., Jr., *Biochemistry 20,* 7449-7457 (1981).
42. Thomas, G. J., Jr., Prescott, B. and Day, L. A., *J. Mol. Biol. 165,* 321-356 (1983).
43. Thomas, G. J., Jr., Li, Y., Fuller, M. T. and King, J., *Biochemistry 21,* 3866-3878 (1982).
44. Verduin, B. J. M., Prescott, B. and Thomas, G. J., Jr., *Biochemistry 23,* 4301-4308 (1984).
45. Lafleur, L., Rice, J. and Thomas, G. J., Jr., *Biopolymers 11,* 2423-2437 (1972).
46. Chou, C. H., Thomas, G. J., Jr., Arnott, S. and Campbell-Smith, P. J., *Nucl. Acids Res. 4,* 2407-2419 (1977).
47. Chou, C. H. and Thomas, G. J., Jr., *Biopolymers 16,* 765-789 (1977).
48. Morikawa, K., Tsuboi, M., Takahasi, S., Kyogoku, Y., Mitsui, Y., Iitaka, Y. and Thomas, G. J., Jr., *Biopolymers 12,* 799-816 (1973).
49. Arnott, S. and Chandrasekaran, R., personal communication. See also: Arnott, S., Smith, P. J. C. and Chandrasekaran, R., In *CRC Handbook of Biochemistry and Molecular Biology,* 3rd. Ed., *Vol. 2,* Fasman, G. D., Ed., CRC Press, Cleveland (1976).
50. Zimmerman, S. B., Cohen, G. H. and Davies, D. R., *J. Mol. Biol. 92,* 181-192 (1975).
51. Day, R. O., Seeman, N. C., Rosenberg, J. M. and Rich, A., *Proc. Natl. Acad. Sci. U.S.A. 70,* 849-853 (1973).
52. Rice, J., Lafleur, L., Medeiros, G. C. and Thomas, G. J., Jr., *J. Raman Spectrosc. 1,* 207-215 (1973).
53. Fish, S. R., Chen, C. Y., Thomas, G. J., Jr. and Hanlon, S., *Biochemistry 22,* 4751-4756 (1983).

Biomolecular Stereodynamics IV, Proceedings of the Fourth Conversation in the Discipline Biomolecular Stereodynamics, State University of New York, Albany, NY, June 04-09, 1985, Eds., Ramaswamy H. Sarma & Mukti H. Sarma, ISBN 0-940030-18-7, Adenine Press, ©Adenine Press 1986.

Homogeneous Double-Helix-Stability in Individual Genes

Akiyoshi Wada and Akira Suyama
Department of Physics, Faculty of Science
The University of Tokyo, Bunkyo-ku, Tokyo, Japan

Abstract

The local unfolding of DNA double helix is a biologically crucial process, because the opening of the coding region is required for the genetic processes of duplication, transcription, and recombination. Thus, the stability distribution in the double helix constrains the evolution of genetic information, so that the positional correlation between the distribution of the physical stability and genetic divisions may be revealed.

In the present report, we discuss this physico-genetic coupling in DNA. We present in the form of a review, the basic idea and material proofs of the correlation between the genetic structure and the local double-helix-stability which are found in many DNAs of biological species.

Finally we report a newly found fact that the average (G+C)-content at the third letter (wobble letter) of codons negatively correlates with that of the first-plus-second letters. This evidence strongly indicates that the third letter which is free from the protein functional constraints counterbalances the (G+C)-content change produced by the other letters so that a homogeneous (G+C)-content distribution may be attained in a gene. We believe that the double-helix homostabilizing propensity in individual genes has been created because of the desirable function of smoothly opening the double helix and/or correct and efficient translation, by producing codon-anticodon interactions with nearly the same energy, over the whole stretch of a gene, and that this propensity is produced by the proper use of the wobble site in codons.

Introduction

The main purpose of this article is to collect and analyze in a systematic way the basic ideas and material proofs of the correlation between the genetic structure and the local double-helix-stability which is found in DNAs of many biological species. These ideas and much evidence have sprung up and have been pursued in a number of laboratories in the last ten years (1-22) and they offer plausible, although not yet definite, roles of the molecular physical property of the DNA double helix in genetic processes.

Together with affirmative data, critical analysis by citing exceptional cases will be given in this article. We believe that it is necessary to exhibit the whole data which we have, as objectively as possible, in order to build up a correct image, because the correlation is not an ideal one and its biological significance is still indistinct. We do not intend to affirm that the correlation is perfect nor that it plays a decisive role in DNA functions. We simply wish to present comprehensible evidence that the physical stability of the double helix plays a part in the genetic structure of DNA.

In the following, first, a short introduction to the genetic role of local unfolding of double helix is given. Then we describe briefly the algorithm of the calculation of helix-coil transition of the DNA of a given base sequence. We investigate the double-helix stability distribution along the DNA molecular thread, which is represented by a local vulnerability to double-helix-unfolding perturbations such as thermal and ligand binding. Then we analyze the correlation between this stability distribution and units of genetic structure. Finally, we discuss the origin of the correlation in terms of whether a unique physical stability distribution is created for better biological function or living economy, or whether the base sequence that is required as a biological signal or as protein coding eventually gives a characteristic stability distribution as a result.

Genetic significance of the local unfolding of the DNA double helix

In a number of genetic processes the local unwinding of the DNA double helix is an essential requirement to expose a base sequence which is usually hidden inside regularly folded nucleotide backbone chains (23,24). In transcription, opening of the double helix by RNA polymerase allows access to the coding region in DNA for the successive copying of the genetic message into RNA (25). In replication, prior to fairly complex molecular processes mediated by several enzymic proteins, the DNA double strand must be opened first in order to be covered with single-strand DNA binding proteins (26). In recombination, the formation of heteroduplex structure by homologous DNA molecules must also involve unwinding of the double helix (27). Thus the local helix-coil transition that is a pure molecular physical process should be closely coupled with the genetic events. Usually, this type of coupling, or "resonance", is expected to occur between units having similar characteristic length.

In Fig. 1 characteristic lengths of genetic units are compared with the cooperative lengths of events of double helix unfolding. *Base pairing* and *base-base stacking* are very local but play fundamental roles in stabilizing the double helical folding of nucleotide chains. The energy of stabilization by these units originates in the hydrogen bonding between bases and the van der Waals contact of stacked base pairs, so that the energy directly depends upon the base sequence. It is, however, hard to disrupt any single unit of hydrogen bonded or stacked bases because they are tightly packed into a regularly folded polynucleotide backbone chain. Here the nucleotide backbone chains have a long-range effect on the stability. If disruption occurs, it will impose a stress on the backbone chain, which has a cooperative effect on the stability of the neighboring structures (8,23,28,29). Thus, the unwinding of the DNA double helix takes place in highly cooperative fashion, as can be seen in the phenomena of *transient opening* (16,30-33) and *cooperative melting* (1-8) with longer characteristic lengths as shown in Fig. 1.

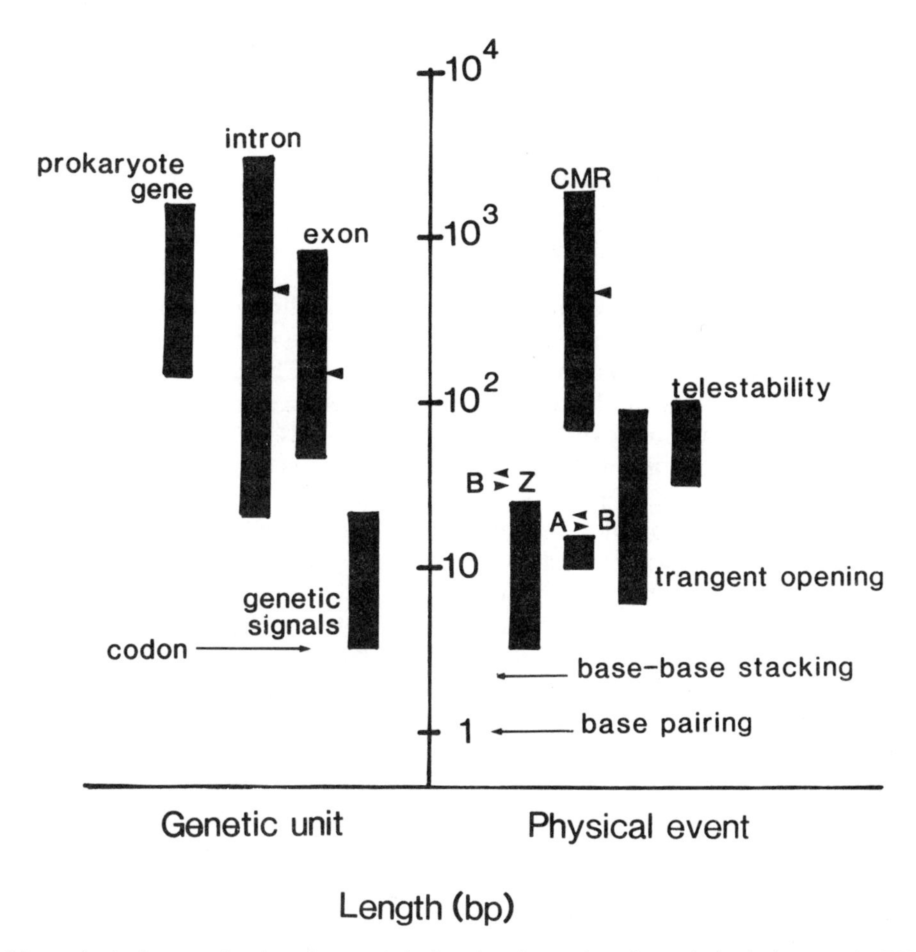

Figure 1. A diagram showing characteristic lengths of genetic units and physical events in DNA. Triangles indicate approximate mean values.

The essential feature in this long-range correlated helix-coil transition of DNA is that the local stability variation of the double helix, which comes from the base sequence, is modified by a tendency toward formation of a macroscopic intra-molecular phase through a long-range cooperative nature. The cooperative nature cannot be attributed to each nucleotide residue separately, so that neither transient opening nor cooperative melting has a direct correlation with local G+C content. Instead, virtual blocks with different double helix stability—that is, cooperatively melting regions—can be assigned along a DNA molecule. In this regard, we must note that the long-range physical stability distribution covers an entirely different aspect from the base sequence, and that the highly unique double-helical structural feature which Watson and Crick discovered provides this difference.

The characteristic length of the blocks of these long-range events range from ten to a hundred in the case of transient opening, and from a hundred to several thousands in cooperative melting. The former range overlaps with the length of genetic signals which interact with double-helix-opening proteins; and the latter range is just comparable to the length of genetic units, i.e. genes, exons (34,35) and introns (35).

In the following discussion we confine ourselves to the cooperative melting problem and its relation to DNA genetic structures.

Calculation of cooperatively melting region (CMR)

The vulnerability of a DNA double helix to an unwinding perturbation can be delineated in the framework of the statistical mechanical theory of helix-coil transition by the use of molecular thermodynamic parameters. The parameter values at which calculated melting profiles (i.e., optical-density change vs. temperature curve) best reproduce observed ones were established by Gotoh (16). This set of parameters has been used for calculating melting curves and has proved to be reliable (17-19,21). Recently, Vologodskii et al. (36) correctly pointed out in their theoretical study that the DNA melting profiles depend on only eight invariants that are linear combinations of ten original stacking parameters which Gotoh determined empirically. This means that, in principle, it is impossible to determine all of the ten parameters from the melting profiles. Even though the physical means of the parameters are somewhat uncertain, they are nevertheless very useful in the analysis of DNA helix-coil transition such the prediction of its stability distribution. Details of the calculation and parameters have been given elsewhere (8,16), and further discussion and quantitative assessments will also be given in a future publication (37).

Once the universal parameters are determined, double-helix stability distribution can be plotted along a DNA molecule by computer work for any DNA of which the whole base sequence is given (4,5,8,16-22). In the following, the stability distribution is plotted on the axis of base numbers, 1 to n, from 5′ to 3′ terminus. Stability of the i-th base pair is expressed in terms of the temperature Tm, giving it a probability of half-helix and half-coil.

Although temperature is not a physiological parameter, the Tm has simply been used in our works as a measure of double helix stability. It has a physiological meaning because CMR has also been shown to be a unit of ligand unfolding such as that by single-strand binding proteins at such physiological conditions as described in a previous report (38). Thus, the thermal stability distribution revealed by a series of CMR can be regarded as an unwinding map by DNA binding functional proteins.

Ionic strength changes both the stability of the double helix and the cooperativity of the helix-coil transition of DNA so that the pattern of stability distribution is altered to some extent (39). However, the location of major stability-jump is not altered by ionic strength change (39,40). Kalambet et al. (40) made a very detailed analysis of the location of CMR by using electron microscopic method along with theoretical examination. They proved that the position of CMRs on the DNA and the order of their melting are conserved independent of ionic strength. This means that the unique features of stability distribution, as a sequence of CMRs calculated by a set of parameters at a specific solution condition, reveal the general and inherent characteristics of a DNA.

Conciliatory nature of CMR with respect to DNA genetic structure

The characteristic features of DNA genetic structure correlated to CMR are as follows.

1. The terminus of a CMR frequently coincides with the terminus of a gene, i.e. protein coding region and exon (4-22). The coincidence has been proved to be of high confidence at the level of significance of 0.001 ~ 0.05 by statistical analysis of the positional correlation of the termini of genes and CMRs in six viral DNAs (Fig. 2) (18). In addition to these DNAs, the coincidence is evident in many cases, including DNA of T7 phage (Fig. 3) (22) and globin genes (Fig. 4a) (19).

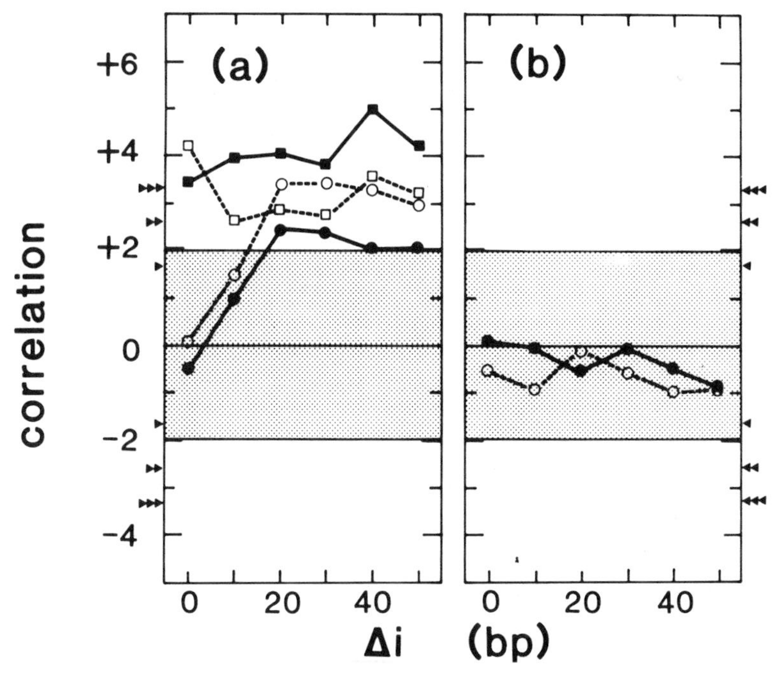

Figure 2. Positional correlation between termini of genetic divisions and CMRs. (a) Correlation between significant boundaries of cooperatively melting regions (SBCMRs) and boundaries of real protein coding regions; (b) Correlation between SBCMRs and boundaries of mere open reading frames. Degree of correlation is plotted against the extent Δi base pairs of DNA segment within which the coincidence between two kinds of the boundaries is examined. The correlation is not statistically significant in the shaded area at the level of significance 0.05. The other levels of significance are indicated at ordinate by ▶ (0.1), ▶▶ (0.01), and ▶▶▶ (0.001). Calculation is performed in the cases where the cooperativity parameter $\sigma = 10^{-3}$ (broken line) and $\sigma = 10^{-5}$ (solid line) for the three phage DNAs (ϕX174, G4, and fd DNAs) (○,●) and the three tumor virus DNAs (SV40, BKV, and polyoma DNAs) (□,■). Refer to our previous report (Ref. 18) for further details.

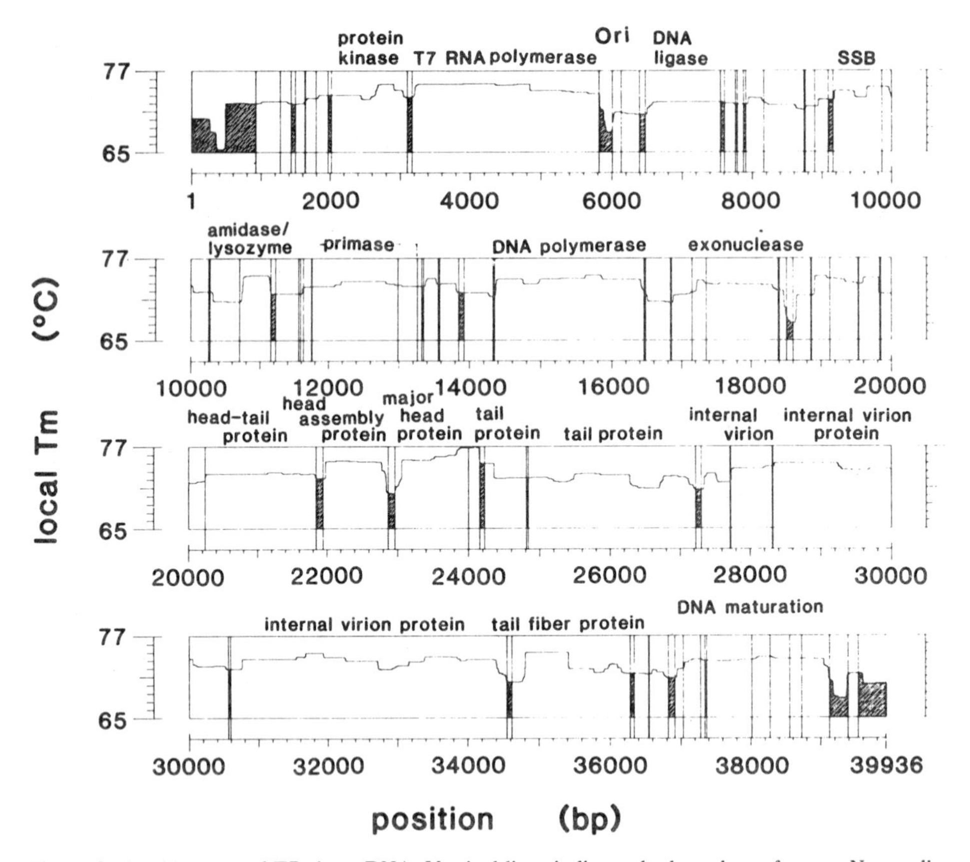

Figure 3. Stability map of T7 phage DNA. Vertical lines indicate the boundary of genes. Non-coding regions are shown by shadowing. Products are indicated for only long genes.

2. On the other hand, the positional correlation between the CMR terminus and that of the open reading frames which are defined by initiation (ATG and GTG) and termination codons (TAA, TAG, and TGA), but do not code proteins, does not exist, as shown in the result of statistical analysis and compared with the result of protein coding genes in Fig. 2 (18). This comparative analysis validates the significant correlation in the case of protein coding regions described in the above section.

3. The range of distribution of the length of CMR is comparable to that of exons and a little shorter than those of prokaryote genes and introns (see Fig. 1).

4. Since the termini of CMR and genetic regions coincide as stated above, a stretch of a unit of CMR often positionally agrees perfectly with one gene (22) or one exon (19) (Figs. 3 and 4). This can be regarded as evidence that each gene carries a tendency to make its double-helix stability uniform at around its own characteristic stability value which has been defined by some unknown condition (21,22).

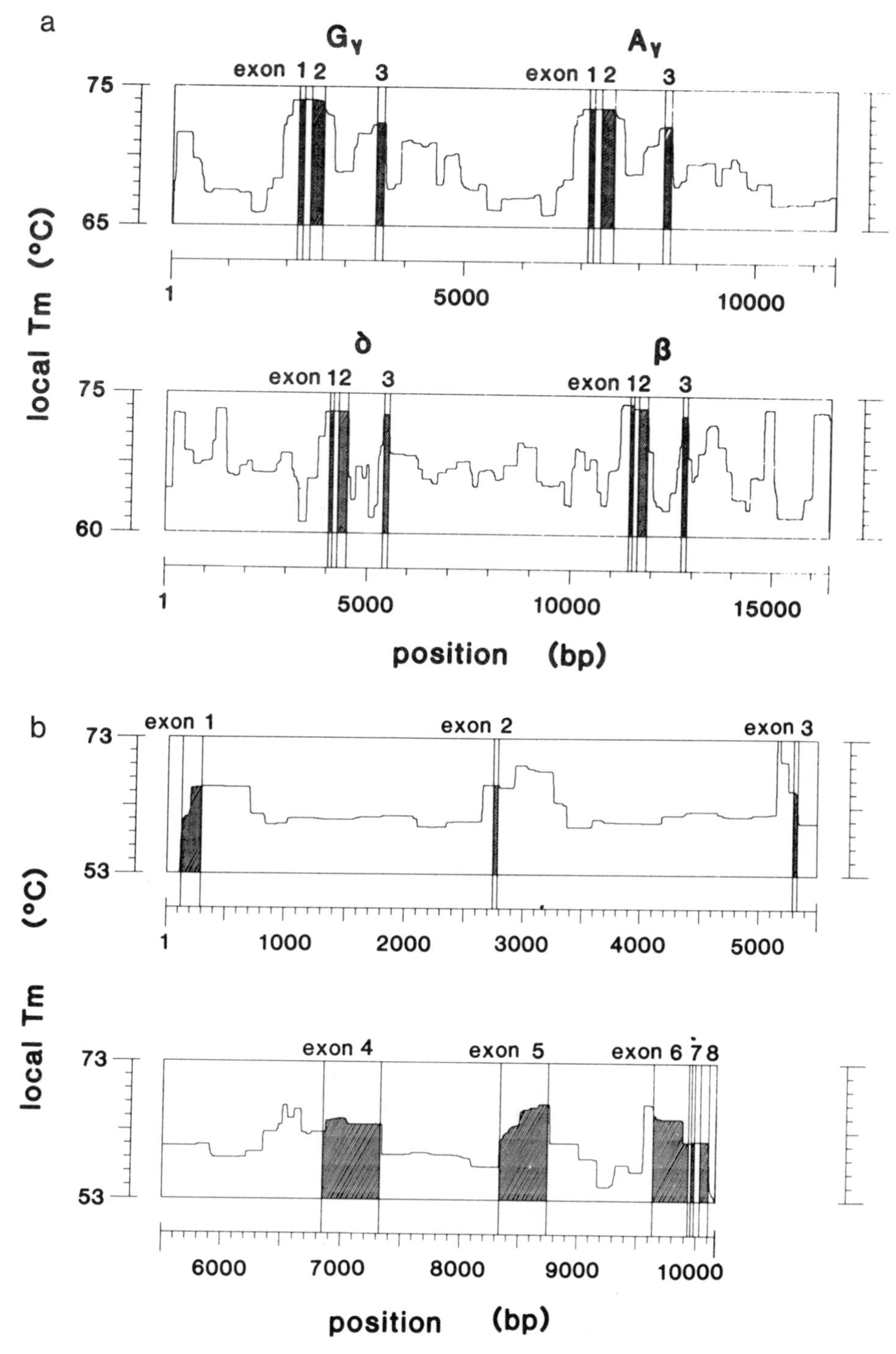

Figure 4. Stability maps of DNA having exon-intron structure. a) Globin gene (Human G_γ, A_γ, δ, and β) DNA; b) Mitochondria (yeast cytochrome oxidase) DNA.

5. The related exons joined by introns which have to be spliced after transcription seem to have an equivalent characteristic stability (Fig. 4a) (19). The stability of mitochondrial genes fluctuates much more than that of the nucleus. Even so, this equality of exon stabilities can be seen, as shown in the typical example of the cytochrome oxydase gene of Yeast mitochondria in Fig. 4b.

6. A gene tends to include several CMRs when it becomes longer; thus several stability-jumps appear in one gene (21,22) (Fig. 3). Mitochondrial DNAs show exceptionally large stability-fluctuation in coding regions (Fig. 5). This special feature might originate in the extraordinally high base-substitution-frequency of this DNA.

7. Intervening sequences, even shorter ones, are found to have many stability jumps (6,17-22) (Figs. 3 and 4). Statistical analyses quantitatively demonstrate this "noisy" feature (21). In other words, analysis of the stability fluctuation in both protein-coding and non-coding regions reveals that the stability fluctuates greatly in the non-coding region, and the homostabilizing propensity seems to prevail in the protein-coding region (21,22).

8. On the other hand, boundary regions of protein coding and non-coding regions show great fluctuation of stability (21). Stability particularly fluctuates at the sides of the protein coding region; in contrast to the quiet features of the interior part of the protein-coding region, a rather noisy part exists at their edges (21).

9. Introns also have a stability-fluctuation, as in the example of the second intron of globin genes shown in Fig. 4a (19). The first intron has uniform stability in all β-globin genes examined. This uniform stability might be due to a hidden role of this region in the DNA (19).

10. As has already been pointed out by many researchers, the origin of replication (Ori) is an AT-rich region, located at or near the deep stability trough, as shown in the case of T7 phage DNA in Fig. 3 (22). It is also found that the region where an "eye" is formed at the initial stage of T7 DNA replication (41)—that is, a region about 3kb in length running from the *Ori* site in the downstream direction—forms a low-stability basin. The upstream side of the *Ori* site, to which the replication is not directed, forms a high-stability block. Similar stopper-like stability humps are observed at the origin of the complementary strand replication of phages G4 and fd. This evidence suggests that the stability pattern seems to control the direction and range of the initial step—i.e., "eye" formation—in DNA replication.

11. It is found that the sequences that code for homologous proteins have a similar pattern of double-helix stability distribution (19). Mitochondrial DNAs present one typical example (42). The stability maps of DNA in three mammalian species are compared in Fig. 5. The average (G+C)-contents of

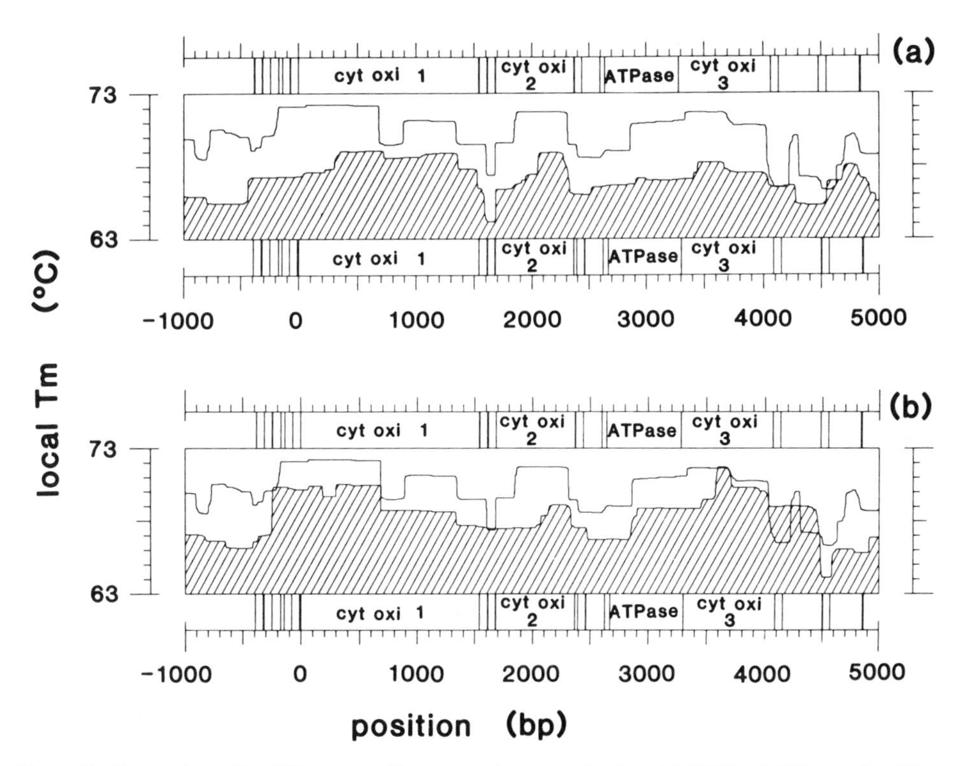

Figure 5. Comparison of stability maps of human and mouse mitochondria DNAs. (a) Thermal stability maps of cytochrome oxidase coding regions of human (solid line) and mouse (solid line with a shadow) mitochondria; (b) Those of human (solid line) and bovine (solid line with a shadow) mitochondria. The genetic maps of cytochrome oxidase coding regions of human are drawn just below the thermal stability map; Those of mouse and bovine just above the thermal stability map in (a) and (b). The first base of the coding region for cytochrome oxidase subunit 1 is numbered 0 in the position scale.

these Human, Mouse, and Bovine DNAs are 46.2%, 39.9%, and 38.6% respectively, so that the stabilities of Human DNA and those of the other two mammalians differs by about 3°C in temperature units. In spite of this difference in over-all stability, and of the high extent of base substitution, it is quite remarkable that the patterns of stability distribution are conserved fairly well in the protein-coding region. This indicates that some functional constraint acts to conserve the stability distribution as a characteristic pattern during the evolution of genetic information by DNA bse substitution.

12. Introns and intervening sequences are found, in many cases, to be low stability regions.

Discussion

The correlation between physical property and the biological structure of DNA described in the previous section may be discussed with respect to its origin and its role in biological functions. These two problems can be combined into one when we introduce the mechanism of natural selection and biological evolution. In this case, the new question arose which is the cause and which is the result; we will discuss the problem from this evolutionary viewpoint.

Even though we realize that the following arguments are speculative, we nevertheless wish to present them in the hope of stimulating discussion on this important subject.

The characteristic features of the stability distribution may be resolved into the following two points: *gene homostabilizing propensity* and *stability discontinuity at the genetic boundary*. These features seem to have evolved due to their functional applications to the smoothness of the unwinding of the double helix in a gene, and to termination of the unwinding at the end of a gene. Furthermore, the gene homostability implies that the codon-anticodon interactions have nearly the same energy over the whole length of a gene. This homogeneous codon-anticodon interaction is an essential requisite for the correct and efficient translation. The evidence that the exons which are spliced have almost the equal stability may come from this functional requirement.

As a matter of fact, the rate of unwinding of the DNA double strand has been found to be strongly dependent upon the local heterogeneity of G+C content as well as its averaged (G+C)-content (43). It is therefore quite plausible that the uniform stability in each of the CMRs and their correlation with the locations of genes have been created for the sake of smoothness of "zipper-opening" of a genetic unit. The evidence that exon regions to be spliced have similar stabilities suggests that this matching of stability is a requisite for the splicing or for some post-splicing process.

On the other hand, the stability jump at the boundary of a gene can be suspected to work as a stopper of unfolding, or as punctuation in the reading of a genetic sentence. The stability distribution around the *Ori* position of T7 phage DNA, together with its unfolding behavior at replication, may give support to this speculation. Wells et al. (24) pointed out that many of the stronger terminators have a (G+C)-rich block followed by an (A+T)-rich region in which RNA is actually terminated. This kind of base sequence is often found at the up-step stability-boundary of CMR, as described in a previous report (21). Furthermore, the evidence that the pattern of stability distribution, but not the average stability, has been conserved as shown in mitochondrial and globin gene DNAs strongly suggests that the pattern itself plays some unknown role in DNA function.

Although it appears that the larger intercistronic regions have the potential for secondary-structure (24), we do not see that the symmetrical sequence which forms a hairpin structure produces the stability jumps which we have observed. It is probable that the primary arrangement of nucleotides is more significant than the secondary structure (24).

It has already been shown on many occasions that local double-helix stability correlates with biological function. For instance, substances that destabilize the double-helix enhance m-RNA initiation (Ref. 24 and references therein). Bensimhon et al. (20) pointed out that promoter recognition and initiation of transcription by the eukaryotic enzyme machinery do not depend on any fixed base sequence, i.e. definite keywords or consensus sequences, but are most probably signaled by collective properties of base pair sets of finite size. They speculated that DNA stability provides relevant collective signals which are the ultimate target sensed, recognized and used by the transcriptional machinery. Tachibana and Ishihama (private communication) recently found that the theoretically calculated opening potential of the DNA segment in promoter sites correlates positively with the experimentally determined rate of open-complex formation between the site and RNA polymerase. Husimi and Shibata (44) pointed out a negative correlation between the CMR stability and the frequency of base substitution. Thus it is shown in many aspects that genetic processes are affected by double-helix stability. In other words, the double-helix stability exercises some genetic functional constraint on the DNA base substitution during the evolution of a gene.

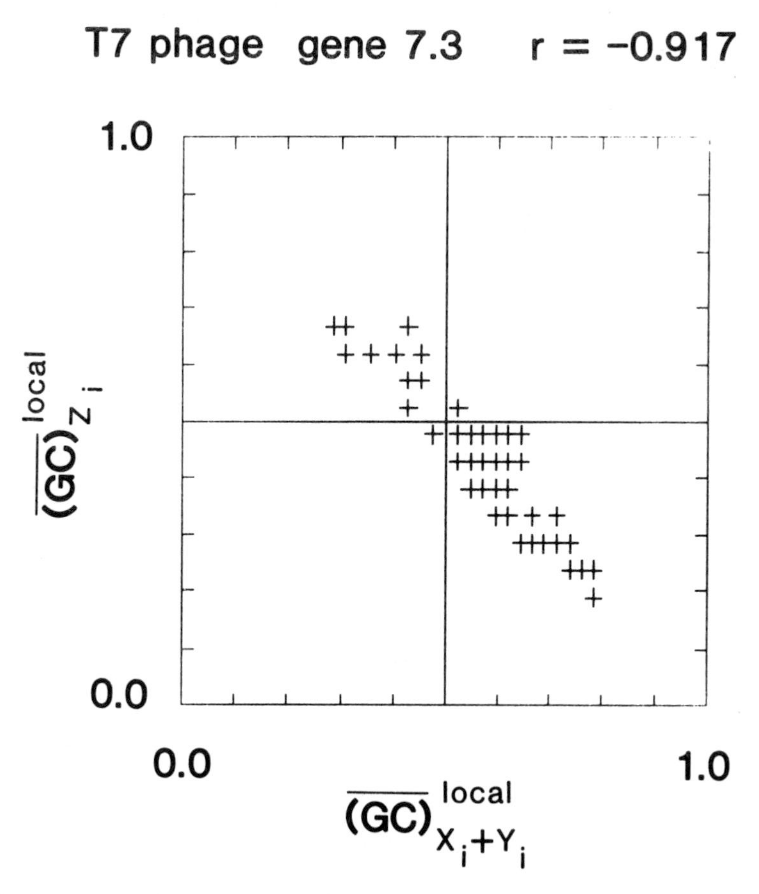

Figure 6. A typical diagram showing the correlation between locally averaged (G+C)-content at first-plus-second letters and third letters of codons in a gene.

A constraint based on a broad physical nature is supposed to be very weak, so that it can be overwhelmed by other stronger constraints such as that of amino-acid changing substitution. However, synonymous (or "silent") substitutions at the third letter of codons are regarded as free from any strong constraint, and seem to contribute to adjusting the local (G+C)-content of DNA. In this regard, it is interesting to study how the frequency of the usage of G and C letters at the 1st and 2nd sites of a codon correlates with that of the 3rd letter (45). The prime object of this examination is to see whether or not the synonymous substitution of the third letter, which is free from the constraint with respect to protein functions, counterbalances the stability change due to the substitution of the 1st and 2nd letters in maintaining homogeneous stability of a gene.

The examination is made by looking at the correlation of the (G+C)-content of the first and second letters of codons XYZ, $(G+C)_{XY}$, with that of their third letters, $(G+C)_Z$ (45). Fig. 6 shows a typical result exhibiting the correlation between locally averaged (G+C)-content $\overline{(G+C)}^{local}_{XY}$ and $\overline{(G+C)}^{local}_{Z}$ in a gene. The average is carried out for each contiguous short segment (63 base pairs in the present study) in a gene. A *negative correlation*—i.e., when the $\overline{(G+C)}^{local}_{XY}$ is low (high) the $\overline{(G+C)}^{local}_{Z}$ in the same segment is high (low)—is clearly observed in this diagram. This is shown to be a general trend of genes which can be described as follows.

In Fig. 7 is plotted a histogram of the correlation coefficients of the above type diagrams obtained in many genes examined, which are listed in the legend. Thus the negative correlation is shown to be a characteristic feature of the gene. The implication of this result is quite clear: the synonymous base substitution at the third letter of codons does work to counterbalance the (G+C)-content change at the first and second letters which was produced by some protein functional requirement. In other words, the third letter actively participates in producing a homogeneous (G+C)-content distribution of gene.

Next, we examine the correlation between (G+C)-content of one whole gene $\overline{(G+C)}^{gene}$ and the $\overline{(G+C)}^{gene}_{Z}$ which are averaged for the same gene, instead of for the local short segments. It is interesting to find in this study that, contrary to the negative correlation in the case of local (G+C)-content mentioned above, a *positive correlation* is found; in other words, when average (G+C)-content of a gene is low, $(G+C)^{gene}_{Z}$ in the same gene is also low, and *vice versa.* A further examination demonstrates that the correlation of (G+C)-content of a gene with that of 1st and 2nd letters of codons in the same gene is relatively weak. Thus this result indicates that the third-letter position of codons is the site where a wide variety of (G+C)-content is allowed. Protein functional constraint reduces the freedom for the variable use of bases at the 1st and 2nd sites of the codon.

In concluding the above two examinations, we may state that there is no detectable tendency among genes to take some common value of (G+C)-content; each gene seems to have its own characteristic value (46,47). At the local base-sequence level in each gene, however, the 3rd letter-position, which has no protein-functional constraint, plays a role in homogenizing its (G+C)-content distribution.

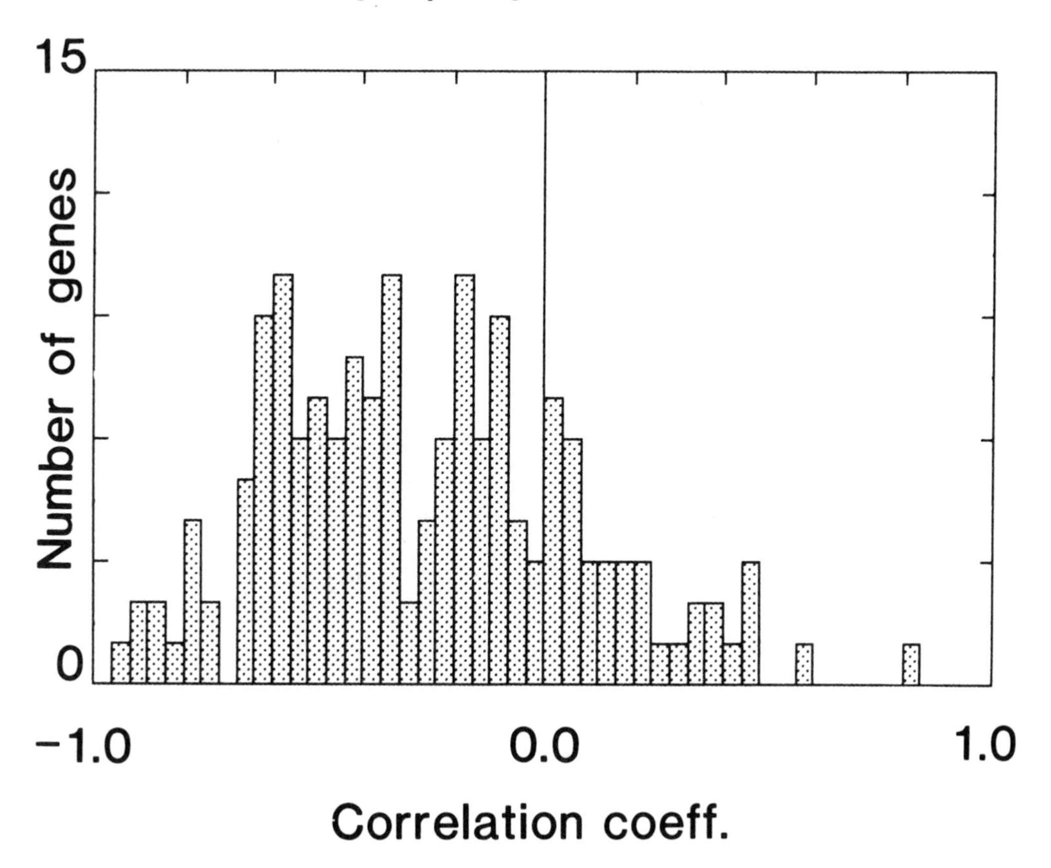

Figure 7. Histogram showing the distribution of the correlation coefficients of many genes, an example of which is shown in Fig. 6. The genes used in these statistics are 161 genes from following species: Phages ϕX174, G4, fd, T7, and lambda[a]; E. coli ribosomal proteins S8 and L6 genes [b]; Mycoplasma ribosomal proteins S8 and L6 genes[b]; Staphylococcus aureus plasmid pT181[a]; and Staphylococcus aureus plasmid pC194[a]. (a: GenBank release 29, b: A. Muto, Y. Kawauchi, F. Yamao, and S. Osawa, Nucleic Acids Res. 12, 8209-8217 (1984).

There could be an elucidation for the counterbalancing effect, that is, nature chooses the codon-anticodon interaction of medium strength, and avoids the codons of extremely high or low (G+C)-content (48). This, however, may not be the case, because it has been found that a number of genes have extremely high and low (G+C)-content. Also the positive correlation between the (G+C)-contents of the first-plus-second letters and the third letter of codons which are averaged over one whole gene may not be elucidated by the fact that the intermediate codon-anticodon interaction energy is preferred in the translation process.

Thus, undoubtedly, the homogeneous (G+C)-content distribution in a gene in fact is closely related to the gene-homostabilizing propensity, although it is very hard to speculate which is the cause and which the result. The authors, however, tend to believe now that homostabilization is the cause and the homogeneous (G+C)-content distribution is the result, simply because we could not find a reason why a homogeneous (G+C)-content should be a primary requirement.

We dedicate this paper to Dr. Paul Doty on the occasion of his 65th birthday.

This work was supported by a grant-in-aid from the Ministry of Education, Science, and Culture, Japan.

References and Footnotes

1. Michel, F., Lazowska, J., Faye, G., Fukuhara, H., and Slonimski, P.P., *J. Mol. Biol. 85,* 411-431 (1974).
2. Vizard, D.L. and Ansevin, A.T., *Biochemistry 15,* 741-750 (1976).
3. Reiss, C. and Arpa-Gabarro, T., *Progress in Molecular and Subcellular Biology* (Hahn, F.E., ed.) *Vol. 5.* Berlin. Springer-Verlag (1977).
4. Ueno, S., Tachibana, H., Husimi, Y., and Wada, A., *J. Biochem. 84,* 917-924 (1978).
5. Tong, B.Y., and Battersby, S.J. *Nucleic Acids Res. 6,* 1073-1079 (1979).
6. Wada, A., Ueno, S., Tachibana, H. and Husimi, Y., *J. Biochem. 85,* 827-832 (1979).
7. Hardies, S.C., Hillen, W., Goodman, T.C., and Wells, R.D. *J. Biol. Chem. 254,* 10128-10134 (1979).
8. Wada, A., Yabuki, S., and Husimi, Y., *CRC Crit. Rev. Biochem. 9,* 87-144 (1980).*
9. Horn, G.T. and Wells, R.D. *J. Biol. Chem. 256,* 2003-2009 (1981).
10. Hillen, W., Goodman, T.C., Benight, A.S., Wartell, R.M., and Wells, R.D. *J. Biol. Chem. 256,* 2761-2766 (1981).
11. Hillen, W. and Unger, B. *Nucleic Acids Res 10,* 2685-2700 (1982).
12. Hillen, W. and Unger, B. *Nature 297,* 700-702 (1982).
13. Klein, R.D. and Wells, R.D. *J. Biol. Chem. 257,* 12954-12961 (1982).
14. Klein, R.D. and Wells, R.D. *J. Biol. Chem. 257,* 12962-12969 (1982).
15. Goodman, T.C., Klein, R.D., and Wells, R.D. *J. Biol. Chem. 257,* 12970-12978 (1982).
16. Gotoh, O. *Adv. Biophys. 16,* 1-52 (1983).*
17. Suyama, A. and Wada, A., *Nucleic Acids Res., Symposium Series No. 11,* 165-168 (1983).
18. Suyama, A. and Wada, A. *J. Theor. Biol. 105,* 133-145 (1983).
19. Wada, A. and Suyama, A. *J. Phys. Soc. Jpn. 12,* 4417-4422 (1983).
20. Bensimhon, M., Gabbaro-Arpa, J., Ehrlich, R., and Reiss, C. *Nucleic Acids Res. 13,* 4521-4540 (1983).
21. Wada, A. and Suyama, A. *J. Biomol. Struct. Dynm. 2,* 131-148 (1984).
22. Wada, A. and Suyama, A. *"Molecular Basis of Cancer", part A,* pages 37-46, Ed. Rein, R., Alan R. Liss, Inc., New York (1984).
23. Wells, R.D. Blakesley, R.W., Burd, J.F., Chan, H.W., Dodgson, J.B., Hardies, S.C., Horn, G.T., Jensen, D.F., Larson, J., Nes, I.F., Selsing, E. and Wartell, R.M. *CRC Crit. Rev. Biochem. 4,* 305-340 (1977).*
24. Wells, R.D., Goodman, T.C., Hillen, W., Horn, G.T., Klein, R.D., Larsen, F.E., Muller, V.R., Nevendorf, S.D., Panayotatos, N., Stirdivant, S.M., *Progr. Nucl. Acid Res. Mol. Biol. 24,* 167-267 (1980).*
25. Chamberlin, M.J. in *"RNA Polymerase"* Eds. Losick, R. and Chamberlin, M.J. p. 159, Cold Spring Harbor Lab., New York (1976).*
26. Stayton, M.M., *Cold Spring Harbor Symp. Quant. Biol. 47,* 693-700 (1982).
27. Fisher, M. *Nature 299,* 105-106 (1982).*
28. Wartell, R.M. and Burd, J.F., *Biopolymers 15,* 1461-1479 (1976).
29. Wartell, R.M. *Nucleic Acids Res. 4,* 2779-2797 (1977).
30. Teitelbaum, H. and Englander, S.W. *J. Mol. Biol. 92,* 55-78 (1975).
31. Teitelbaum, H. and Englander, S.W. *J. Mol. Biol. 92,* 79-92 (1975).
32. Lukashin, A.V., Vologodskii, A.V., Frank-Kamenetskii, M.D. and Lyubchenko, Y.L. *J. Mol. Biol. 108,* 665-682 (1976).
33. Gotoh, O. and Tagashira, Y. *Biopolymers 20,* 1043-1058 (1981).
34. Blake, C. *Nature 306,* 535-537 (1983).
35. Naora, H. and Deason, N.J. *Proc. Natl. Acad. Sci. USA 79,* 6196-6200 (1982).
36. Vologodskii, A.V., Amirikyan, B.R., Lyubchenko, Y.L. and Frank-Kamenetskii, M.D. *J. Biomol. Struct. Dynm. 2,* 131-148 (1984).

37. Suyama, A. and Wada, A. to be published.
38. Tachibana, H. and Wada, A., *Biopolymers 21,* 1873-1885 (1982).
39. Tachibana, H., Ueno-Nishio, S., Gotoh, O. and Wada, A. *J. Biochem. 92,* 623-635 (1982).
40. Kalambet, Y.A., Borovik, A.S., Lyamichev, V.I. and Lyubchenko, Y.L. *Biopolymers 24,* 359-377 (1985).
41. Romano, L.J., Tamanoi, F., and Richardson, C.C. *Proc. Natl. Acad. Sci. USA 78,* 4107-4111 (1981).
42. Wada, A. and Suyama, A. to be published.
43. Suyama, A. and Wada, A. *Biopolymers 23,* 409-433 (1984).
44. Husimi, Y. and Shibata, K., *J. Phys. Soc. Jpn. 53,* 3712-3716 (1984).
45. Wada, A. and Suyama, A., FEBS Lett. *188,* 291-294 (1985).
46. Sueoka, N., *J. Mol. Biol. 3,* 31-40 (1961).
47. Freese, E., *J. Theor. Biol. 3,* 82-90 (1962).
48. Grosjean, H., and Fiers, W., *Gene 18,* 199-209 (1982).

*Article of review style where a number of original works are cited.

Biomolecular Stereodynamics IV, Proceedings of the Fourth Conversation in the Discipline Biomolecular Stereodynamics, State University of New York, Albany, NY, June 04-09, 1985, Eds., Ramaswamy H. Sarma & Mukti H. Sarma, ISBN 0-940030-18-7, Adenine Press, ©Adenine Press 1986.

Evolution and Functional Significance of the Bias in Codon Usage

R.D. Blake, Philip W. Hinds, Scott Earley, A.L. Hillyard and Gary R. Day
Department of Biochemistry, University of Maine
Orono, ME 04469, USA

Abstract

DNA sequences from *E. coli, S. cerevisiae,* mouse and human have been analyzed for the distribution of codon frequencies. Almost 300 gene sequences from *E. coli* have been characterized according to the nature and level of average codon use. The distribution is broad and exhibits some bimodality; with about half the number of genes using an average of only 36 codons and the remainder about 42 codons. There is a high correlation between the levels of codon bias, the levels of tRNAs and the abundance of protein products, indicating that the biased pattern is exploited by the cell for the production of widely different levels of gene product. This relationship is especially striking in genes involved in the production of components for transcription and translation. Overall, the genes for these processes generate about five-fold more protein than the average gene, and use about five fewer codons.

The codon bias in yeast sequences is even more pronounced. One large family of sequences uses an average of 38 codons and another just 29 codons. Bias levels in this entire latter group, some which use only 27 codons, far exceed those of any other sequence examined thus far. Sequences in this highly biased group are dominated by genes for enzymes of glycolysis and fermentation.

Mouse and human sequences indicate similar broad distributions of average codon use with few examples of sequences with the extraordinarily high levels of codon bias found in yeast and *E. coli.* The bias in the mouse genome has remained conserved to a remarkable degree in the human genome, indicating a sensitivity or resistance to change are important factors in the preservation of the biased codon pattern in the more complex eukaryotic cell.

Introduction

Reports by Grantham (1-3) and his coworkers in Lyon as well as some from our own and other laboratories (4-7), have shown that the frequencies synonymous codons are used for a particular amino acid are often far from that expected from a random choice. Our Table I illustrates the codon bias in four widely different species. Suggestions have been made to explain the bias, and some have even gained a certain popularity. Unfortunately, as we have improved and broadened the statistical basis for the bias in different species, we find that most of these earlier explanations cannot be sustained. In the light of present evidence we now believe that biased codon patterns evolved to fill the need for widely different levels of gene product.

Results

Table I indicates frequencies of codon occurrence in sequences from *E. coli,* yeast, mouse and human. We have put our initial focus on this assortment of organisms for in-depth study because the combined database is especially large. At the present time the database for these four species alone totals well above one million base pairs which is a large fraction of the complete GENBANK DNA sequence electronic database. Almost one third or three hundred thousand base pairs from these four species are reading frame or coding sequences.

Table I
Frequencies of Synonymous Codon Usage

amino acid	codon	E. coli	yeast	mouse	human	(expected)
ALA	GCA	0.23	0.11	0.19	0.14	0.25
	GCG	0.32	0.03	0.15	0.11	
	GCC	0.21	0.26	0.30	0.43	
	GCT	0.25	0.60	0.35	0.32	
ARG	AGA	0.04	0.72	0.29	0.28	0.17
	AGG	0.02	0.08	0.23	0.27	
	CGA	0.04	0.02	0.09	0.09	
	CGG	0.04	0.01	0.12	0.14	
	CGC	0.33	0.03	0.19	0.18	
	CGT	0.53	0.14	0.08	0.05	
ASN	AAC	0.68	0.68	0.57	0.63	0.50
	AAT	0.33	0.32	0.43	0.38	
ASP	GAC	0.51	0.47	0.54	0.61	0.50
	GAT	0.50	0.53	0.46	0.39	
CYS	TGC	0.54	0.18	0.52	0.53	0.50
	TGT	0.46	0.82	0.48	0.47	
GLN	CAA	0.28	0.86	0.23	0.26	0.50
	CAG	0.72	0.14	0.77	0.74	
GLU	GAA	0.71	0.85	0.32	0.39	0.50
	GAG	0.29	0.16	0.68	0.61	
GLY	GGA	0.06	0.05	0.30	0.21	0.25
	GGG	0.09	0.04	0.21	0.21	
	GGC	0.39	0.08	0.33	0.41	

	GGT	0.46	0.83	0.16	0.17	
HIS	CAC	0.59	0.58	0.54	0.52	0.50
	CAT	0.41	0.42	0.46	0.48	
ILE	ATA	0.05	0.09	0.13	0.11	0.33
	ATC	0.58	0.41	0.47	0.62	
	ATT	0.37	0.51	0.40	0.27	
LEU	TTA	0.09	0.19	0.04	0.04	0.17
	TTG	0.10	0.58	0.10	0.12	
	CTA	0.02	0.08	0.06	0.06	
	CTG	0.62	0.05	0.46	0.48	
	CTC	0.08	0.02	0.23	0.23	
	CTT	0.09	0.07	0.12	0.08	
LYS	AAA	0.74	0.33	0.40	0.34	0.50
	AAG	0.26	0.67	0.60	0.66	
MET	ATG	1.00	1.00	1.00	1.00	1.00
PHE	TTC	0.58	0.63	0.61	0.61	0.50
	TTT	0.42	0.37	0.39	0.39	
PRO	CCA	0.18	0.68	0.27	0.18	0.25
	CCG	0.61	0.03	0.13	0.11	
	CCC	0.08	0.07	0.22	0.37	
	CCT	0.13	0.22	0.38	0.33	
SER	AGC	0.23	0.05	0.21	0.31	0.17
	AGT	0.07	0.08	0.10	0.11	
	TCA	0.10	0.11	0.11	0.10	
	TCG	0.12	0.04	0.05	0.05	
	TCC	0.23	0.28	0.27	0.24	
	TCT	0.25	0.43	0.26	0.20	
THR	ACA	0.09	0.16	0.21	0.22	0.25
	ACG	0.16	0.6	0.15	0.09	
	ACC	0.50	0.36	0.44	0.42	
	ACT	0.25	0.42	0.20	0.28	
TRP	TGG	1.00	1.00	1.00	1.00	1.00
TYR	TAC	0.56	0.67	0.63	0.59	0.50
	TAT	0.44	0.33	0.38	0.41	
VAL	GTA	0.21	0.07	0.07	0.06	0.25
	GTG	0.27	0.09	0.55	0.52	
	GTC	0.14	0.36	0.22	0.27	
	GTT	0.37	0.48	0.16	0.15	
STOP	TAA	0.77	0.72	0.25	0.39	0.33
	TAG	0.02	0.17	0.04	0.21	
	TGA	0.21	0.10	0.71	0.41	

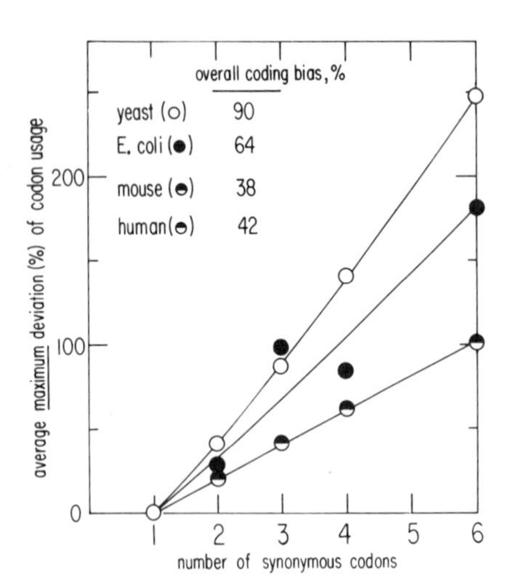

Figure 1. Variation in average maximum level of bias (percent) in the usage of synonymous codons for a given amino acid with the number of synonymous codons. Results are plotted for yeast (○), *E. coli* (●), mouse and human (◒). (Results for mouse and human are essentially coincident).

As shown in Figure 1, the bias is greatest in yeast and least in mouse and human sequences. This figure shows a good correlation to exist between the maximum level of bias and the number of synonymous codons. Thus, those amino acids with six codons have an average maximum bias of 250% in yeast sequences. The bias then falls off with decreasing numbers of synonymous codons at a rate that differs for each of the four species.

Examination of Table I indicates some distinct biases, particularly for amino acids with largest numbers of synonymous codons. Thus *E. coli* uses the codon CGT for arginine almost 53% of the time or three times above the frequency that would be expected from a random choice among the synonymous codons. The bias for arginine is even stronger in yeast where a different codon, AGA, is used 4½ times more often than expected. The bias for this amino acid is considerably less in mouse and human, where three different codons are preferred at about the same level in both species. Thus, the bias is both quantitatively and qualitatively different among the four species. We would also note that with one small exception codon preferences show no apparent systematic or overall preference for the wobble base. There are strong biases for and against the same wobble base within and among different organisms. The small exception is that an AC in the second and third base is preferred over an AT 15 times out of 16 occurrences in our table.

There is one further observation we can note at this time. The bias in the mouse genome has remained conserved to a remarkable degree in the human genome, indicating that vulnerability to survival or resistance to change have been important factors in the preservation of the biased codon pattern in the more complex eukaryotic cell.

One question that arises here concerns the extent amino acid composition effects the level of codon bias. We find the effect to be negligible. Variations in amino acid composition from one protein to another are very small. Variations in amino acid compositions in proteins from different species are just as small, as can be seen in Table II. The level of correspondence between amino acid frequencies in proteins from widely different species is very high (Table III) and at about the same level as we see among individual globular proteins, consequently we can rule out amino acid composition as a significant factor in the evolution of the bias.

Table II
Amino Acid Frequencies

	amino acid	aggregate* of 314 sequences	E. coli (53)	yeast (31)	mouse (44)	human (54)
1.	ALA	0.086	0.096	0.086	0.082	0.071
2.	GLY	0.084	0.079	0.073	0.073	0.060
3.	LEU	0.074	0.097	0.083	0.098	0.111
4.	SER	0.072	0.055	0.072	0.061	0.081
5.	VAL	0.066	0.078	0.076	0.070	0.063
6.	LYS	0.066	0.054	0.075	0.051	0.060
7.	THR	0.061	0.049	0.059	0.054	0.054
8.	GLU	0.060	0.068	0.061	0.066	0.068
9.	ASP	0.055	0.056	0.058	0.051	0.049
10.	PRO	0.052	0.042	0.041	0.052	0.044
11.	ARG	0.049	0.054	0.041	0.060	0.050
12.	ILE	0.045	0.061	0.059	0.036	0.040
13.	ASN	0.043	0.040	0.048	0.040	0.037
14.	GLN	0.039	0.043	0.031	0.041	0.044
15.	PHE	0.036	0.035	0.040	0.039	0.045
16.	TYR	0.034	0.028	0.035	0.031	0.028
17.	CYS	0.029	0.009	0.010	0.021	0.029
18.	HIS	0.020	0.021	0.022	0.029	0.023
19.	MET	0.017	0.026	0.020	0.022	0.027
20.	TRP	0.013	0.009	0.011	0.023	0.013

*Globular protein database (*Atlas of Protein Sequence and Structure*, Vol 5, Sup 3, 1979 (Dayhoff, M.O., Ed.).

Table III
Pairwise Correlation Coefficients for Amino Acid Frequencies from Different Species

	E. coli	yeast[a]	(314)[b]	mouse	human
E. coli					
yeast[a]	0.912				
(314)[b]	0.902	0.931			
mouse	0.940	0.868	0.900		
human	0.859	0.852	0.846	0.919	

[a]S. cerevisiae
[b]amino acid frequencies in 314 globular proteins (from: *Atlas of Protein Sequence and Structure*, Vol 5, Sup 3, 1979 (Dayhoff, M.O., Ed.).

A more interesting question concerns the level of variability in the codon pattern among sequences from the same species. For this question we focused initially on sequences from *E. coli* since it has among the largest representation in the sequence database. We find the correspondence in the levels of bias among all sequences from *E. coli* to be very high. The correlation coefficient rarely falls below .85 for any pair of sequences from *E. coli,* but approaches zero for most heterologous pairs, for example, between an *E. coli* sequence and one from a different bacteria or from yeast.

We are dealing with just under 10% of the *E. coli* genome in our present sequence database, nevertheless there are strong indications that the biased codon pattern we observe is distributed fairly uniformly throughout the genome. This question is important to the interpretation we give to the existence of the bias, therefore, we have made a special effort to find other indications that the limited sequence database is a good representation of total genomic DNA. First, as just indicated, we find good pair-wise correspondence of the bias in any two sequences from *E. coli.* Secondly, we obtain excellent correspondence between dinucleotide frequencies determined by direct count in the sequence database and nearest-neighbor frequencies determined by the Josse-Kornberg chemical analysis of total genomic DNA (8-10). In Figure 2 these are plotted against one another as the difference between observed and expected frequencies so as to exaggerate any small variations, and yet the correspondence remains surprisingly good. The dashed line running diagonally in this figure represents the line of perfect correspondence. (The standard error for the chemical results is larger than the greatest deviation of any point from the dashed line).

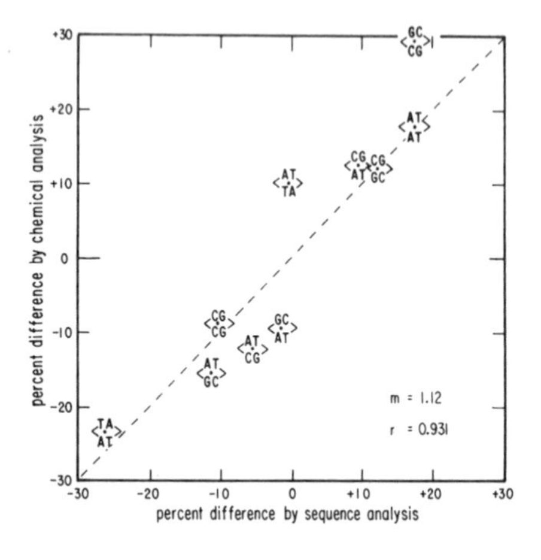

Figure 2. Variation of the difference (percent) between observed and expected frequencies of the ten unique nearest-neighbor base pairs determined by computer count in *E. coli* sequences (241,592 bp), with the same difference for frequencies determined by Josse-Kornberg chemical analysis (8-10). The dashed line represents perfect correspondence.

A third indication that the pattern we've found is distributed uniformly in the genome is obtained from the excellent correspondence between the distribution of base compositions in the sequence database and in total genomic DNA. The distribution for genomic DNA illustrated in Figure 3 by the continuous line was obtained by high resolution melting (7,11), and reflects the distribution of G+C composition within DNA segments of about 150 bp in length. We obtain an average G+C content of 51.1% which compares well with the consensus value of 50.7% in the literature. The mode occurs at 52.2% which is identical with that of reading frame sequences. The distribution is skewed slightly for total genomic DNA and has a standard deviation of ±4% G+C while the distribution for the sequence database has a standard deviation of ±6% when seen through a window of 150 bp in size. We interpret these several observations as indicating the generality of sequence results from only 10% of the genome to the genome as a whole.

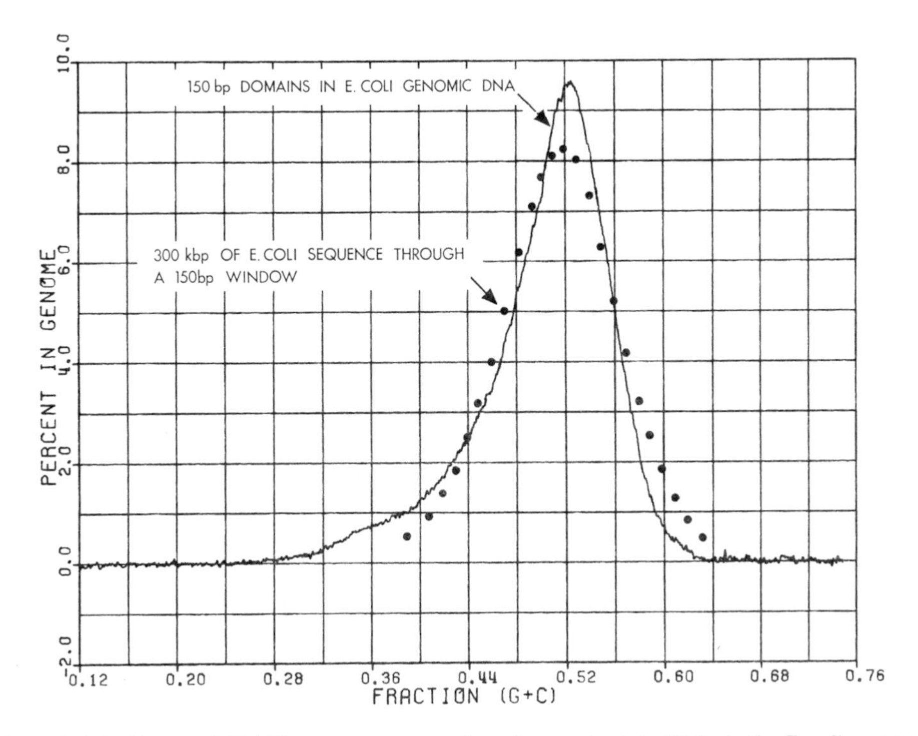

Figure 3. Distribution of (G+C) content over stretches of approximately 150 bp in the *E. coli* genome (solid line) (7,12). The closed circles represent the distribution of (G+C) content in 300 kbp of *E. coli* sequences when viewed through a 150 bp window.

We have characterized a large number of gene sequences from *E. coli* according to the nature and level of average codon preference. We obtain an average codon probability for a sequence by assigning a probability to each triplet equal to the relative overall frequency of use as a codon and then dividing by the number of triplets. Characterization of gene sequences this way may mask any local variations in codon preference within genes, and which may be important for temporal control of translation. However, the average of preferences allows for distinctions to be made between sequences. A distribution was then determined for the variation in average codon probability in both reading and non-reading frame sections of each sequence, and the results illustrated in Figure 4. The average codon probability of triplets in reading frame segments is represented by the clear bars in this figure, and appears to be bimodal in distribution with about 40% of sequences having an average probability of 0.028 and the remaining 60% with values near 0.024. Sequences with the higher value use an average of just 36 codons while those in the lower group use 42.

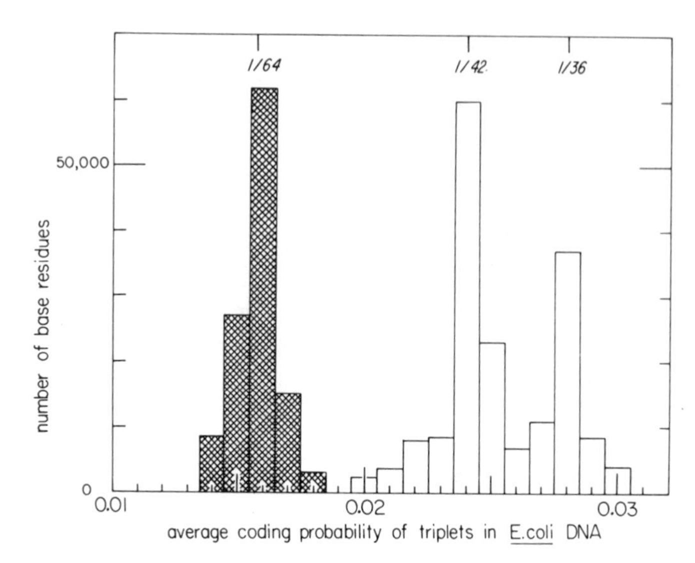

Figure 4. Distributions of average codon probabilities in reading frame (clear bars) and nonreading frame (crosshatched bars) segments of *E. coli* sequences.

The average codon probability of triplets in nonreading frame segments, represented by the crosshatched bars in this figure, form a normal distribution with a mode at 0.0160, which is almost the numerical equivalent of 1/64 and therefore corresponds with the random occurrence of all triplets.

Figure 5 shows the distribution of average codon probabilities for both reading and nonreading-frame triplets in yeast sequences. Nonreading-frame segments, represented by the crosshatched bars in this figure, have average codon probabilities that cluster with a mean of 0.0155, corresponding almost precisely to the random occurrence of triplets (1/64). The standard deviation is somewhat larger than it was for *E. coli* sequences, and about three-fold larger than expected from a random sequence; indicating perhaps, the presence of functional or vestigial sequence elements in nonreading frame segments (6).

A distinct distribution (or distributions) for reading frame segments is not yet apparent, perhaps indicating the population of yeast sequences may still be too small. The database is less than one-half of that for *E. coli,* while the genome is about 2 times larger. There seems to be one large block of sequences with an average codon probability near 0.026 (1/38) and another block at 0.035 (1/29). The codon bias levels in this latter group far exceed any found in any single sequence from *E. coli* or any other species that we have examined so far. Indeed, all sequences from yeast indicate a significantly higher bias, with the average sequence using just 35 codons. Those at the extreme upper end of the distribution scale, at 0.037, are using an average of only 27 codons, which is remarkably close to the absolute minimum number.

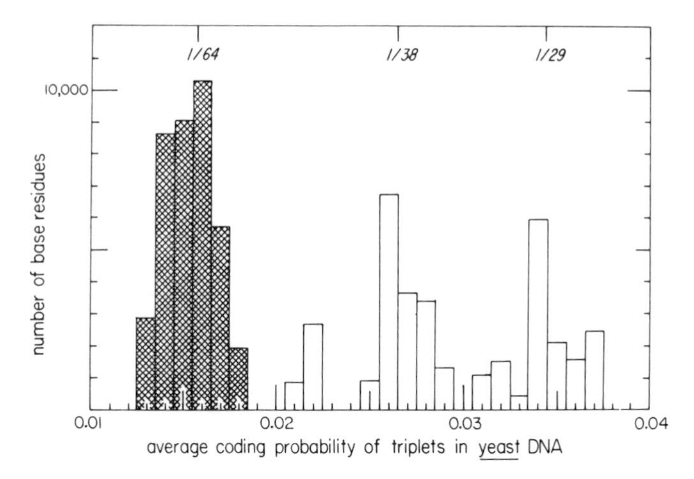

Figure 5. Distributions of average codon probabilities in reading frame (clear bars) and nonreading frame (crosshatched bars) segments of yeast sequences. (The average codon probabilities of triplets in introns is slightly above random (0.017); and is not shown here because the present representation is too small).

Analysis of mouse and human sequences indicate similar broad distributions of average codon probabilities, with few examples of sequences with the extraordinarily high levels of codon bias found in yeast and *E. coli.* A more meaningful interpretation of the bias in mouse and human should be possible as more sequences are added to the database.

In summary, we find sequences from the rapidly dividing *E. coli* cell to exhibit a very characteristic pattern of preferences among the synonymous codons. This pattern corresponds to an average level of preference for only 39 codons, and it extends throughout the genome. In yeast, a primitive eukaryotic cell with approximately 1000 times greater internal volume than *E. coli,* but with only a 3-fold longer generation time, the average level of preference is for only 34 codons. Some sequences even use as few as 27 codons. Again, the biased codon pattern is distributed throughout all sequences from yeast that have been examined thus far.

The codon bias in mouse and human sequences is less pronounced than in *E. coli* and yeast. While the pattern of codon usage is highly conserved between mouse and human, it is as different from that of yeast as the latter is from *E. coli.* The depth of sequence representation in the present database is too low to warrant further comparisons at this time, nevertheless, there are indications the codon pattern in the more primitive cellular forms vary quite widely. We find little or no correspondence between any of the existing patterns and codon frequencies in various bacterial sequences from the database. Certainly, patterns in bacterial species with widely different base composition will differ. The continuous curves in Figure 6 represent the distributions of G+C compositions over domains of 150 bp in length in the genomes of *Cl. perfringens, E. coli* and *M. lysodeikticus* (12). The points in this figure represent the overlay of three normal distribution curves with modes of 30.6, 52.0 and 68.5% G+C, each with the same standard deviation (±4% G+C) as that found in both reading and nonreading frame sequences from *E. coli* (Figure 3). There is essentially no overlap of base compositions in sequences from these three species indicating their codon patterns must vary widely from one another. Assuming the amino acid composition in proteins of these species is similar to those in Table II, we find that *Cl. perfringens* and *M. lysodeickticus* will be "genetic code-limit" organisms (13), and, therefore, will have very large and widely different codon biases.

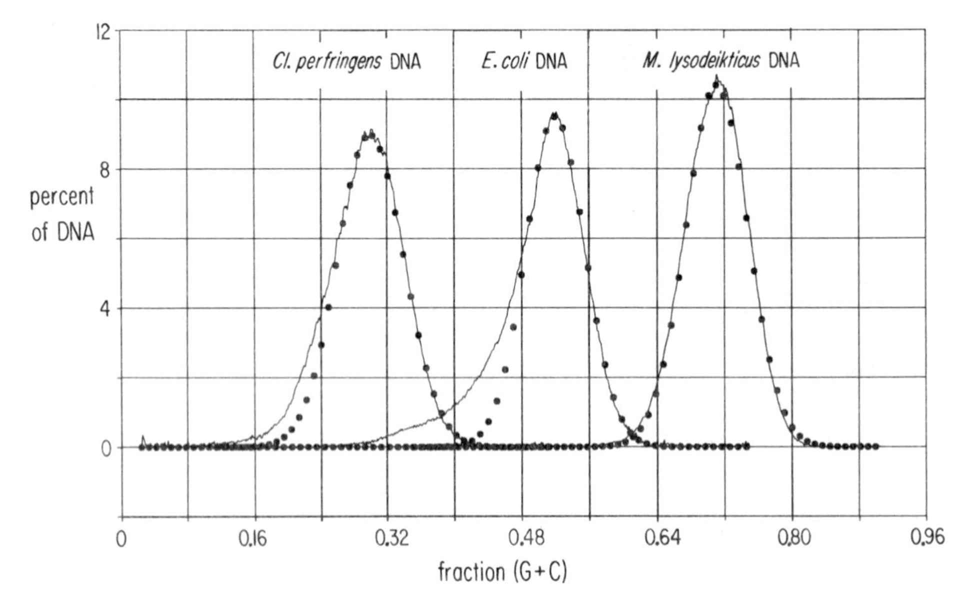

Figure 6. Distributions of (G+C) content over stretches of approximately 150 bp in the genomes of *Cl. perfringens*, *E. coli* and *M. lysodeikticus* (solid lines). The closed circles represent normal distribution curves with modes of 30.6, 52.0 and 68.5% G+C, each with the same standard deviation (±4% G+C) as that found in both reading and nonreading frame sequences from *E. coli* (cf. Figure 3).

Discussion

The principal remaining question concerns the functional significance of biased codon patterns. Given the variety of patterns found among different species, it seems unlikely that factors intrinsic to the interaction between codon and anticodon are responsible for the bias.

We believe a strong case can be made for the concept that biased patterns are exploited by the cell for the production of widely different levels of gene product. We find a high correlation between the level of codon bias, the tRNA population and the abundance of protein product. To achieve a maximum rate of expression, we imagine the levels of specific codons will converge to meet those of their cognate tRNAs, although the cause and effect of this relationship remains a question. A strong correlation between the abundance of different leucine tRNAs determined by Ikemura (14-16) and the pattern of usage among the six leucine codons in both *E. coli* and yeast can be seen in Figure 7. The ordinate in this figure represents the frequency of specific leucine tRNA species in either *E. coli* or yeast while the abscissa represents the frequencies of occurrence in coding regions of the corresponding codons for those same leucyl tRNAs. We note that one particular codon is used in *E. coli* about 60% of the time for leucine, which has six synonymous codons, while a different codon is used about 60% of the time in yeast. Concurrently, the corresponding tRNAs for these codons make up 60% of the total leucyl tRNA present in the cells of each species. The dashed line represents the case where perfect correspondence would exist between codon usage and cognate tRNA abundance. The physical consequence of this correlation will be an enhancement in both rate and fidelity of incorporation of leucine into protein. tRNAs for other amino acids show similar correlations with codon frequencies, although lacking in the same level of quantitation.

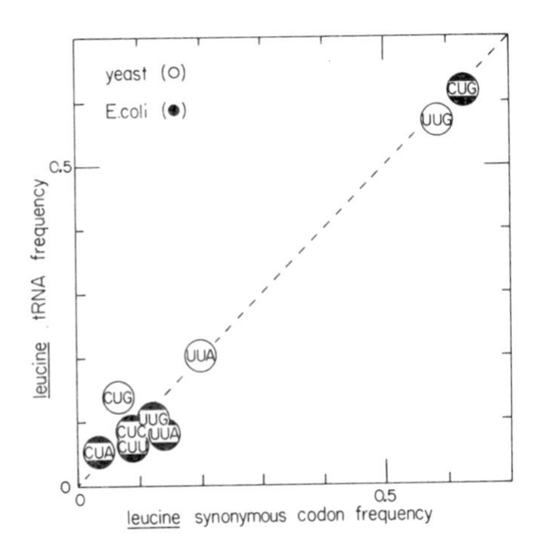

Figure 7. Variation in the relative frequency of different leucyl tRNAs with the relative frequency of occurrence in coding sequences of the corresponding codon. There are six synonymous codons for leucine. Results for *E. coli* are denoted by the closed circles, while the open circles represent yeast results.

We also note the existence of a correlation between the level of codon bias in a gene, and the level of gene expression. Table IV lists fifteen sequences from *E. coli* with the highest average codon probability. The last column in this table gives the approximate numbers of proteins produced by each sequence. These seem to vary, yet their average output exceeds 50,000 copies. This contrasts with the output of sequences at the lower end of the scale, which is often just a few molecules per sequence. Thus, a general consequence of the correlation between codon and cognate tRNA levels is increased abundance of gene product. Genes with a high codon bias almost invariably produce, or have the capacity to produce, extraordinarily high levels of protein, while the converse is found among sequences with a low bias.

This table of sequences of highest codon probability is particularly rich in genes involved in the production of components for transcription and translation, denoted in Table IV by an asterisk. Thus, the S1 ribosomal protein gene, with the highest average codon probability yet found in *E. coli* leads to the synthesis of approximately 150,000 copies of this largest ribosomal constituent in each log phase cell. Such levels are characteristic of most products of the roughly 100 genes involved in the synthesis of RNA and protein constituents of transcription and translation (17). The genes for these constituents comprise only 5% of the *E. coli* genome, yet their products make up almost 40% of the dry cell mass. Again, the average codon probability for elongation factor Tu is very high and constitutes 10% of the total cellular protein. It has been said to be the single most abundant protein in the log phase cell. Similarly, the average codon probability of triplets in genes for a number of the ribosomal proteins all exceed 0.027(1/36). This high value corresponds with the observation that the rates of synthesis of most ribosomal proteins in exponentially

Table IV
E. coli Sequences of Highest Average Codon Probability

	gene product	average codon probability	length, bp	approximate number in E. coli cell
*	ribosomal protein S1	0.030	1671	150,000
*	tufA, elongation factor Tu	0.030	1185	60,000
*	β' subunit, RNA polymerase	0.029	4224	7000
*	tufB, elongation factor Tu	0.029	1185	20,000
	recA, recombination-repair	0.029	1062	—
	α subunit, ATP synthetase	0.028	1542	—
	β subunit, ATP synthetase	0.028	1383	—
*	β subunit, RNA polymerase	0.028	3423	7000
*	ribosomal protein, L11/L1	0.028	1134	20,000
*	ribosomal protein, L7/L12	0.028	366	40,000
	outer membrane lipoprotein	0.027	237	720,000
*	ribosomal protein, L10	0.027	498	20,000
	outer membrane protein II	0.027	1041	10,000
	SSb, helix destabilizing protein	0.027	537	500
*	tryptophanyl-tRNA synthetase	0.026	1005	2000

*denotes involved in transcription/translation.

growing cells are all very high and coordinately regulated. Furthermore, it has been observed that the synthesis of the subunits of RNA polymerase are coupled in an undefined way to the synthesis of ribosomal proteins at both transcriptional and translational levels (17). We note in this table the averge codon probability in genes for the β' and β subunits of this enzyme, required for the synthesis of large amounts of RNA for translation, are also very high. Values for the other subunits of RNA polymerase are also high. The occurrence of large codon biases throughout these structural genes for the enzymatic machinery of transcription and translation then provides the basis for a simple, autogenous mechanism for their coordinate synthesis, and which operates as follows: The level of RNA polymerase in the cell affects the level of specific tRNAs according to the gene frequencies of the latter. Since the levels of specific tRNAs affect the levels of translation, the expression of those genes with large codon biases will be disproportionately amplified.

Further correspondence between codon bias levels and translation rate or abundance of gene product can be found throughout this table. The outer membrane lipoprotein gene has an average codon probability of 0.027(1/36), and correlates very well to an extraordinary translational level of 720,000 molecules/cell (18). The very high level of codon bias in the RNA polymerase subunit genes contrasts with a somewhat lower level in the DNA polymerase I gene while the gene product of the latter is present in considerably lower abundance (19). Similarly, a very low level of codon bias in the tryptophan repressor gene correlates with the very low abundance of gene product found in the cell (20).

Table V lists the top eleven sequences of highest average codon probability in yeast, which is 10% of the present yeast database. Interestingly, eight of these sequences code for enzymes involved directly in glycolysis and fermentation, which is what this organism has been selected for over the past several millennia.

Table V
Yeast Sequences of Highest Average Codon Probability

	gene product	average codon probability	
*	enolase-I	0.037	(1/27)
*	enolase-II	0.037	
*	3-phosphoglycerate kinase	0.037	
	histone H2A-I	0.036	(1/28)
*	glyceraldehyde-3-P dehydrogenase-I	0.035	(1/29)
	ribosomal protein L3	0.035	
*	alcohol dehydrogenase-I	0.035	
*	pyruvate kinase	0.034	
	histone H2A-II	0.034	
*	glyceraldehyde-3-P dehydrogenase-III	0.034	
*	glyceraldehyde-3-P dehydrogenase-II	0.034	

*denotes involved in glycolysis/fermentation.

These observations indicate that biased patterns exist to fill the need for different levels of gene product. A codon bias offers a primitive, mass-action form of control over the levels of gene expression, therefore, we imagine it would have originated shortly after the code itself. Indeed, we note that the bias reflects the widest differences among the most divergent species.

Selective pressure would favor the bias so long as existing rates of codon-anticodon interaction during translation were unable to meet the demand for more and different gene product. Of course, the bias would be subject to the counter-balancing pressure for more random codon usage among the synonymous codons, nevertheless, we expect the level of bias for a particular amino acid to exhibit some correlation with the frequency that that amino acid occurs in protein. Figure 8 shows that such a correlation exists. In this figure the maximum frequency a synonymous codon is used is plotted against the frequency that the same amino acid is found in proteins.

While there is some scatter the data clearly supports the existence of a correlation. Since this plot represents an average over all sequences from *E. coli* a scatter would be expected. A scatter will arise in any case if there are unequal numbers of tRNA genes for each amino acid. The dashed line at the bottom of the figure with the low slope represents the position that would be expected if there were no bias while the upper dashed line corresponds to that for maximum bias, in which only 20 codons are used. The same plot for the yeast codon pattern (Figure 9) shows a steeper slope, and perhaps a slightly better correlation.

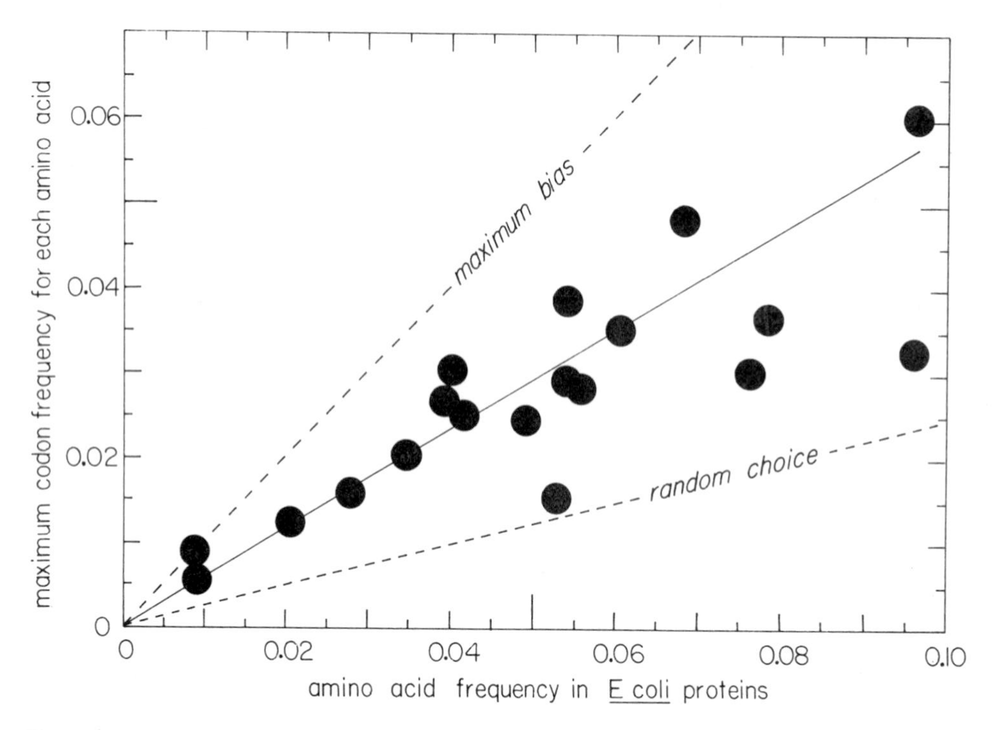

Figure 8. Variation in the maximum codon frequency for each amino acid in *E. coli* reading frame sequences with the frequency that the same amino acid occurs in *E. coli* proteins. The dashed line labeled 'random choice' represents the variation in the reciprocal of the number of synonymous codons for each amino acid with amino acid frequency. The dashed line labeled 'maximum bias' represents the line of correspondence for the extreme case when the relative frequency of occurrence of just one synonymous codon for each amino acid is 1.0.

A few incidental consequences of the codon bias deserve brief comment. First, it is clear that what we have previously labeled as neutral or silent mutations (21) may, in fact, produce significant physiological or developmental effects. In 1975 David Smith observed that a series of mutant messengers for globins had greatly reduced

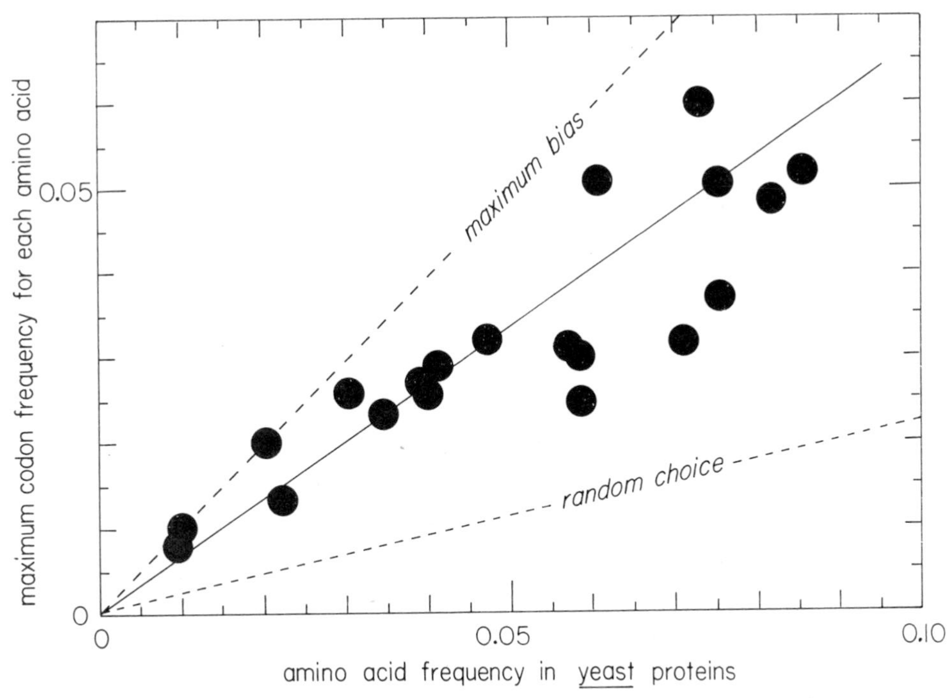

Figure 9. Variations in the maximum codon frequency for each amino acid in yeast sequences with the frequency that that amino acid occurs in yeast proteins. The variations represented by the dashed lines were determined as described in the legend to Figure 8.

translation rates (22). He pinned the cause of this kinetic effect to the unavailability of tRNAs for the mutant codons. For the same reason we may expect that the expression of recombinant sequences in either *E. coli* or yeast will generally be less than optimal, particularly if the sequences are derived from higher eukaryotic cells.

Acknowledgement

This work was supported in part by BRSG S07RR07161 awarded by the Biomedical Research Support Grant Program, Division of Research Resources, NIH.

References and Footnotes

1. Grantham, R., Gautier, C., Gouy, M., Mercier, R. and Pave, A., *Nuc. Acids Res. 8,* r49-r62 (1980).
2. Grantham, R., Gautier, C. and Gouy, M., *Nuc. Acids Res. 8,* 1893-1912 (1980).
3. Grantham, R., Gautier, C., Gouy, M., Jacobzone, M. and Mercier, R., *Nuc. Acids Res. 9,* r43-r74 (1981).
4. Bennetzen, J.L. and Hall, B.D., *J. Biol. Chem. 257,* 3026-3031 (1982).
5. Day, G.R. and Blake, R.D., *Nuc. Acids Res. 10,* 8323-8339 (1982).
6. Hinds, P.W. and Blake, R.D., *J. Biomol. Struct. Dyn. 2,* 101-118 (1984).
7. Blake, R.D. and Hinds, P.W., *J. Biomol. Struct. Dyn. 2,* 593-606 (1984).

References and Footnotes

1. Grantham, R., Gautier, C., Gouy, M., Mercier, R. and Pave, A., *Nuc. Acids Res. 8,* r49-r62 (1980).
2. Grantham, R., Gautier, C. and Gouy, M., *Nuc. Acids Res. 8,* 1893-1912 (1980).
3. Grantham, R., Gautier, C., Gouy, M., Jacobzone, M. and Mercier, R., *Nuc. Acids Res. 9,* r43-r74 (1981).
4. Bennetzen, J.L. and Hall, B.D., *J. Biol. Chem. 257,* 3026-3031 (1982).
5. Day, G.R. and Blake, R.D., *Nuc. Acids Res. 10,* 8323-8339 (1982).
6. Hinds, P.W. and Blake, R.D., *J. Biomol. Struct. Dyn. 2,* 101-118 (1984).
7. Blake, R.D. and Hinds, P.W., *J. Biomol. Struct. Dyn. 2,* 593-606 (1984).
8. Josse, J., Kaiser, A.D. and Kornberg, A.,*J. Biol. Chem. 236,* 864-875 (1961).
9. Russell, G.J., McGroch, D.J., Elton, R.A. and Subak-Sharpe, J.H. *J. Molec. Evolution,* 277-292 (1973).
10. Setlow, P. in *"Handbook in Biochemistry and Molecular Biology",* 3rd Ed. (Fasman, G.D., ed) Vol. II, CRC Press, Cleveland, OH (1976).
11. Yen, S.W. and Blake, R.D., *Biopolymers 19,* 681-700 (1980).
12. Blake, R.D. and Hydorn, T.G. *J. Biochem. Biophys. METHODS 11,* 307-316 (1985).
13. Woese, C.R. and Bleyman, M.A., *J. Molec. Evolution 1,* 223-229 (1972).
14. Ikemura, T. in: *"Genetics and Evolution of RNA Polymerase, tRNA and Ribosomes"* (S. Osawa, H. Ozarki, H. Uchida and T. Yura, eds.) Univ. of Tokyo Press, Tokyo, 519-573 (1980).
15. Ikemura, T., *J. Mol. Biol 146,* 1-21 (1981).
16. Ikemura, T., *J. Mol. Biol. 151,* 389-409 (1981).
17. Zubay, G., in *Cell Biology,* (L. Goldstein, D.M. Prescott, eds.) Vol. 3, Academic Press, N.Y. pp. 153-214 (1980).
18. Nakamura, K. and Inouye, M., *Cell 18,* 1109-1117 (1979).
19. Joyce, C.M., Kelley, W.S. and Grindley, N.D.F., *J. Biol. Chem. 257,* 1958-1964 (1982).
20. Singleton, C.K., Roeder, W.D., Bogosian, G., Somerville, R.L. and Weith, H.L., *Nuc. Acids Res. 8,* 1551-1560 (1980).
21. Jukes, T.H., *Naturwissenschaften 67,* 534-539 (1980).
22. Smith, D.W.E., *Science 190,* 529-535 (1975).

Biomolecular Stereodynamics IV, Proceedings of the Fourth Conversation in the Discipline Biomolecular Stereodynamics, State University of New York, Albany, NY, June 04-09, 1985, Eds., Ramaswamy H. Sarma & Mukti H. Sarma, ISBN 0-940030-18-7, Adenine Press,

On the Conformational Properties of 5S RNA

Neocles B. Leontis, Partho Ghosh and Peter B. Moore
Department of Chemistry
Yale University
New Haven, CT 06511

Abstract

Recent NMR data on the sensitivity of 5S structure to alterations in ionic conditions and temperature will be presented. These results will be discussed from the standpoint of trying to understand why it is that only about half the secondary structure believed to exist in 5S RNA has so far been identified by NMR spectroscopic methods. This situation contrasts strikingly from that which prevails in tRNA where all the secondary structure elements postulated on the basis of sequence comparisons have been confirmed by NMR (e.g. Heerschap, A., Haasnoot, C.A.G. & Hilbers, C.W., *Nucleic Acids Res. 11,* 4501, 1983).

Introduction

Of the abundant nucleic acids the least understood are the ribosomal RNAs. We are beginning to have precise models for their secondary structures, but have little information about their tertiary structures (for a review see 1). Of the details of their chemical and physical properties we are largely ignorant. This is unfortunate because there is a growing sense among those who study ribosome structure and function that the catalytic activity of the ribosome in protein synthesis depends crucially upon the participation of its RNAs.

Because of its small size, 5S ribosomal RNA has long attracted the attention of those seeking a simple, tractable model for rRNAs in general. The view has often been taken that 5S RNA is qualitatively similar to its larger cousins and that its interactions with ribosomal proteins are typical. In addition, the 5S RNA nucleoprotein complex occupies a conspicuous position on the 50S ribosomal subunit close to the location of the peptidyl transferase site (for review see 2) and ribosomes lacking 5S RNA are unable to support protein synthesis (3). Thus as well as being useful as a model, 5S RNA and its protein complexes are of interest for their own sake.

5S RNA was the first RNA other than tRNA to be sequenced (4); hundreds of 5S sequences are known today (5). These sequences combined with the results of enzymatic and chemical probings of the structure have led to a secondary structure model for 5S RNA in its native conformation, now generally accepted (for a review

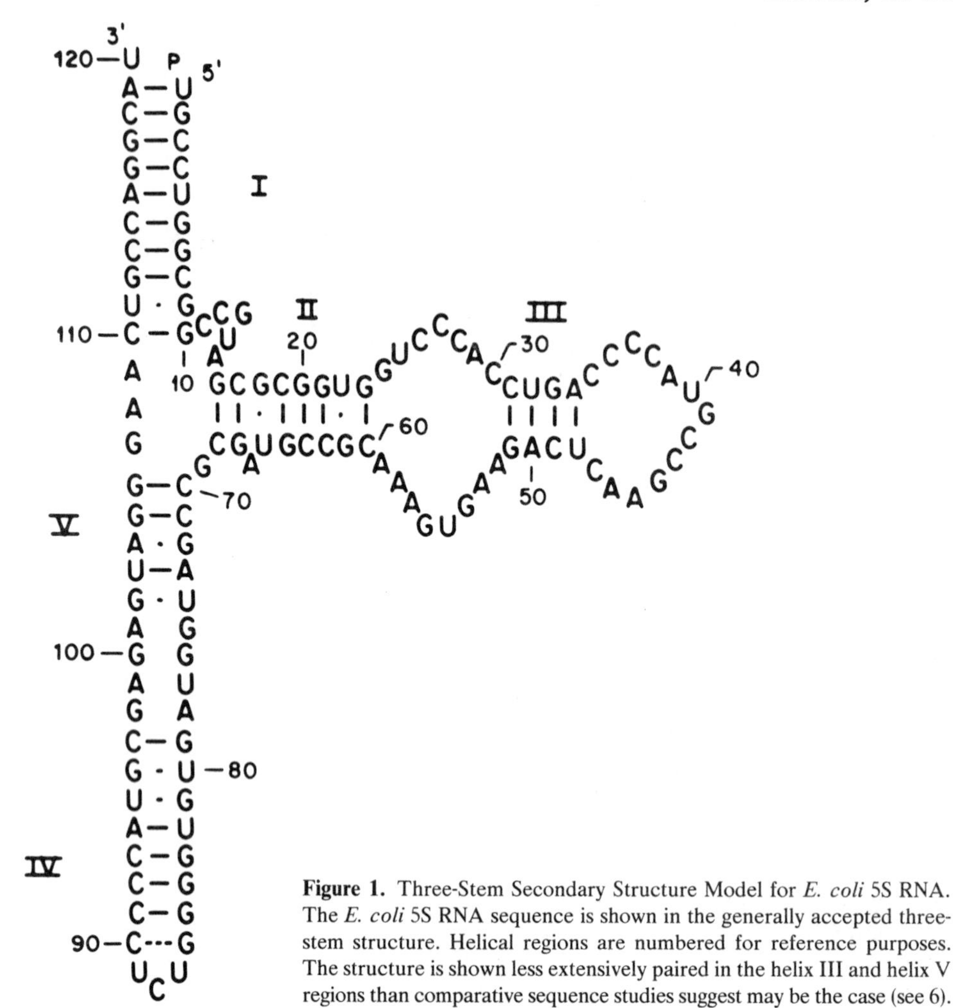

Figure 1. Three-Stem Secondary Structure Model for *E. coli* 5S RNA. The *E. coli* 5S RNA sequence is shown in the generally accepted three-stem structure. Helical regions are numbered for reference purposes. The structure is shown less extensively paired in the helix III and helix V regions than comparative sequence studies suggest may be the case (see 6).

see 6). The canonical model is shown in Figure 1 in its *E. coli* version. The details of the hydrogen bonding in some parts of the structure remain uncertain, and it is still unclear what the tertiary structure of the molecule is.

A few years ago we set out to examine the 5S from *E. coli* and its complexes with ribosomal protein by NMR, believing that techniques had evolved to the point that NMR could contribute substantially to our understanding of this system. The purpose of this paper is to examine the NMR behavior of 5S RNA, contrasting it with that of the tRNAs, the one class of RNAs whose structures and properties are reasonably well understood.

Materials and Methods

RNAs. Unless otherwise noted, the 5S RNA used in these studies is the product of the 5S overproducing strain HB101/pKK5-1 (7). The methods used for growth of this strain and purification of 5S RNA have been described elsewhere (8). 5S fragment is prepared from intact 5S RNA by digestion with RNase A. The RNA is dissolved in 0.1 M KCl, 5 mM $MgCl_2$, 50 mM tris-borate, pH 7.8, at 20 OD_{260nm}/ml. RNase A is added at a level of 10 μg/ml and digestion carried out at 0°C for 45 min. The reaction is stopped by addition of SDS and the product recovered by phenol extraction followed by purification on Sephadex G-75 columns (8).

NMR. The spectra shown below were obtained on the spectrometers of the Northeast Regional NMR facility, using either a Bruker WM500 spectrometer or the 490 MHz spectrometer constructed by the staff of the facility. Where water suppression was necessary a (45°-t-45°) pulse sequence was used (9). Unless otherwise stated, samples were dialyzed into 0.1 M KCl, 4 mM $MgCl_2$, 5 mM cacodylic acid, pH 7.2, 5% D_2O, and brought to concentrations around 1 mM by ultrafiltration. A small amount of dioxane was included in each sample as a chemical shift reference. Dioxane is assumed to have a chemical shift of 3.741 ppm relative to the methyl resonance of 3-(trimethylsilyl)-1-propanesulfonic acid, independent of temperature.

Results

5S RNA

Shortly after our NMR work began it became clear that in order to reach our final objective, an understanding of 5S RNA/protein complexes, we would have to understand 5S RNA first. Our efforts in this direction have concentrated on the ^1H spectrum of 5S RNA in the region downfield of H_2O between 9 and 15 ppm.

The downfield resonances in the ^1H spectrum of a nucleic acid arise from imino protons belonging to the nucleotide bases (10,11). In free nucleotides these protons exchange readily with H_2O protons in a base catalyzed process. Under the buffer conditions we use (5 mM cacodylate, pH 7.2), the exchange rate for GN(1) or UN(3) imino protons would be expected to be of the order of 5×10^4 sec^{-1} (12,13). In general, because of exchange, imino resonances are detectable in ^1H spectra only in nucleic acid samples dissolved in H_2O. Furthermore, even in H_2O, when their exchange rates exceed a few hundred per second, their line widths become so large as to make them essentially undetectable. Thus the only imino protons observed downfield are those protected from solvent exchange, usually by participation in hydrogen bonding. To a first approximation the downfield spectrum of a nucleic acid reflects its hydrogen bonded structure.

It is well known that as molecules increase in size two factors conspire to make their spectra difficult to interpret. First, the number of atoms contributing resonances increases, leading to spectral complexity, and second, line widths increase, compounding the difficulty of resolving resonances. It is an unfortunate fact that while the downfield spectra of tRNAs are often quite well resolved (eg. 14) those of the larger 5S RNAs seldom are. Figure 2 shows spectra for 5S RNA at 500 MHz in D_2O and in H_2O. (For another example of a 5S downfield spectrum, see 15). Clearly,

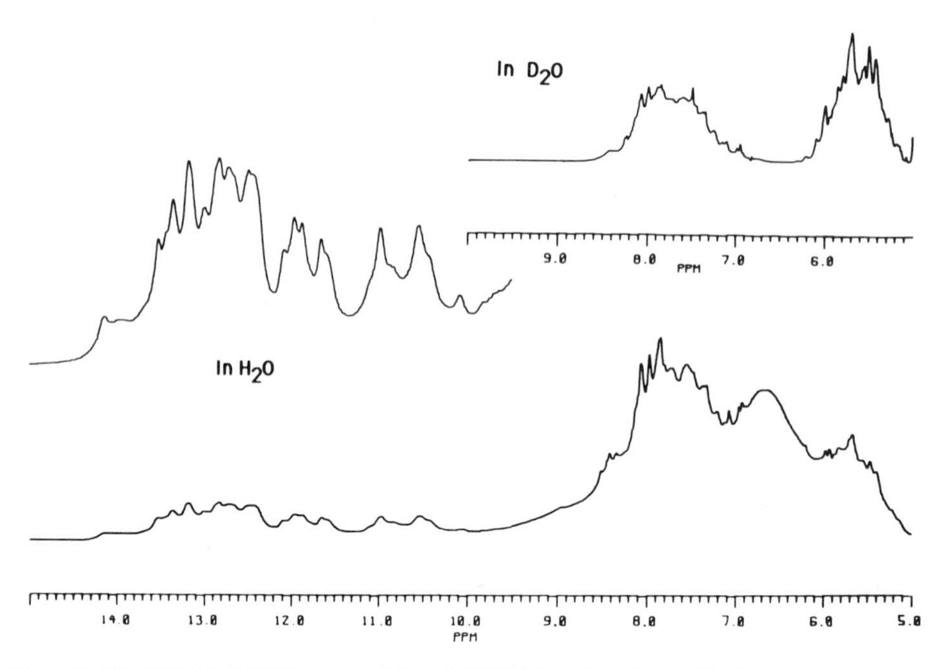

Figure 2. The 500 MHz NMR Spectrum of *E. coli* 5S RNA from 5 to 15 ppm. The lower spectrum is that of 5S RNA from 5 to 15 ppm collected under standard conditions (see Materials and Methods) in 95% H_2O-5% D_2O using a (45°-t-45°) observe pulse sequence to suppress the H_2O resonance. The spectrum in the upper right is of the same material collected in 100% D_2O.

over much of the spectrum one has an envelope to examine, not a set of resolved resonances. This fact makes the study of this molecule by NMR a major challenge.

5S fragment

In all likelihood, this project would have foundered were it not for the fact that light digestion of 5S RNA with RNase A leads to the production in high yield of a resistant fragment (16). Fragment encompasses bases (1-11, 69-120) in the parent molecule and the majority of most preparations are broken at bases 87, 88, 89 (see Figure 1). Fragment has a tRNA-like spectrum as one would anticipate from its tRNA-like molecular weight, 20,000 (see Figure 3). Further, its downfield spectrum is a subset of the downfield spectrum of 5S RNA, suggesting that the cleavages which produce it leave its internal structure intact (9).

Assignment of the Fragment Spectrum

In the late 1970s, Redfield and his colleagues demonstrated that the nuclear Overhauser effect (NOE) could be used to assign resonances in the downfield spectra of nucleic acids (17,18). This discovery provided spectroscopists interested in nucleic acids with a general method for making downfield assignments for the first time, and techniques which work for tRNA are usually helpful for 5S fragment.

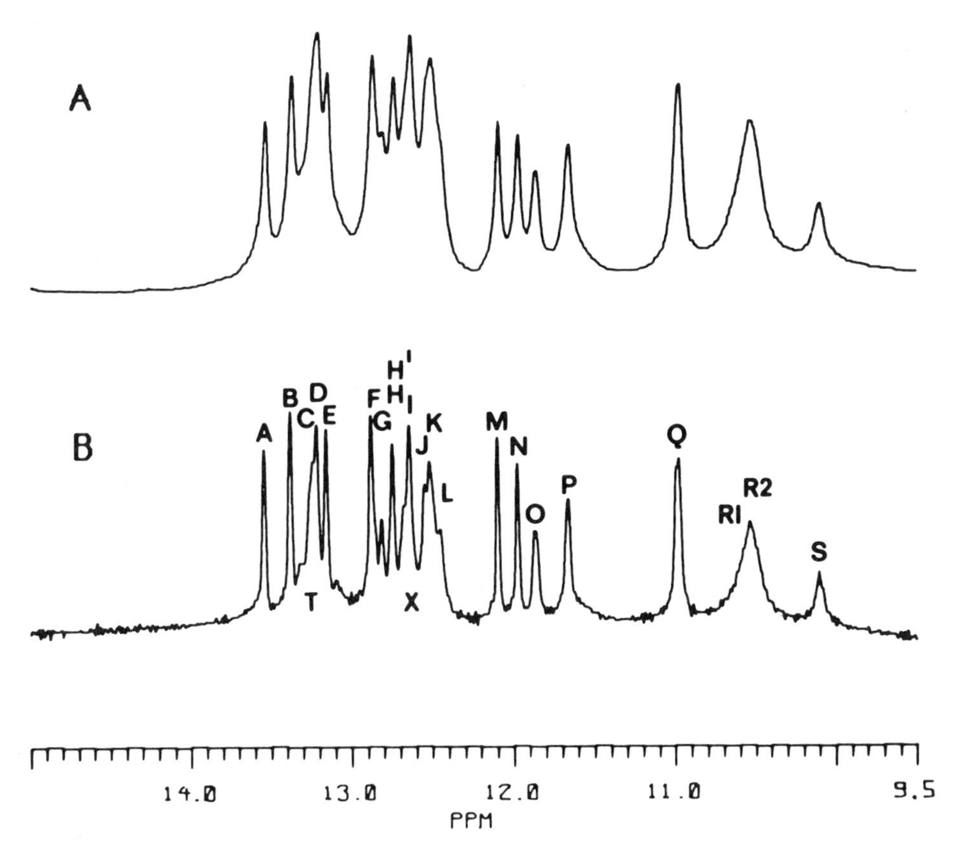

Figure 3. The 500 MHz Proton Spectrum of *E. coli* 5S Fragment. Spectrum A is the downfield spectrum of a 1 mM sample of fragment in 95% H_2O-5% D_2O plotted without resolution enhancement. Spectrum B is the same spectrum as A resolution enhanced with the convention for naming resonances given.

Figure 3 shows the downfield spectrum of fragment (top) and a resolution-enhanced version of the spectrum (bottom) giving the convention for naming resonances.

NOE experiments done in this region of the spectrum rapidly identified two helical segments in the molecule, one which clearly arises from helix I (Figure 1), resonances J, C, F, B, M, E, H, and S, and a second which comes from helix IV, resonances I, N, A, P, Q, O, and D, (see Figure 4) (8). Unassigned at this point were resonances G, K, L, H′, R1, R2, T and X, most of which must come from bases between helices I and IV, the helix V region.

The alignment of the J, C, F, ... series of resonances in helix I was also indeterminant at this stage because that stem is palindromic around its central AU (resonance B). To resolve this ambiguity the distances from the imino proton of resonance B (the UN(3) of U5) to its neighbors on either side, M and F, were estimated from the kinetics of the development of the NOE's between them. The F proton is closer to the B proton than the M proton. This observation is consistent with right-handed

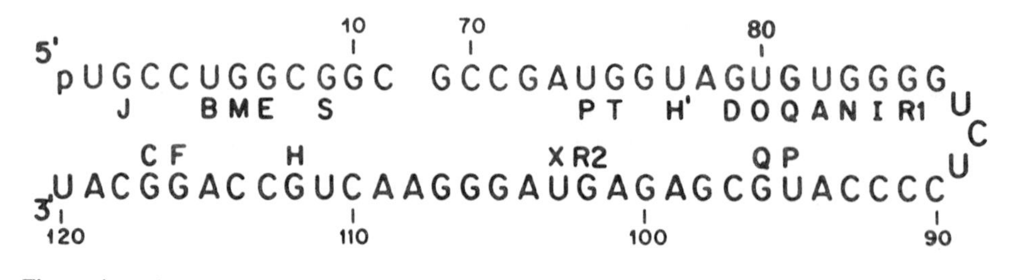

Figure 4. Assignments of Resonances in 5S Fragment. The sequences of fragment is given in the canonical secondary structure. The assignment of resonances to base imino protons are shown (see text for details).

helical geometry only if the resonances are assigned as shown in Figure 4. Ring current shift calculations also strongly favor this alignment (19).

In the past few years it has been demonstrated that examination of the effects of ^{15}N labelling of a nucleic acid on its ^{1}H downfield spectrum can assist in making assignments (20,21,22,23). ^{15}N is a spin ½ nucleus. An imino proton's ^{1}H resonance will split into a 80-90 Hz doublet if the N to which it is bonded is ^{15}N. Irradiation at ^{15}N frequencies causes these doublets to collapse to singlets. ^{15}N chemical shifts are dominated by covalent chemistry (24). For example, all UN(3) imino doublets decouple over a narrow range of ^{15}N frequencies which does not overlap with the range that decouples GN(1)H's. Uniformly ^{15}N labelled 5S and 5S fragment were studied by one or two dimensional methods to obtain $^{1}H/^{15}N$ chemical shift correlations (25,26; Jarema, M., Redfield, A.G. & Moore, P.B., unpublished results). The chemical nature of all the imino protons in the fragment downfield spectrum was thus established. The $^{15}N/^{1}H$ experiments revealed the existence of two previously unrecognized UN(3) resonances, H′ and X, both with chemical shifts near 12.7 ppm (25). Data to be presented elsewhere suggests that H′ should be assigned to U77, X assigns to U103, and that a thermally sensitive GU (R2, P) is (G102, U74) (19; Jarema, M., Redfield, A.G. & Moore, P.B., unpublished results). At this stage, only resonances G, K. L remain unassigned. All three are GN(1) imino protons. By elimination, all three must belong to helix V.

Assignments in 5S RNA. Almost all the NOEs demonstrable in fragment have also been seen in intact 5S RNA (27) showing that resonances that correspond in chemical shift in the two molecules represent corresponding protons in the two sequences. It is a striking fact that almost no NOEs can be detected in intact 5S beyond those found in fragment, even though many other imino resonances are visible in its spectrum (27). Only one GU base pair could be detected.

Utilization of Assignments. While assignments are useful in and of themselves because they test models of the secondary structure, the real purpose in making them is to obtain markers within the molecule which can be used to follow what happens when the molecule is perturbed. Some of the applications made of the assignments we have are described below, the ones most relevant to the issue of the conformational dynamics of 5S RNA.

It has been known for many years that several conformations of 5S RNA exist in addition to the native, or A conformation. The best known is the B form, which has an altered electrophoretic mobility (28) and equilibrates remarkably slowly with A form under conditions where the two coexist (29). More recently Kao and Crothers discovered a fast, low temperature melting transition in 5S RNA (30). The conformation of 5S on both sides of this transition is A form by electrophoretic criteria. Thus there are no fewer than 3 conformations to be considered for 5S RNA, and an obvious question to ask is what the structural relationships are between them.

Ionic Effects

One of the first observations we made on 5S RNA by NMR was that its spectrum is sensitive to Mg^{+2} ion concentration (31,9). This sensitivity is not confined to the fact that the absence of Mg^{+2} predisposes the structure to transform itself slowly

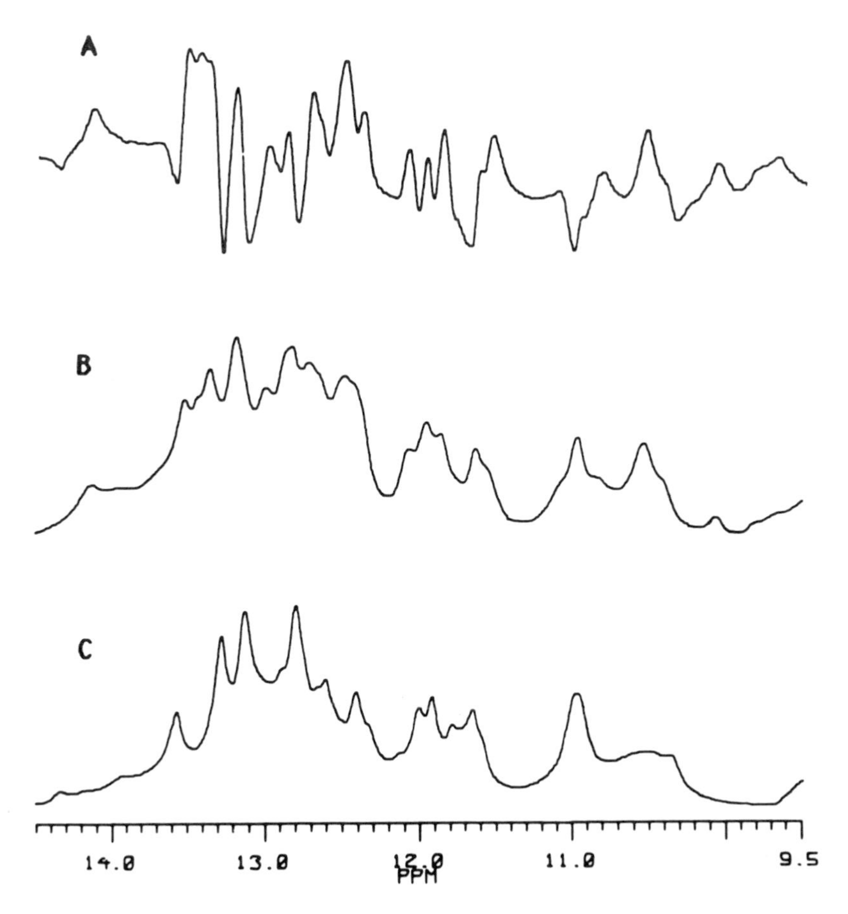

Figure 5. The Effect of Mg^{+2} on the Downfield Spectrum of 5S RNA at 303 K. Spectrum A: Spectrum B-Spectrum C. Spectrum B: ~1 mM sample in 2 mM $MgCl_2$, 0.10 M KCl, 5 mM cacodylate pH 7.2, 0.10 mM EDTA. Spectrum C: ~1 mM sample in identical buffer except $MgCl_2$ is replaced by 2 mM EDTA.

into the B form, which it does (28). There are also changes induced in the spectrum of 5S RNA "instantaneously" and reversibly by withdrawal or addition of Mg^{+2} ion. Part of the interest in these fast changes lies in the fact that the Kao and Crothers transition is also magnesium-sensitive (30). Figure 5 (b and c) shows 5S RNA in both its high and low Mg^{+2} forms.

The Kao-Crothers transition has a number of characteristics which make it unlike most nucleic acid transitions (30). First, high magnesium ion concentration or high monovalent cation concentration favor the high temperature form of the molecule. Second, high pH favors the high temperature form. Third, the transition is more sensitive to monovalent catons than to divalent cations.

A Mg^{+2} titration study was done on 5S RNA under standard buffer conditions to ascertain the approximate midpoint of the NMR transition. At pH 7.2 in 5 mM cacodylate, 0.1 M KCl at 30°C, the high Mg^{+2} spectrum is seen at concentrations of Mg^{+2} ion above 1 mM. The downfield spectrum of 5S in .03 mM Mg^{+2} is identical to that in EDTA. The changes in chemical shifts and disappearances of resonances seen as the Mg^{+2} ion concentration drops do not all occur in parallel (Ghosh & Leontis, unpublished data). Thus there is no reason to believe the NMR-detected changes are cooperative.

Starting with the molecule at the midpoint, pH, temperture and monovalent cation concentration were varied to see whether the molecule would respond spectrally in the way observed by Kao and Crothers for their optical transition. Figure 6 shows a typical result. When the monovalent cation concentration is pushed up five fold, the spectrum changes very little. If anything, it becomes more like the low Mg^{+2} spectrum (e.g. note the loss of intensity on the part of resonance R). Were the NMR spectrum responding like the Kao-Crothers transition, this change in ionic condition should have been equivalent to an increase in temperature of 35°C, and the H form appearance should have been restored. Negative results were obtained using other perturbants as well. The changes one sees in the spectrum of 5S RNA when it is deprived of Mg^{+2} ion for the most part are not reflections of the Kao-Crothers phenomenon.

In an effort to specify the structural changes taking place upon the removal of Mg^{+2}, both 5S and 5S fragment were examined in the presence and absence of Mg^{+2} (Figures 5, 7). The difference spectra shown (Fig. 5a and Fig. 7a) are quite similar. The negative and positive features in the difference spectrum of fragment can all be found at identical chemical shifts in the whole molecule difference spectrum. Thus even in low Mg^{+2} fragment remains a useful model for the behavior of the parent molecule. Furthermore, since the fragment differences account for a large fraction of the whole molecule differences much of what happens to the structure of 5S RNA when Mg^{+2} is removed must take place within the fragment region. In this part of the molecule resonances A, B, M, N, O and P undergo obvious changes in chemical shift. Resonances O and P broaden and there are substantial changes in the H, I, J, K, L region of the spectrum. S and R are largely lost, and G merges with F. NOE experiments are still underway to establish which connectivities remain in

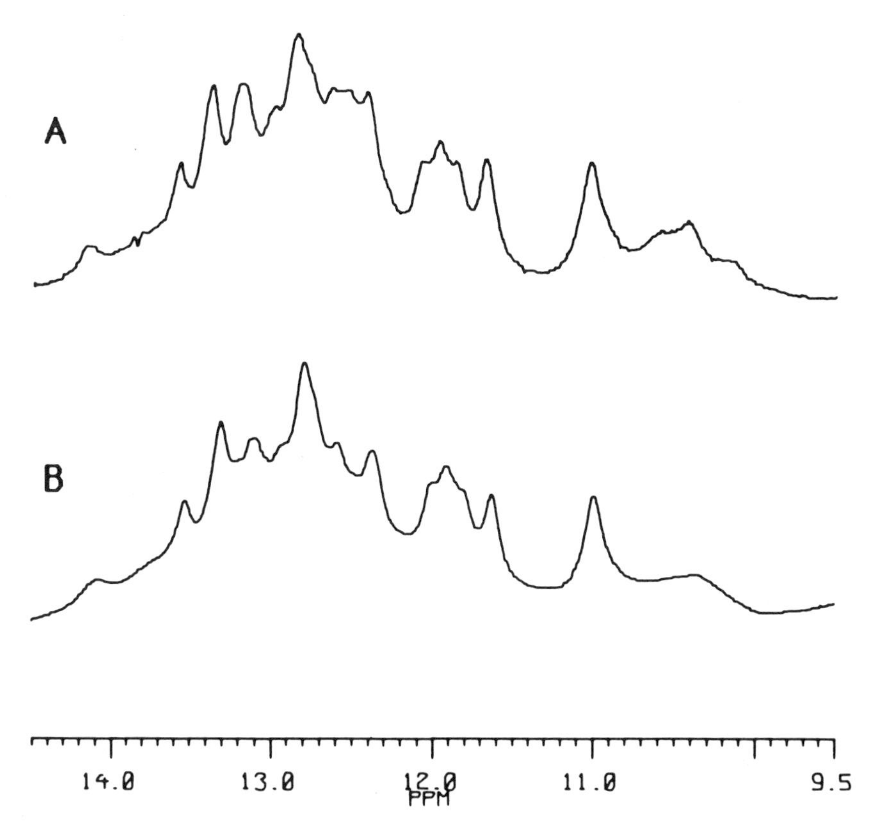

Figure 6. The effect of Monovalent Cation Concentration on the Downfield Spectrum of 5S RNA at 303 K. A: ~1 mM sample in 5 mM HEPES at pH 7.86, 0.3 mM $MgCl_2$, 0.10 M KCl. B: Same as A except 0.50 M KCl.

fragment in low Mg^{+2}. It is clear that helices I and IV are still basically intact. The resonances which have been "lost" come from the interior end of helix I, the GU's at the interior end of helix IV and from helix V.

Two additional observations should be made about the low Mg^{+2} state. First, this state has been difficult to characterize using fragment preparations broken at the 87, 88, 89 loop. The strand from base 69 through base 87 dissociates after several hours at 30°C in low Mg^{+2} giving rise to new resonances in the spectrum as well as the weakening of old ones (unpublished data). Full characterization of the low Mg^{+2} state depends on the use of preparations containing species whose loops are intact. Preparations of this kind have only recently become available. Second, L25 binds to 5S and to fragment in the absence of Mg^{+2}. This result implies that no serious breakdown of helix IV structure occurs when Mg^{+2} is removed. Helix IV is the part of 5S RNA most strongly affected when L25 binds (27). It is interesting to note that a recent chemical modification study done on the 5S RNA from *B. stearothermophilus* identified helix V as the region most destabilized by Mg^{+2} deprivation (32), consistent with the NMR data.

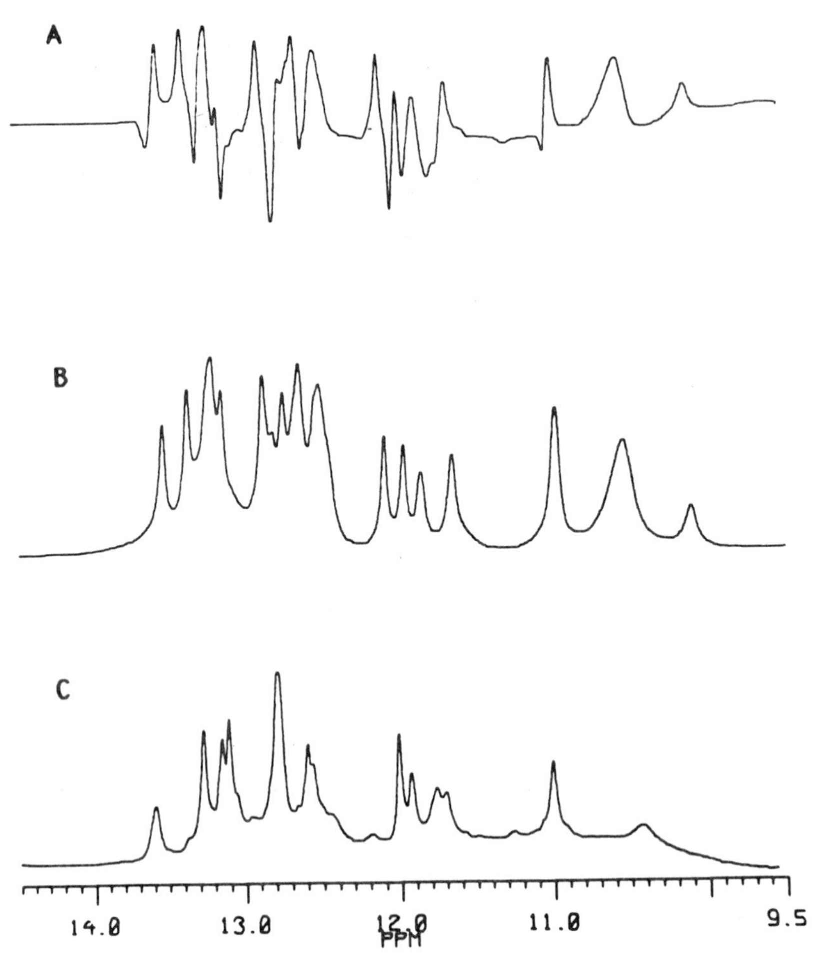

Figure 7. The Effect of Mg^{+2} on the Downfield Spectrum of 5S Fragment at 303 K. Spectrum A: Spectrum B-Spectrum C. Spectrum B: ~1.5 mM sample in same buffer as 5B. Spectrum C: ~1.5 mM sample in same buffer as 5C.

Some conclusions about 5S structure can be drawn from these data which will be of use when we compare 5S RNA with tRNA. Helix V clearly has a stability which is marginal at room temperature. Its central GU is only fully apparent by NMR criteria at 10°C in the presence of Mg^{+2}. The removal of Mg^{+2} ion at room temperature destroys all its contributions to the downfield spectrum. (The poor stability of this part of the molecule may well play a role in the A to B transition which on current theory requires the IV-V helix to open so that base pairing can be established involving the loop at base 40 with bases around 80 (33). Despite its poor stability, helix V gives some NOEs. Indeed, the assignments we have in this part of the molecule are based on that fact. A structure does not have to have manifestly high thermal stability in order that it give NOEs.

Temperature

The classic method of perturbing nucleic acids is heating. We are now in the midst of an extensive study of the response of 5S RNA to heat. While the data are not yet in hand necessary to answer all the questions which interest us, enough are available to permit us to draw some useful conclusions.

Figure 8 shows what happens to the downfield spectrum of 5S RNA as temperature rises in the presence of Mg^{+2}. Figure 9 is similar to Figure 8 except that the sample used has no Mg^{+2} in it. What is interesting about these two experiments is their similarity. If one displaces the two melting series making temperature x in the absence of Mg^{+2} correspond to temperature (x + 15°C) in the presence of Mg^{+2} good correspondence is seen. Clearly the primary effect of Mg^{+2} on 5S melting behavior is that it changes T_m. This finding suggests that it is inappropriate to regard the Mg^{+2} induced changes in the downfield spectrum of 5S RNA as indicative of a new conformational state. It is more likely that what is being observed is the same fundamental conformation at two different effective temperatures.

Calorimetric experiments recently initiated show that the midpoint of the 5S melt in the absence of Mg^{+2} (pH 7.2, 0.10 M KCl) occurs in roughly 70°C. The downfield spectrum of 5S is completely featureless at 70°C under these conditions showing that the NMR melt preceeds the thermodynamic melt as has long been recognized.

The similarity of the downfield melt of 5S RNA in the presence and absence of Mg^{+2} is reassuring. Heating of RNAs in the presence of Mg^{+2} is known to lead to chain cleavage, and indeed, our samples are not intact after melting runs such as those shown in Figure 9. The material heated in EDTA is intact, however. When heated samples are cooled, the original spectrum is restored. The fact that the high temperature Mg^{+2} spectra are so similar to the corresponding spectra taken in EDTA suggests that degradation is not contributing seriously to those spectra.

Figure 10 is a melting series for fragment in the presence of Mg^{+2}. Examination of this series reveals that fragment melting is not cooperative by NMR criteria. Comparison of the fragment spectra in this series with the fragment spectrum taken in the absence of Mg^{+2} at 30°C (Figure 7c) suggests again that Mg^{+2} withdrawal amounts to a displacement in T_m of about 15°C. The aspects of the Mg^{+2}-free spectrum which are not fully reproduced by heating are the chemical shift changes which may in part represent small adjustments in secondary structure due to the altered ionic melieu.

In the fragment spectrum at 70° one can see several resonances stoutly resisting melting, resonances B, C, E, F, and M. The same resonances make conspicuous contributions to the downfield spectrum of intact 5S RNA at the same temperature, but there are other, non-fragment, resonances still visible in the intact molecule's spectrum, notably those upfield of F. Clearly there are hydrogen bonded structures in helices II and III which are not only more stable than helix V, but comparable in stability to helices I and IV.

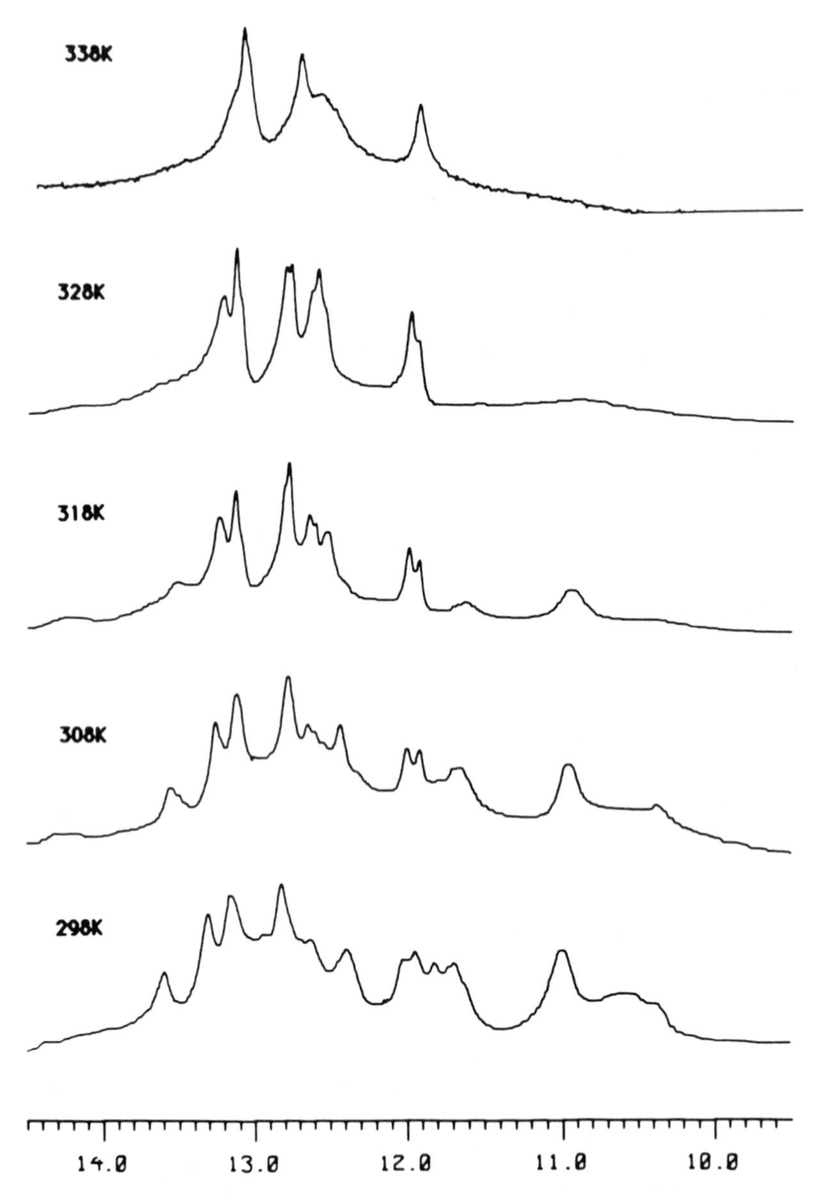

Figure 8. The Dependence of the Downfield Spectrum of 5S on Temperature in the Presence of Mg^{+2}. Spectra were taken on a 1 mM sample of 5S RNA in 5 mM cacodylate pH 7.2, 0.10 M KCl, 2 mM $MgCl_2$ at a series of temperatures. Spectra are shown at 10 K intervals starting at 303 K.

Relaxation Measurements

In an effort to examine the melting process more closely, a series of longitudinal relaxation measurements have been done on the downfield spectrum of fragment in Mg^{+2} as a function of temperature.

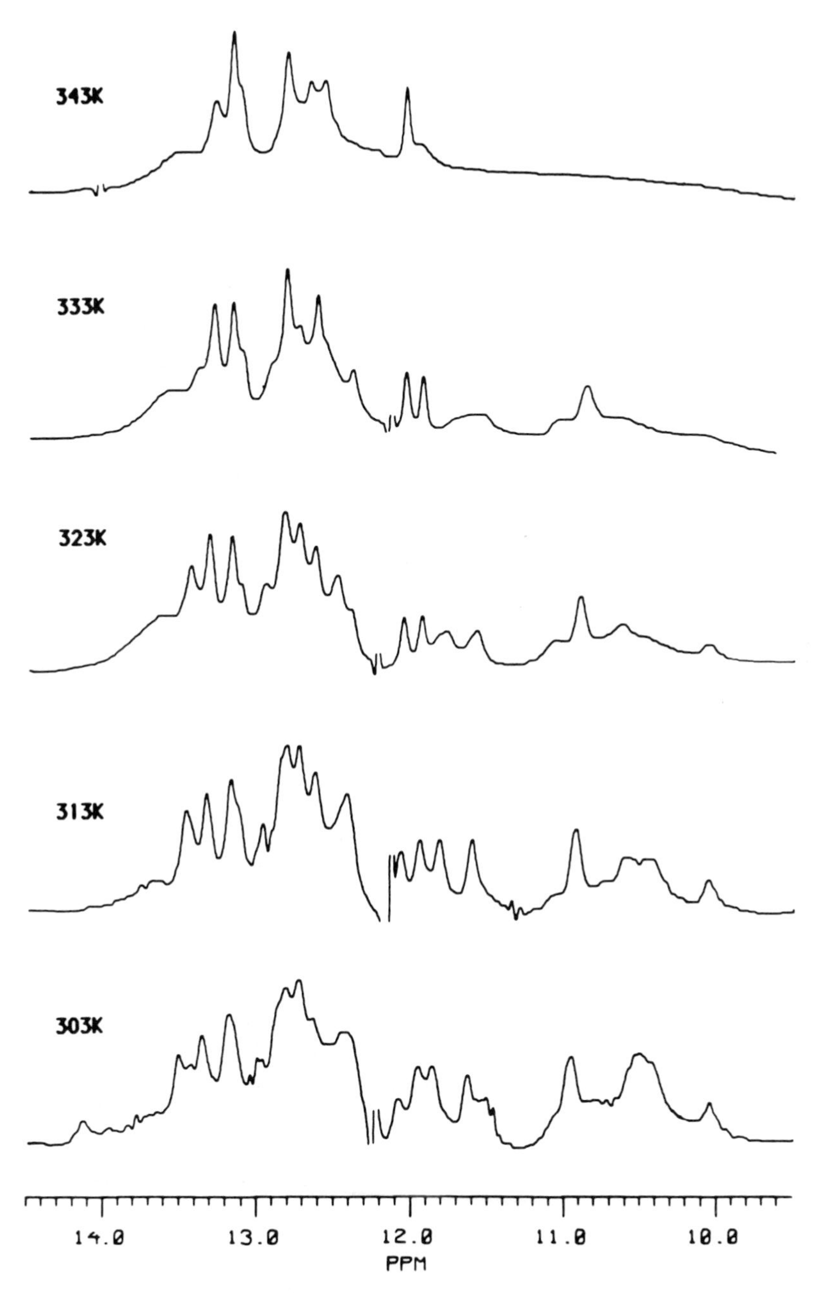

Figure 9. The Thermal Dependence of the Downfield Spectrum of 5S RNA in the absence of Mg^{+2}. This figure shows the results of an experiment similar to that depicted in Figure 8 except that the solvent contained 2 mM EDTA instead of 2 mM $MgCl_2$. Spectra are shown for every 10 K starting at 298 K.

What one wishes to obtain from T_1 measurements is evidence of the processes which relax individual imino proton resonances. To obtain this information for molecules like 5S RNA it is advisable to measure T_1's by saturation recovery

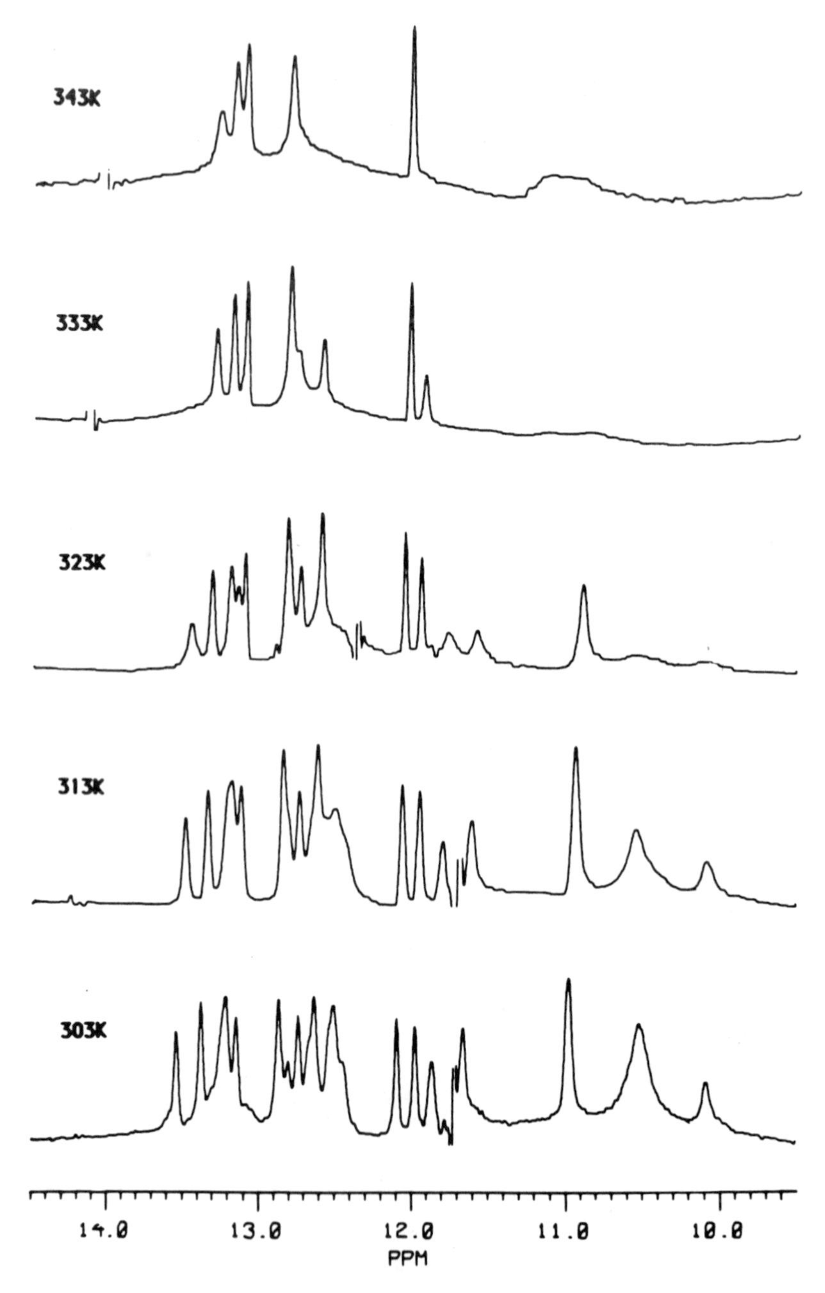

Figure 10. The Thermal Dependence of the Downfield Spectrum of 5S Fragment in the Presence of Mg^{+2}. See the legend for Figure 8.

methods because at short times the kinetics of saturation recovery are not affected by spin diffusion (34). Limitations in the instrument time available to us, however, have compelled us to make these measurements by inversion recovery methods, a more economical technique, but one which potentially suffers from spin diffusion (35). A pulse sequence (90°-t-90°) was used to invert the downfield region, leaving H_2O protons undisturbed. This pulse sequence by its nature perturbs the magnetization of ribose protons and pyrimidine C5 protons very little because they resonate close to H_2O. The aromatic protons are substantially disequilibrated by this pulse sequence, however, and the entire downfield region is inverted. In principle, the kinetics we measured in these experiments are affected to some degree by spin diffusion effects, although less so than they would have been had a totally non-selective pulse been used for inversion. In order to find out how reliable the data are, the T_1's of 4 resonances in the fragment downfield spectra were measured at 313 K by saturation recovery. Within error the results were the same as those obtained at the same temperature by inversion recovery. We feel confident, therefore, that the data shown in Table I are reasonably accurate.

Table I shows the longitudinal relaxation times for the resonances in helix IV as a function of temperature. There is a tendency for T_1's to increase with temperature at low temperatures in this molecule. The same phenomenon has been seen in other polynucleotides (e.g. 35) and is due to the decrease in correlation time as solvent viscosity decreases. As the temperature increases further, T_1's start to drop,

Table I

T_1's versus temperature in helix IV in 5S fragment. T_1's were measured on a fragment sample dissolved in 0.1 M KCl, 4 mM $MgCl_2$, 5 mM cacodylic acid, pH 7.2, 5% D_2O at several temperatures. Resonances in the imino proton region were inverted using a twin pulse sequence (9) with the offset in the imino region and times adjusted to produce a null at the H_2O frequency. At each temperature, the spectrum was observed at 10 different times following inversion, starting at 1 msec and extending on to several hundred msec, as required. Intensities were plotted semilogarithmically against time and the initial slope of the decay estimated by eye. For temperatures above 293 K, most plots were linear for at least 50 msec. At 293 K or below, in some cases, curvature could be observed after 30 msec. Based on repetitive processing of the same data, the standard error estimated for the values given is 15%. Times are quote in msec. Entries marked with a dash imply that the resonance in question is not detectable at the temperature in question. "N.D." means "not determined," usually because of spectral overlaps.

Base Resonance	G79 D	U80 O	G81 G96 Q	U95 P	U82 A	G83 N	G84 I	G85 R1
283K	86	110	91	83	63	104	110	45
293K	88	108	127	85	72	93	90	14
298K	112	94	106	83	88	106	138	11
303K	104	69	80	64	88	124	104	–
308K	166	49	74	49	93	139	135	–
313K	100	22	49	30	51	142	N.D.	–
318K	80	15	35	20	27	145	134	–
323K	65	9	20	10	19	116	125	–
328K	25	7	10	6	11	102	120	–
333K	N.D.	–	5	–	–	70	100	–
338K	–	–	–	–	–	28	18	–

finally reaching a regime where log ($1/T_1$) plotted against 1/T becomes linear. This Arrhenius behavior is ascribed to breathing of the bases which permits solvent exchange (36). The activation energies we calculate from the Arrhenius plots range from 25-40 kcal per mole.

This high activation energy breathing process affects the U imino protons of GU base pairs first. The next class of protons to go are the G iminos of GU's closely followed by the U iminos of AUs. The GC resonances are the most stable, but there is a tendency for GC's at helical termini to melt at low temperature. The T_1's obtained here and their qualitative melting behavior correspond well with what spectroscopists have observed for tRNAs and short DNA oligonucleotides (37,38,39).

The only notable exceptions to this behavior are the R resonances, represented by R1 in Table I. These resonances' T_1s start shortening immediately as temperature rises and the resonances disappear completely at quite low temperatures. A strong case can be made that the R resonances exchange with H_2O at a significant rate at all temperatures. It should be added that resonance G, which is still unassigned acts like a normal GC with respect to its T_1 behavior. Resonances K and L, also unassigned, are not quite as stable as O and P by this criterion.

Discussion

The RNAs most intensively studied by NMR are the tRNAs. NOE methods have made it possible to assign spectra reliably and NOE connectivities in the imino region are readily demonstrated corresponding to the full helical secondary structure of tRNA both in the presence and absence of Mg^{+2} (40,41,42,43,14). Operationally the contrast between tRNA and 5S RNA is striking. About half the downfield resonances in 5S RNA give no NOEs to other imino resonances. So far it has been possible to demonstrate the existence of only part of the putative secondary structure of the molecule. One of the things we would like to understand is what this failure means.

A priori one can identify four factors which influence the detection of NOEs. They are: (1) correlation time, (2) time-averaged geometry, (3) exchange processes, and (4) accidents related to placement of resonances in the spectrum. The last possibility is the most trivial of the four, and can be dealt with first. Helix II in the non-fragment part of 5S consists of 8 base pairs with a bulged A between the second and third pair (see Figure 1). The uninterrupted secondary structure goes (GU)$(GC)_3$(GU)(GC). The GU's of this helix should contribute resonances in the 11-12 ppm region. Using the ring current shift rules of Robbilard (44), for example, one estimates that the GC's resonances of this structure should be spread over 0.8 ppm from 12.7 to 13.5 ppm. Were this structure A type helix, there should be no problem following its NOEs due to spectral overlaps.

There exist data which speak to the other possible causes for our failing to get NOEs. The nuclear Overhauser effect has been used to estimate the correlation time of the uracil residues in *E. coli* 5S RNA. 5S RNA was heavily substituted with

5-fluorouracil and ^1H to ^{19}F NOEs measured. The sign of the effect indicated a correlation time in excess of 10^{-8} sec for most, if not all, of 5S uracils (45). Moreover, one can estimate the rigid body correlation time for 5S RNA from its molecular volume. It should be of the order of $2\text{-}3\times10^{-8}$ sec. Correlation times have to be around 3×10^{-10} sec in order for the NOE to fall to zero at 500 MHz. Thus at 500 MHz all ^1H-^1H NOEs should be cross-relaxation driven, and have a negative sign (46). It is unlikely NOEs are lost in 5S RNA because correlation times accidentally reduce their magnitudes to zero. It would appear that geometrical or dynamical problems are the only likely sources for NOE failure.

The dipolar coupling on which NOEs depend has an r^{-6} dependence. By the time the distance between a pair of protons exceeds 5Å, NOEs become so weak that it is not practical to measure them in tRNA or 5S RNA. In a regular nucleic acid helix, distances between imino protons in neighboring base pairs vary with sequence, but should not exceed 5Å. Obviously helices I and IV meet all criteria for NOEs and with difficulty, some structure in helix V can also be detected. Why helix V is "better" than helices II and III, however, is unclear.

In the correlation time regime being dealt with here, the magnitude of the NOE one sees between a pair of protons is governed by the ratio of the cross-relaxation rate between the two to the overall rate at which the proton manifesting the NOE is relaxing (for review see 52). An imino proton able to transfer its magnetization to other protons with high efficiency will give weak NOEs. An obvious process to suspect as a cause for such losses of magnetization is exchange with solvent.

^3H exchange results on 5S RNA were reported in 1981 by Ramstein and Erdmann (47). Their studies identified about 100 protons exchanging with half times of roughly 3 minutes and a class of protons exchanging with half times of about 0.6 hours, this group consisting of 19 to 25 depending on Mg^{+2} concentration. Their results suggest that if one were to transfer protonated 5S RNA into D_2O and quickly acquire downfield spectra that the exchange of D for H in the imino protons of 5S RNA might be followed in real time. Experiments of this kind have been done successfully on tRNAs (48,49,50). We performed such an experiment on 5S RNA and the result was completely negative. By the time the first spectrum could be recorded, about 5 minutes, no imino resonances were visible. We conclude that none of the slowly exchanging protons in 5S RNA detected by ^3H methods are imino protons, and that imino proton exchange rates in 5S RNA exceed .005 sec^{-1} at 30°C under our standard buffer conditions.

The T_1 experiments we have done give us some indications about exchange in fragment. Only resonances R1 and R2 have relaxation rates which accelerate rapidly with temperature over the whole range of temperatures examined. These resonances are also anomolously broad at room temperature supportive of the view that exchange is an important component in their relaxation. It is to be emphasized that the upfield component of R, R2, nevertheless gives NOEs at room temperature, and at reduced temperature and/or pH shows itself to be the GN(1) of the GU base pair in helix V. Clearly appreciable exchange does not necessarily prohibit NOEs. Equally obviously, a case can be made that the reason more NOEs are not seen in helix V is that the structure is only marginally stable. Exchange may be the answer for helix V. It is interesting to note that by the criterion of sensitivity to small molecule reagents and insensitivity to single strand nucleases, helix V is a helix. It is not cleaved by double strand nucleases, however.

Our knowledge of the relaxation properties of intact 5S RNA is limited at this point in part because it is hard to interpret inversion recovery data obtained on intact 5S RNA because the downfield spectrum is so badly overlapped. Suffice it to say that the melting data show the existence of imino resonances presumably belonging to helices II and III which are as stable to heating as any in helices I, IV and V (i.e. fragment).

The position we are left with can be summarized as follows. In the one part of 5S RNA where we can document the existence of an important contribution of exchange to relaxation (helix V), and where the structure is unstable also by the criterion of Mg^{+2} sensitivity and temperature, we are able to get some NOEs. We are forced, therefore, to consider the possibility that helices II and III, where we see almost no NOEs, have non-standard geometries.

This is not an entirely comfortable view to take. Helix II is double stranded by the criteria of its resistance to small molecule probes and to single stranded nucleases (51,32). It is also cleaved by double stranded nucleases, something which might be expected to fail if the double strand were of nonstandard geometry. Clearly there is a need for further experimental work to see if this hypothesis can be confirmed or denied.

Whatever the outcome may be it is hard to escape the impression that tRNAs, as a class, are "better built" molecules than 5S RNA. Since there is nothing about their secondary structure notably superior to the secondary structure postulated for 5S RNA, one is led to suspect that it is the tertiary interactions of the molecule which render it so tractable for investigation by NMR.

Acknowledgements

We thank Betty Freeborn and Grace Sun for their assistance with the preparative biochemistry required for this project. NMR data were obtained at the Northeast Regional NMR Facility which is supported by the National Science Foundation (CHE-7916210). We acknowledge the help of Dr. Benedict Bangerter and Mr. Peter Demou in the use of the spectrometers in the facility. This work was supported by a grant from the National Institutes of Health (GM-32206).

References and Footnotes

1. Noller, H.F., *Ann. Rev. Biochem. 53,* 119-162 (1984).
2. Stöffler, G. & Stöffler-Meilicke, M., *Ann. Rev. Biophys. Bioeng. 13,* 303-330 (1984).

3. Dohme, F. & Nierhaus, K.H., *Proc. Natl. Acad. Sci. USA 73,* 2221-2225 (1976).
4. Brownlee, G.G., Sanger, F. & Barrell, B.G., *J. Mol. Biol. 34,* 379-412 (1968).
5. Erdmann, V.A., Wolter, J.H., Haysmans, E., Vandenberghe, A. & DeWachter, R. *Nuc. Acids Res. 12,* 3133-3166 (1984).
6. Delihas, N., Anderson, J. & Singhal, R.P., *Prog. in Nuc. Acid Res. & Mol. Biol. 31,* 161-190 (1984).
7. Brosius, J., Dull, T.J., Sleeter, D.D. & Noller, H.F., *J. Mol. Biol. 148,* 107-127 (1981).
8. Kime, M.J.& Moore, P.B., *Biochemistry 22,* 2615-2622 (1983).
9. Kime, M.J. & Moore, P.B., *FEBS Letters 153,* 199-203 (1983).
10. Kearns, D.R., Patel, D. & Shulman, R.G., *Nature (London) 229,* 338-339 (1971).
11. Kearns, D.R., Patel, D., Shulman, R.G. & Yamane, T., *J. Mol. Biol. 61,* 265-270 (1971).
12. Eigen, M., Angew, *Chem. Int. Ed. Engl. 3,* 1-19 (1964).
13. Leroy, J.L., Bolo, N., Figueroa, N., Plateau, P. & Gueron, M., *J. Biomolec. Struct. Dynam. 2,* 915-939 (1985).
14. Hare, D.R. & Reid, B.R., *Biochemistry 21,* 1835-1842 (1982).
15. Salemink, P.J.M., Raue, H.A., Heerschap, A., Planta, R.J. & Hilbers, C.W., *Biochemistry 20,* 265-272 (1981).
16. Douthwaite, S., Garrett, R.A., Wagner, R. & Feunteun, *J. Nuc. Acids Res. 6,* 2453-2470 (1979).
17. Johnston, P.D. & Redfield, A.G., *Nuc. Acids Res. 4,* 3599-3615 (1978).
18. Johnston, P.D. & Redfield, A.G., *Biochemistry 20,* 1147-1156 (1981).
19. Kime, M.J., Gewirth, D.T. & Moore, P.B., *Biochemistry 23,* 3559-3568 (1984).
20. Griffey, R.H., Poulter, C.D., Yamaizumi, Z., Nishimura, S. & Hurd, R.E., *J. Am. Chem. Soc. 104,* 5801-5811 (1982).
21. Griffey, R.H., Poulter, C.D., Yamaizumi, Z., Nishimura, S. & Hurd, R.E., *J. Am. Chem. Soc. 104,* 5811-5813 (1982).
22. Griffey, R.H., Poulter, C.D., Yamaizumi, Z., Nishimura, S. & Hawkins, B.L., *J. Am. Chem. Soc. 105,* 143-145 (1983).
23. Griffey, R.H., Poulter, C.D., Box, A., Hawkins, B.L., Yamaizumi, Z. & Nishimura, S., *Proc. Natl. Acad. Sci. USA 80,* 5895-5897 (1983).
24. Gonnella, N.C., Birdseye, T.R., Nee, M. & Roberts, J.D., *Proc. Natl. Acad. Sci. USA 79,* 4834-4837 (1982).
25. Kime, M.J., *FEBS Letters 173,* 342-346 (1984).
26. Kime, M.J., *FEBS Letters 175,* 259-262 (1984).
27. Kime, M.J. & Moore, P.B., *Biochemistry 22,* 2622-2629 (1983).
28. Aubert, M., Scott, J.F., Reynier, M. & Monier, R., *Proc. Natl. Acad. Sci. USA 77,* 3360-3364 (1980).
29. Lecanidou, R. & Richards, E.G., *Eur. J. Biochem. 57,* 127-133 (1975).
30. Kao, T.H. & Crothers, D.M., *Proc. Natl. Acad. Sci. USA 77,* 3360-3364 (1980).
31. Kime, M.J. & Moore, P.B., *Nuc. Acids Res. 16,* 4973-4983 (1982).
32. Kjems, J., Olesen, S.O. & Garrett, R.A., *Biochemistry 24,* 241-250 (1985).
33. Christensen, A., Mathiesen, M., Peattie, D. & Garrett, R.A., *Biochemistry 24,* 2284-2291 (1985).
34. Kearns, D.R., *Critical Rev. Biochem. 15,* 237-290 (1984).
35. Chou, S.-H., Wemmer, D.E., Hare, D.R. & Reid, B.R., *Biochemistry 23,* 2257-2263 (1984).
36. Crothers, D.M., Cole, P.E., Hilbers, C.W. & Shulman, R.G., *J. Mol. Biol. 87,* 63-88 (1974).
37. Johnston, P.D. & Redfield, A.G., *Biochemistry 20,* 3996-4006 (1981).
38. Reid, B.R. & Hare, D.R., *CIBA Found. Symp. 93,* 208-225.
39. Pardi, A., Morden, K.M., Patel, D.J. & Tinoco, I., Jr., *Biochemistry 21,* 6567-6574 (1982).
40. Heerschap, A., Haasnoot, C.A.G. & Hilbers, *Nuc. Acids Res. 10,* 6981-7000 (1982).
41. Heerschap, A., Haasnoot, C.A.G. & Hilbers, *Nuc. Acids Res. 11,* 4483-4499 (1983).
42. Heerschap, A., Haasnoot, C.A.G. & Hilbers, *Nuc. Acids Res. 11,* 4501-4520 (1983).
43. Roy, S. & Redfield, A.G., *Biochemistry 22,* 1386-1390 (1983).
44. Robillard, G.T., in *"NMR in Biology,"* Dwek, R.A., Campbell, I.D. & Williams, R.J.P. eds., pp. 201-230 (1977).
45. Marshall, A.G. & Smith, J.L., *Biochemistry 19,* 5955-5959 (1980).
46. Jardetzky, O. & Roberts, G.C.K., *"NMR in Molecular Biology,"* Academic Press, N.Y., Chapter 2 (1981).
47. Ramstein, J. & Erdmann, V.A., *Nuc. Acids Res. 9,* 4081-4088 (1981).

48. Johnston, P.D., Figueroa, N. & Redfield, A.G., *Proc. Natl. Acad. Sci. USA 76,* 3130-3134 (1979).
49. Figueroa, N., Keith, G., Leroy, J.L., Plateau, P., Roy, S. & Gueron, M. *Proc. Natl. Acad. Sci. USA,* 4330-4333 (1983).
50. Leroy, J.L., Bolo, N., Figueroa, N., Plateau, P. & Gueron, M., *J. Biomolec. Struct. & Dynam. 2,* 915-939 (1985).
51. Douthwaite, S. & Garrett, R.A., *Biochemistry 20,* 7301-7307 (1981).

Biomolecular Stereodynamics IV, Proceedings of the Fourth Conversation in the Discipline Biomolecular Stereodynamics, State University of New York, Albany, NY, June 04-09, 1985, Eds., Ramaswamy H. Sarma & Mukti H. Sarma, ISBN 0-940030-18-7, Adenine Press, ©Adenine Press 1986.

The structure of yeast tRNA^{Asp}. A model for tRNA interacting with messager RNA[61]

D. Moras, A.C. Dock, P. Dumas, E. Westhof, P. Romby, J.P. Ebel and R. Giegé
Institut de Biologie Moléculaire et Cellulaire du CNRS
15, rue René Descartes, 67084 Strasbourg Cedex, France

Abstract

The anticodon of yeast tRNA^{Asp}, GUC, presents the peculiarity to be self-complementary, with a slight mismatch at the uridine position. In the orthorhombic crystal lattice, tRNA^{Asp} molecules are associated by anticodon-anticodon interactions through a two-fold symmetry axis. The anticodon triplets of symmetrically related molecules are base paired and stacked in a normal helical conformation. A stacking interaction between the anticodon loops of two two-fold related tRNA molecules also exists in the orthorhombic form of yeast tRNA^{Phe}. In that case however the GAA anticodon cannot be base paired. Two characteristic differences can be correlated with the anticodon-anticodon association: the distribution of temperature factors as determined from the X-ray crystallographic refinements and the interaction between T and D loops. In tRNA^{Asp} T and D loops present higher temperature factors than the anticodon loop, in marked contrast to the situation in tRNA^{Phe}. This variation is a consequence of the anticodon-anticodon base pairing which rigidifies the anticodon loop and stem. A transfer of flexibility to the corner of the tRNA molecule disrupts the G19-C56 tertiary interactions. Chemical mapping of the N3 position of cytosine 56 and analysis of self-splitting patterns of tRNA^{Asp} substantiate such a correlation.

Transfer ribonucleic acids: the adaptor molecules

The existence and function of transfer RNA molecules were predicted by Crick in its adaptor hypothesis (1) before their biochemical discovery by Hoagland et al. (2). Crick's original thought limited the adaptor molecules to the trinucleotides corresponding to the anticodons. In fact these molecules turned out to be much more sophisticated, containing 70 to 90 nucleotides folded in a L-shaped three-dimensional structure (3) exhibiting numerous functions. Their main function, however, remains that of transfering the amino acid in its correct order in the growing polypeptide chain within the ribosome. That crucial step of protein synthesis requires a perfect deciphering of the codon of messenger RNA by the anticodon of tRNA (Fig. 1).

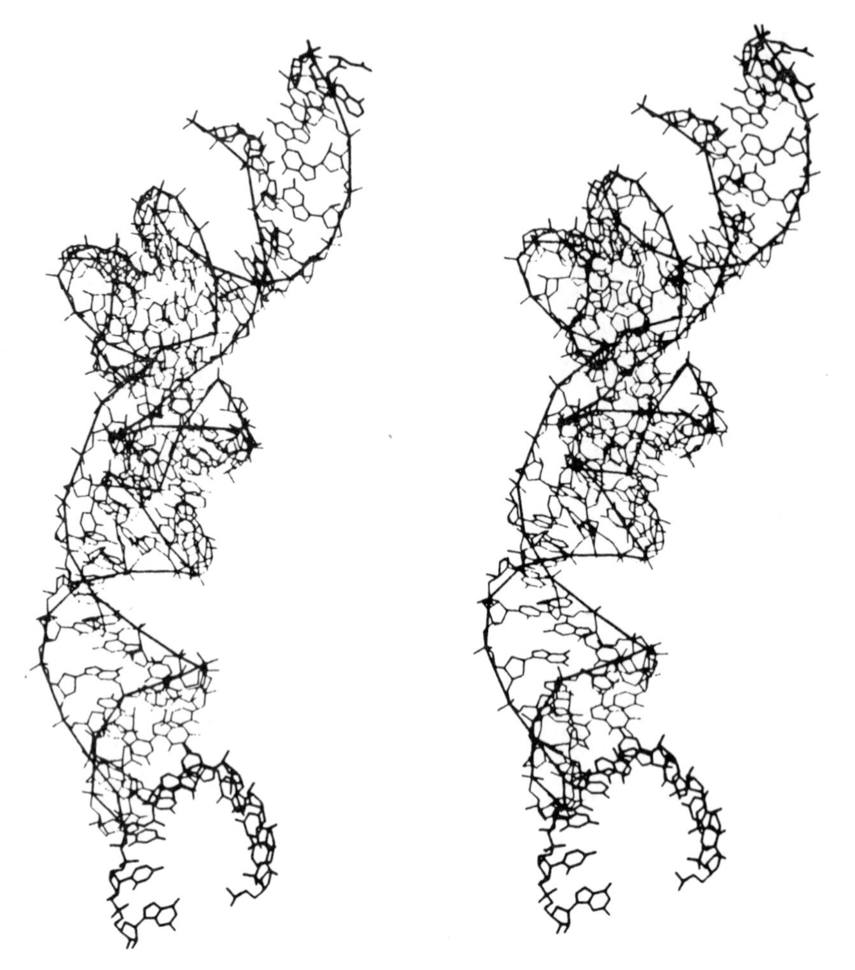

Figure 1. Model of codon-anticodon recognition. The figure represents the crystal structure of yeast $tRNA^{Asp}$ (4,5) interacting with a strand of mRNA. The geometry of the codon-anticodon interaction corresponds to that found in the packing of $tRNA^{Asp}$ crystal, where two molecules interact in a helical fashion through Watson Crick pairing.

Many models have been proposed to explain the molecular mechanism involved in that nucleic acid recognition process. Their goal was to try to bring a solution to the many questions raised by the origin of the degeneracy of the genetic code and the source of the selectivity. The wobble hypothesis of Crick (6) was a major contribution toward the understanding of the biochemical results. The missing triplet hypothesis (7) or the two out of three model (8) are two other significant attempts to the problem. Very soon it appeared that the binding to tRNA of free trinucleotides complementary to the anticodon (9,10) or the association between tRNAs of complementary anticodons is much stronger than expected (11,12). So for instance in 1976, Grosjean, Soll and Crothers reported on the basis of T-jump experiments that the binding constant of yeast $tRNA^{Phe}$ and *E. coli* $tRNA^{Glu}$, which have complementary anticodons, is about six orders of magnitude higher than expected for complementary trinucleotides (13). This prompted the systematic investigation of anticodon-anticodon interactions. In their 1978 study, Grosjean et al. (14) showed striking parallels with the genetic coding rules. In that paper the existence of a relatively stable yeast $tRNA^{Asp}$ duplex, formed through the interaction between its quasi self-complementary GUC anticodon, was mentioned for the first time. The lifetime of that complex was about 20 times shorter than that measured for the complex between yeast $tRNA^{Asp}$ (GUC) and *E. coli* $tRNA^{Val}$ (GAC). The crystal structure of yeast $tRNA^{Asp}$ solved from ammonium sulfate grown crystals (4,5), enabled the visualization of the $tRNA^{Asp}$ dimers. This prompted further investigations of anticodon interactions in solution by T-jump techniques (15,16). The main results of that study concern the stabilization of the $tRNA^{Asp}$ duplexes as well as of the complexes with a *E. coli* $tRNA^{Val}$ species by addition of ammonium sulfate up to 1.6 M (38% of saturation). Salt effects are essentially increasing the association rate constant by one order of magnitude, whereas no incidence on the dissociation constant was noticed.

This paper reports an analysis of anticodon-anticodon interactions in yeast $tRNA^{Asp}$ both in the crystalline state and in solution. Comparisons with other known tRNA structures and especially with that of yeast $tRNA^{Phe}$, refined at a comparable resolution, enables to draw some structure-function relationships. The most important one is the correlation between the association and long range conformational changes occuring in the D and T-loop regions.

Comparison of the crystal structures of yeast $tRNA^{Asp}$ and $tRNA^{Phe}$

The three-dimensional structure of two elongator tRNAs from yeast specific for phenylalanine (17-19) and for aspartic acid (4,5) are known in great details and with a similar accuracy. Both structures have been refined using X-ray diffraction data to a 3 A resolution or slightly better. Moreover for both tRNAs two different crystal forms were analysed and refined. For simplicity, when not specifically stated otherwise, the following description and comparisons will deal with the data of references 20 and 21. A detailed analysis of crystal form A, as well as the comparisons with crystal form B and $tRNA^{Phe}$ will be published later. Figure 2 shows the overall similarities of the two L-shaped molecules, but stresses also the main difference between them, which is the larger opening of the L in $tRNA^{Asp}$. As a consequence the distance between the anticodon and the amino acid accepting-end is significantly larger in $tRNA^{Asp}$ than in $tRNA^{Phe}$.

Other remarkable differences concern a more regular anticodon stem in $tRNA^{Asp}$ and the conformations of the D and T-loop regions. Whereas in $tRNA^{Phe}$ the interaction between these two loops involves a G-C Watson Crick pair between the second conserved G residue (G19) in the D-loop and the conserved C56 in the T loop, this interaction is absent in the structure of $tRNA^{Asp}$. Figure 3 presents the local structures in the two molecules. As a result of the disruption of the G-C base pair, a large part of the D-loop and the tip of the T-loop are labilized. This is reflected by the temperature factor distribution along the nucleotide chain. Although

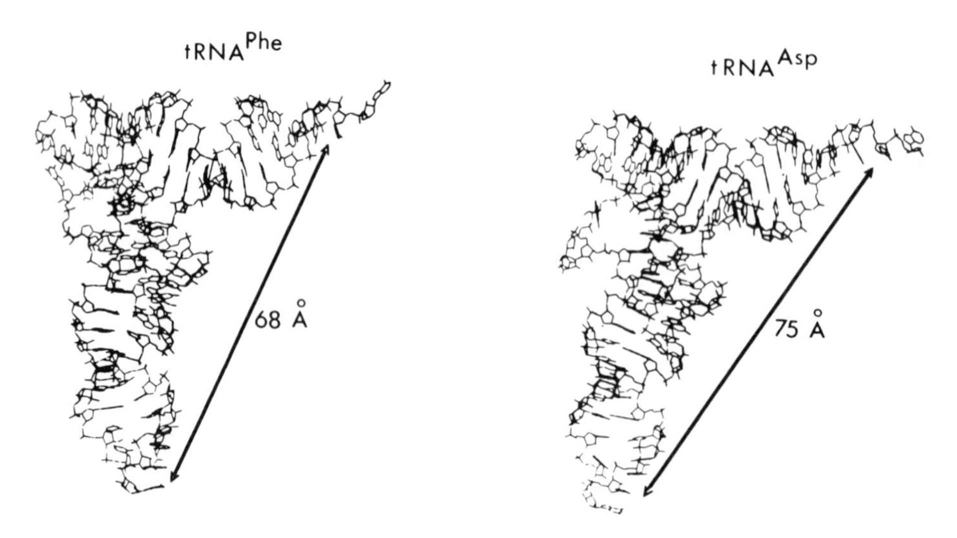

Figure 2. Comparison between the three-dimensional structure of yeast $tRNA^{Phe}$ and $tRNA^{Asp}$. The coordinates of $tRNA^{Phe}$ are taken from (20); those of $tRNA^{Asp}$ corresponds to refined form B (21). The figure illustrates the different overall opening of the L-shape in the two molecules. The distances shown on the picture are measured between phosphorus atoms of residues G34 and A73 or G73 respectively.

these factors, results of the least squares refinement, contain various components, it is legitimate to correlate large B values to flexible parts of the molecules (22,23). Two stereoviews representing the thermal vibrations along the backbone of each tRNA are shown on figure 4. These drawings clearly show the differences between $tRNA^{Phe}$ for which the most rigid part is the corner of the molecule and $tRNA^{Asp}$ for which part of the D-loop and the T-D corner are quite agitated. An opposite situation can be noticed for the anticodon stems. Packing effects cannot be the cause of the differences since a similar overall distribution of thermal factors is observed for the two different crystal forms of yeast $tRNA^{Phe}$.

The anticodon-anticodon interaction

A stereoview of a $tRNA^{Asp}$ dimer, as observed in the crystal structure, is shown in figure 5. The two two-fold related anticodon arms form a continuous helix whereas the two acceptor arms form an angle of 110°. Figure 6 summarizes some informations about the anticodon-anticodon interaction. Figure 6A shows a stereoview of the molecular model fitted in its experimental electron density. The symmetrically related anticodons form a short regular RNA helix. The modified base m1G37 stacks on both sides of the interacting G-C base pairs, enforcing the continuity with the anticodon stem via C38. This interstrand stacking can also modulate the stability of the association. The middle base U35 is the only one in that position which allows tRNA duplex formation. The model clearly shows that replacement of uridine bases by purines (A or G) will induce steric hindrance which in normal conditions would prevent external base pair formation. The model can also explain the absence of interaction between GCC anticodons. In that case the amino group at position 4 would come too close and the resulting repulsion forces would destabilize the duplex.

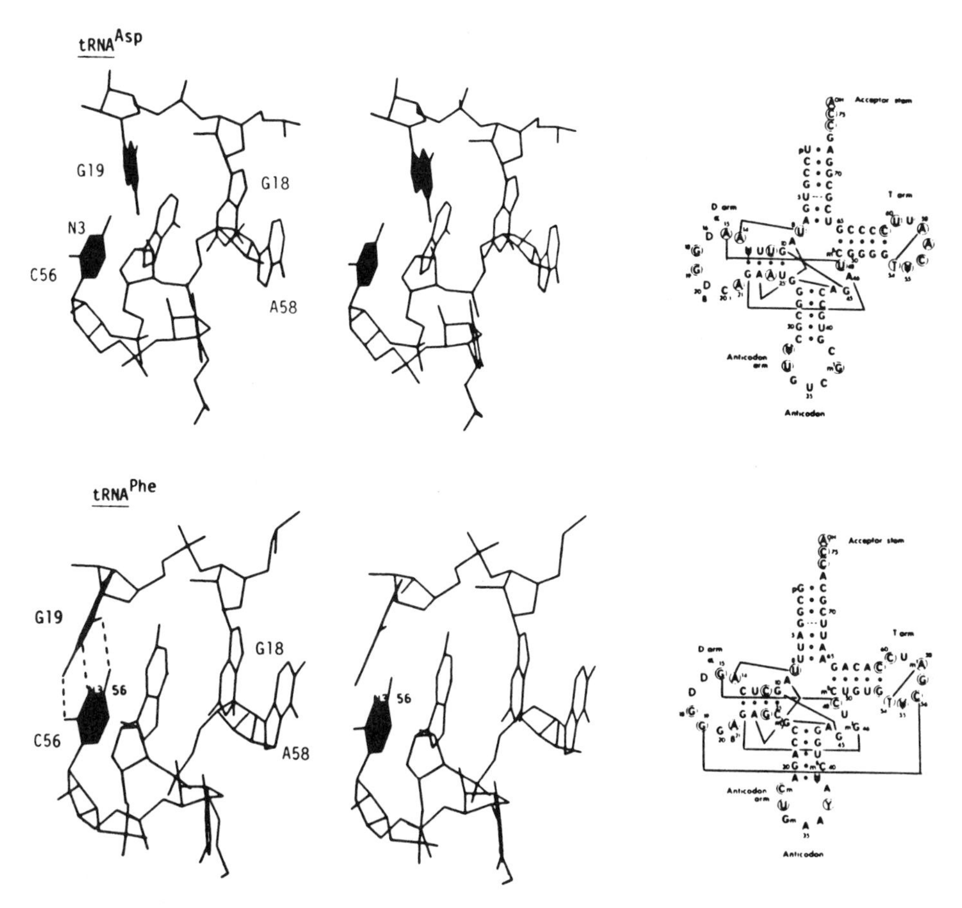

Figure 3. Nucleotide sequences of yeast $tRNA^{Asp}$ (39) and $tRNA^{Phe}$ (40) and structural environment of C56. The stereoviews show the G19-C56 Watson-Crick base pair in $tRNA^{Phe}$, and the changed orientation of G19 in $tRNA^{Asp}$ which prevents this pairing. Notice the intercalation of G18 between ψ57 and A58 in both tRNAs. The nucleotides are numbered according to the sequence of $tRNA^{Phe}$; thus the constant G18 and G19 correspond to the 17th and 18th nucleotide in the $tRNA^{Asp}$ sequence. Conserved and semi-conserved nucleotides in most elongator tRNAs are circled.

With U two possibilities exist: self-pairs formation via two hydrogen bonds ($O_4 \ldots H\text{-}N_3$ and $N_3\text{-}H \ldots O_2$), but that imposes a local crystalline disorder since that association does not obey the two-fold symmetry with the axis in the plane of bases; or no U-U pairing like in figure 6B. One of our heavy atom derivatives, a mercury salt, binds between the two uridines (fig. 6B). The distance between the two symmetry related N3 atoms of U35 is well suited for the binding of the mercury atom.

In order to compare the conformation of the anticodon loop in $tRNA^{Phe}$ and $tRNA^{Asp}$, a superposition of the two models was done using a graphic display. The best fit, presented on figure 6C, shows striking structural homologies. The observed differences in base orientations are at the limit of the significance for that resolution. When fitting the two loops as shown on the figure, the acceptor ends of the two molecules are distant by about 12 Å. Since in the ribosomes both molecules will

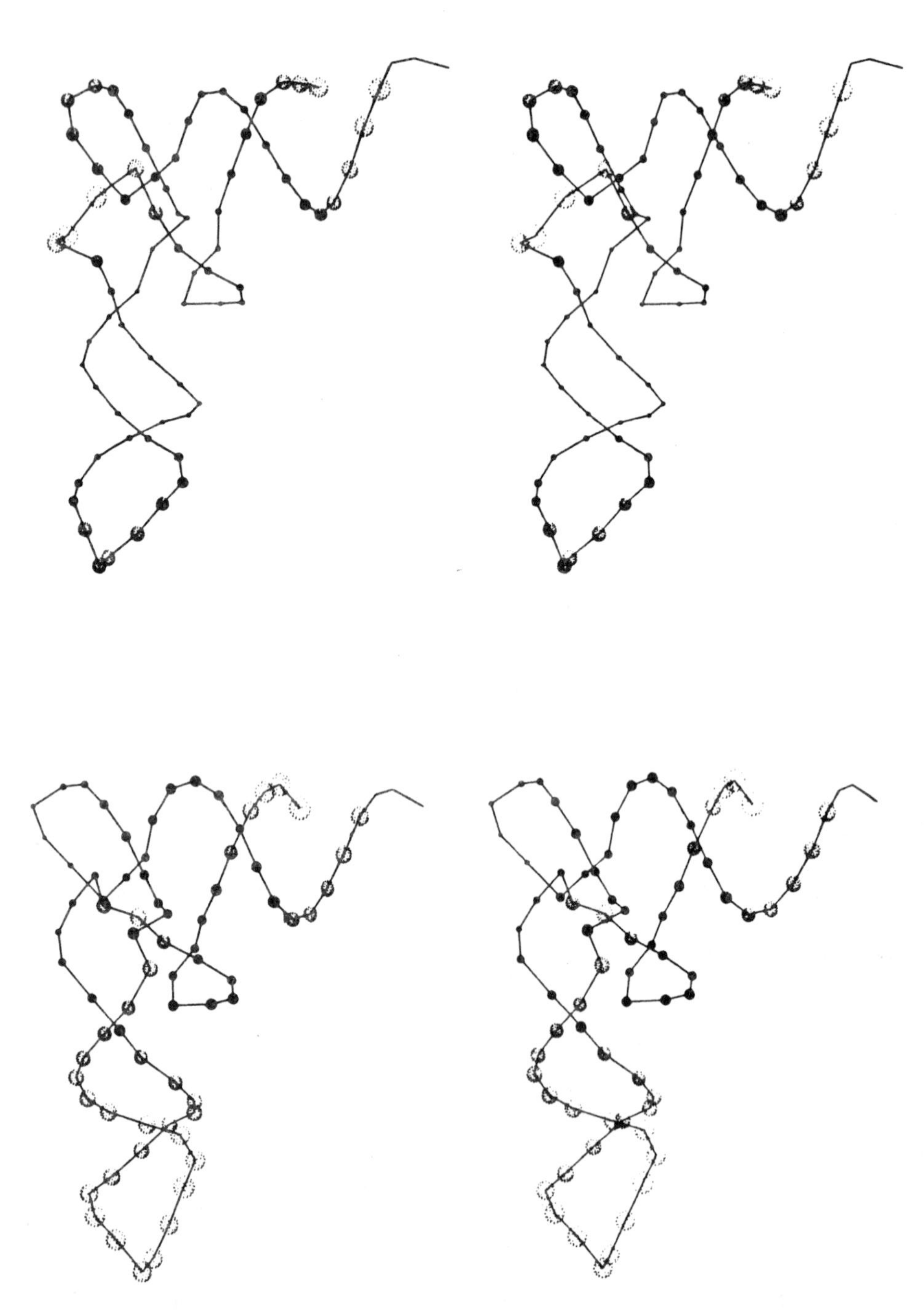

Figure 4. Thermal vibrations of the backbones of $tRNA^{Asp}$ (top) and $tRNA^{Phe}$ (bottom). The size of the spheres located at the positions of the phosphorus atoms are proportional to the averaged B factor of the corresponding residue. Values are taken from references 20 and 21. CCA-end values, not available for $tRNA^{Asp}$ and too much packing dependent for $tRNA^{Phe}$, are not shown.

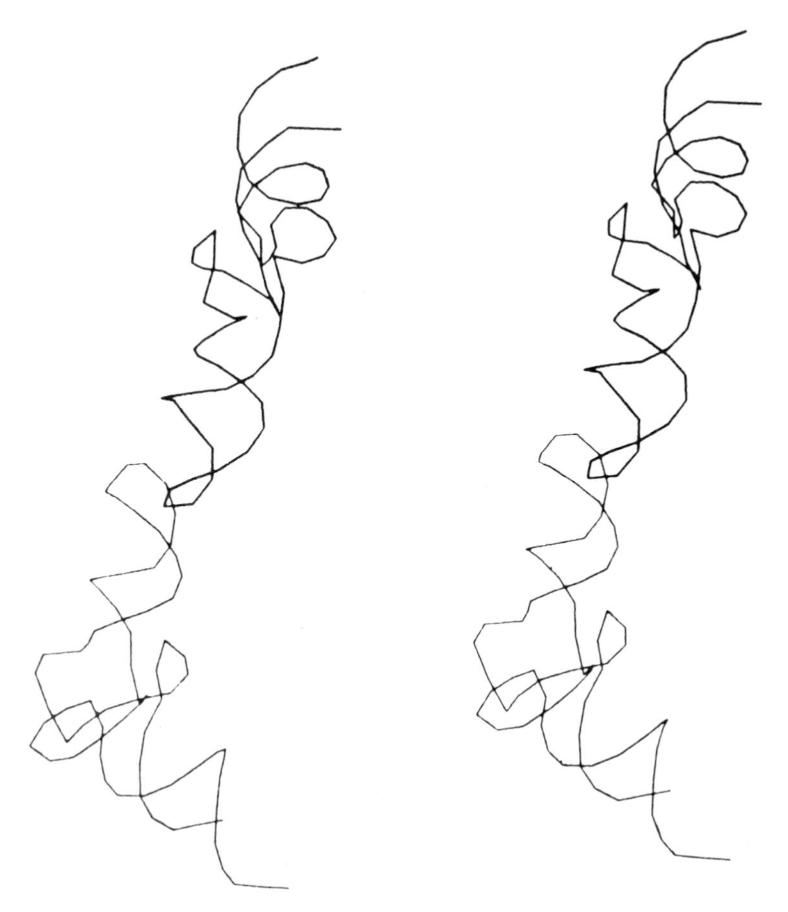

Figure 5. Stereoview of the phosphate backbone of a $tRNA^{Asp}$ duplex. Notice the continuous helix formed by the two anticodon arms and the 110° angle formed by the two acceptor stems.

bind to the messenger RNA in a similar way and deliver the amino acid at the same elongation site it is tempting to think that $tRNA^{Asp}$ represents the model of a bound tRNA whereas $tRNA^{Phe}$ is the free molecule. A corollary of that proposition is that we are looking at two different conformational states of these molecules.

Chemical fragility of $tRNA^{Asp}$ in solution and in the crystal

Due to the presence of the free 2′-hydroxyl residues on their ribose moieties, RNA molecules are intrinsically unstable (24), and under certain circumstances undergo self splitting. Under physiological conditions at neutral pH the extent of such splitting, except for self splicing (25), has to be extremely low, if not zero, otherwise RNAs could not exist. The development of electrophoretic methodologies for the analysis of end-labeled RNAs allowed to detect low levels of such degradations, and it was found that splitting exhibits a preference for pyrimidine-adenine sequences (26,27) and for certain positions in loop regions (e.g. 28) likely because their flexibility

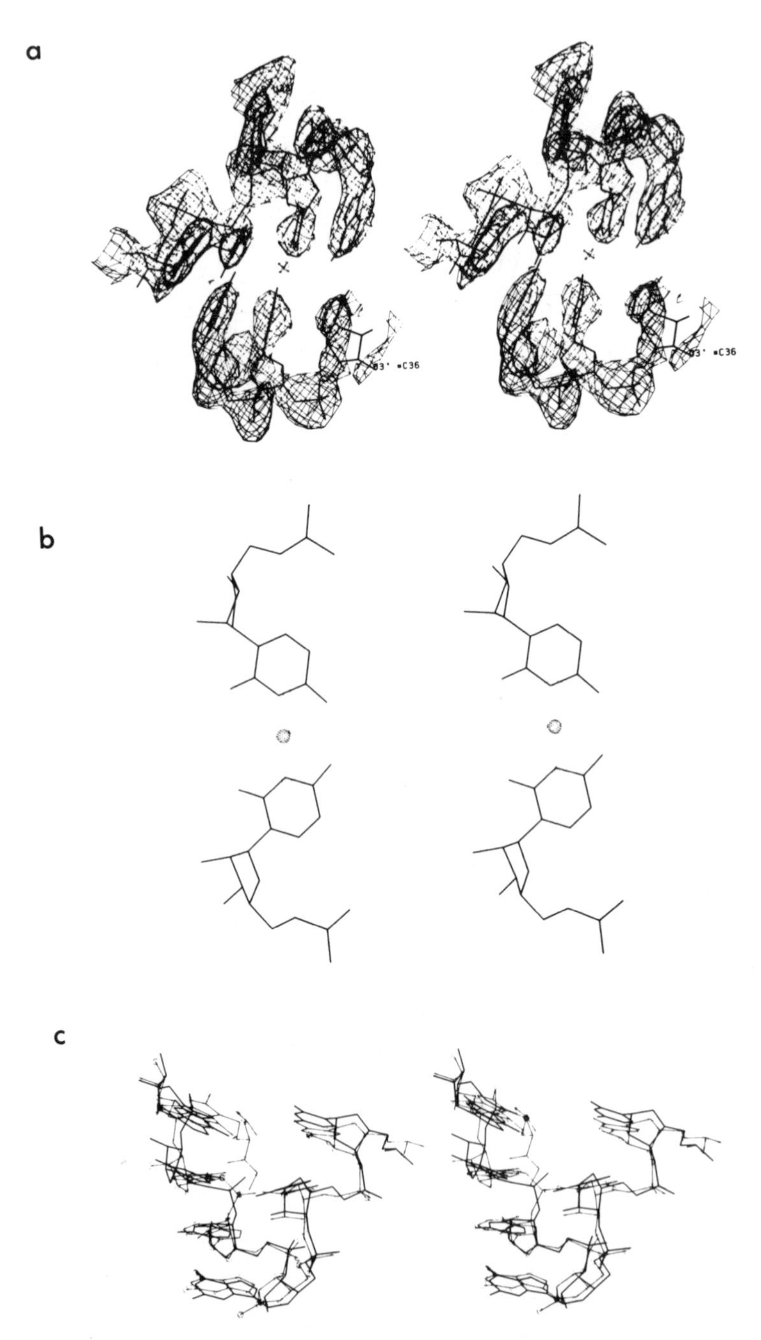

Figure 6. Anticodon-anticodon interaction in yeast tRNA^{Asp} (a) Stereoview of the molecular model in the experimental electron density map. The top includes residues 33 to 37, the bottom are anticodon residues 34 to 36. Notice the perfect interstrand stacking of the modified base m1G37 (top) with G34 (bottom) (b) Orientation of the two middle U35 residues with the position of the mercury atom, used in the MIR structure determination (4) sitting on the two-fold symmetry axis. (c) The superposition of the anticodon loops of tRNA^{Asp} in broad line and tRNA^{Phe} in light line (residues 32 to 37) shows the similarity of the two anticodon conformations.

favors the cyclization mechanism to occur. It was also found that metal ions might cut tRNA chains at particular positions (29-31). Thus a careful analysis of the self-splitting patterns of tRNAs might inform about some of their structural characteristics. Following these lines we have studied the self degradation of $tRNA^{Asp}$ when mixed with a minor species of *E. coli* $tRNA^{Val}$ having GAC anticodon strickly complementary with the GUC anticodon of $tRNA^{Asp}$. Two experimental condition were chosen, (i) in the presence of ammonium sulfate, spermine, $MgCl_2$ and cacodylate buffer at pH 6.8, conditions mimicking the crystallization medium of $tRNA^{Asp}$ and known to favor duplex formation (16), and (ii) in the absence of ammonium sulfate, under conditions where the association constant is lower. The results displayed in Figure 6 show striking differences. In the presence of ammonium sulfate, cuts occur in the anticodon and D-loop; in the absence of the salt, when the duplex is less stable, the cuts in D-loop are practically suppressed. It is interesting to recall that similar experiments conducted on $tRNA^{Asp}$ alone, either in a standard solution or crystallized in the orthorhombic form with ammonium sulfate yield similar results. For the free molecules, cuts mainly occur in the anticodon loop at position U33, G34, U35 and C36. For the tRNA recovered from crystals, they are found predominantly in the D-loop (positions G18, G19 and D20) and confirm the increased flexibility of this part of the loop, when the tRNA is associated (32).

Mapping of the solution structure of tRNAs with chemical probes

Solution structure of tRNAs can be approached by a variety of physical and biochemical techniques (for reviews see ref. 33,34). Structural mapping of RNA structures with chemical probes became in recent years a popular method (35,36), because of the small size of the reagents which allows to have access to discrete conformational features in the tRNA. The principle of the chemical approach derives from the chemical sequencing methodologies of nucleic acids (37,38) and lies in a statistical and low yield modification of them at each potential target in such a way that each molecule undergoes less than one modification. The end-labelled tRNAs are then specifically split at the modified positions and the resulting end-labelled oligonucleotides are analyzed by polyacrylamide gel electrophoresis and autoradiography.

To investigate the effects induced by anticodon-anticodon interactions in tRNAs we studied the accessibility of the N3 position in cytosine to dimethylsulfate (DMS). Because this DMS reaction is sensitive to Watson-Crick G-C pairing it allows to probe the dimerization of $tRNA^{Asp}$ through GUC anticodon associations, as well as the existence or non-existence of the tertiary interaction G19-C56. Figure 7 displays two autoradiograms of 15% polyacrylamide gels on 3′-end labelled $tRNA^{Asp}$. It shows the modification of cytosine residues on both free $tRNA^{Asp}$ and on $tRNA^{Asp}$ mixed in the presence of ammonium sulfate with an equimolar amount of complementary *E. coli* $tRNA^{Val}$ (GAC). On free $tRNA^{Asp}$ (lanes 1 to 3), C36 in anticodon is reactive, as expected under all experimental conditions; C56 on the contrary is completely protected against alkylation under native conditions. The fact that C56 becomes reactive under semi-denaturating conditions, in contrast to other cytosines located in helical regions (e.g. residues 29, 31, 42, 43 and 61 to 64), is an indication

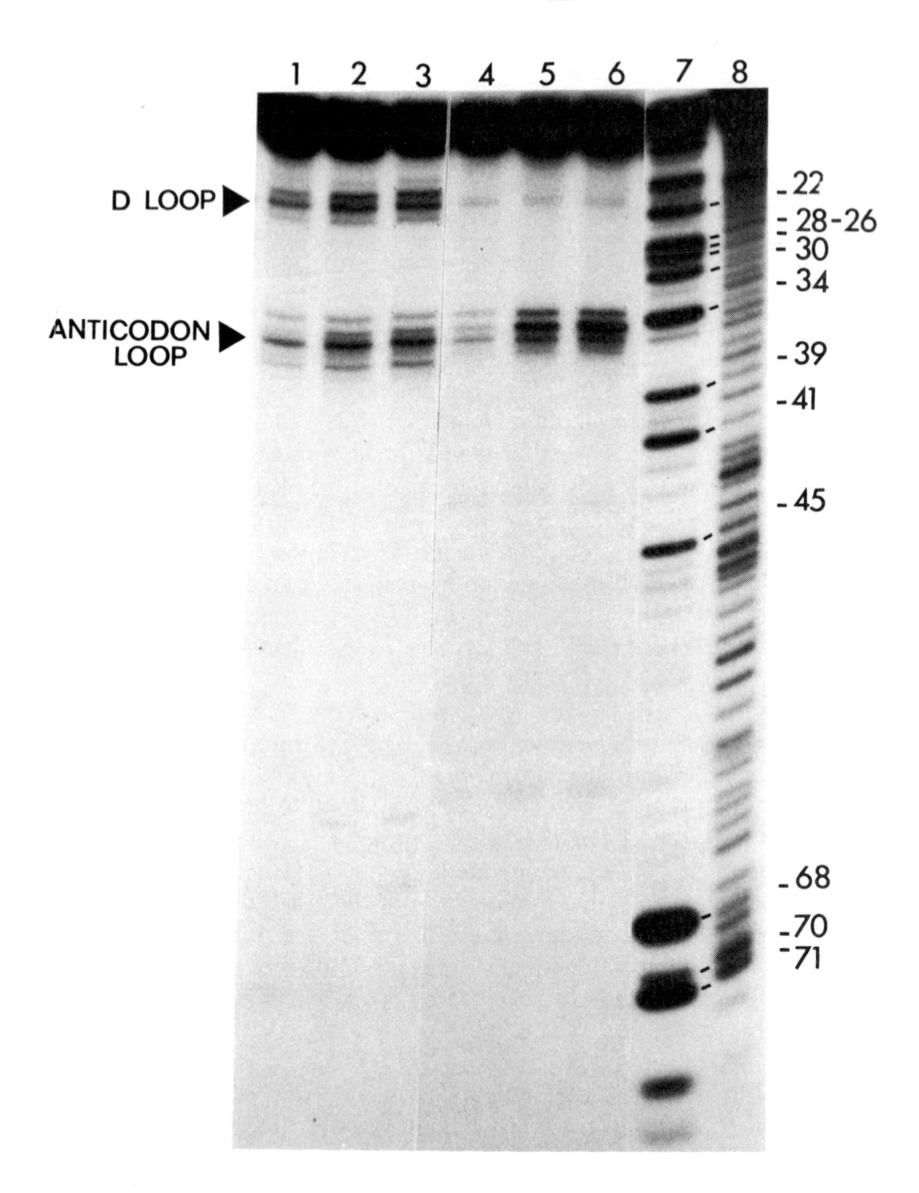

Figure 7. Self splitting patterns in solution of free and duplex tRNA^{Asp}. Autoradiogram of a 15% polyacrylamide gel run in the presence of urea of 3′-end labelled yeast tRNA^{Asp} (GUC) mixed with cold *E. coli* tRNA^{Val} (GAC). Lanes 1 to 37: Duplex between tRNA^{Asp} and tRNA^{Val} formed in the presence of Na cacodylate (pH 6.8) 10 mM, $MgCl_2$ 10 mM, spermine 3 mM, and ammonium sulphate 1.6 M; lanes 4 to 6: same conditions, but without ammonium sulfate; lane 7: partial ribonuclease T_1 digest; lane 8: formamide ladder. 3′-end labelled tRNA^{Asp} was incubated with cold tRNA^{Val} at a final concentration of 20 μM for each tRNA. Lanes 1 and 4: incubation at 4°C for 1 hour; lanes 2 and 5: incubation at 15°C for 4 days and lanes 3 and 6: incubation at 20°C for 17 days. In the presence of ammonium sulphate most molecules are in duplex form; in the absence of the salt the media contains a mixture of free and complexed tRNAs.

for the relative weakness of the G19-C56 bond. When $tRNA^{Asp}$ (GUC) is mixed with the complementary $tRNA^{Val}$ (GAC) (lanes 6 to 16) it is possible to detect anticodon-anticodon complexes. Below 30°C in the presence of ammonium sulfate, it is known from T-jump data (16), that most $tRNA^{Asp}$ molecules interact through their anticodons. This fact is also reflected by the chemical mapping since C36 in the anticodon is protected against alkylation (compare the weak C56 bands, lanes 6 to 11, with the strong bands, lanes 1 to 3). Under these conditions however C56 becomes partially reactive (compare lanes 5, 7 and 11 with lane 1). At higher temperature, above 30°C, C56 remains reactive, but the complex tends to dissociate, as reflected by the increased reactivity of C36 (lanes 12 to 16).

Discussion

For the problem of anticodon-anticodon association the first result of the crystallographic investigation is the close conformational similarity of both anticodon loops, that of the free $tRNA^{Phe}$ molecule and that of the bound $tRNA^{Asp}$. This *experimental* model is very close to the one derived from the nearest neighbour model (49) and, except for the position of the conserved U33, also close to the one proposed by Fuller and Hodgson in 1967 (50). It can easily account for most of the observations made in solution, like the role of the middle base and the importance of the almost always modified adjacent purine 37 (13). The absence of drastic conformational changes at the anticodon loop level stresses the lack of necessity for these perturbations. Furthermore recent NMR studies on model systems, $tRNA^{Asp}$ and a pentadecamer comprising the anticodon loop and stem of yeast $tRNA^{Phe}$, have shown that binding of cognate trinucleotides results only in minor perturbations in the structure of the anticodon loops (51,52).

In addition this paper describes several experimental evidences in favor of the existence of a long range conformational transition between free tRNA and dimeric tRNAs associated through anticodon-anticodon interactions. The crystallographic evidences essentially include the different distribution of temperature factors along the polynucleotide chain of $tRNA^{Phe}$ and $tRNA^{Asp}$ and the disruption of the conserved G-C Watson-Crick base pair at the corner of the molecule. Chemical mapping of cytosines in solution clearly establishes the correlation between anticodon association and C56 accessibility. Analysis of the self-splitting patterns are another proof of the sequentially induced conformational change.

A question remains concerning the importance of the conformational change. From the bulk of our data which combine crystal and solution studies, it is clear that the change, although important by its effect on the accessibility of C56 and the distance between anticodon and acceptor ends, is not as drastic as has been postulated by various authors (53,54,55). Could solvent effects or crystal packing explain the observed differences or suggest some restriction to more important conformational fluctuations? Solvent effects can easily be excluded by our solution studies. In absence of ammonium sulfate, formation of duplexes is hampered and all the observations agree with a more $tRNA^{Phe}$ like structure for $tRNA^{Asp}$. On another

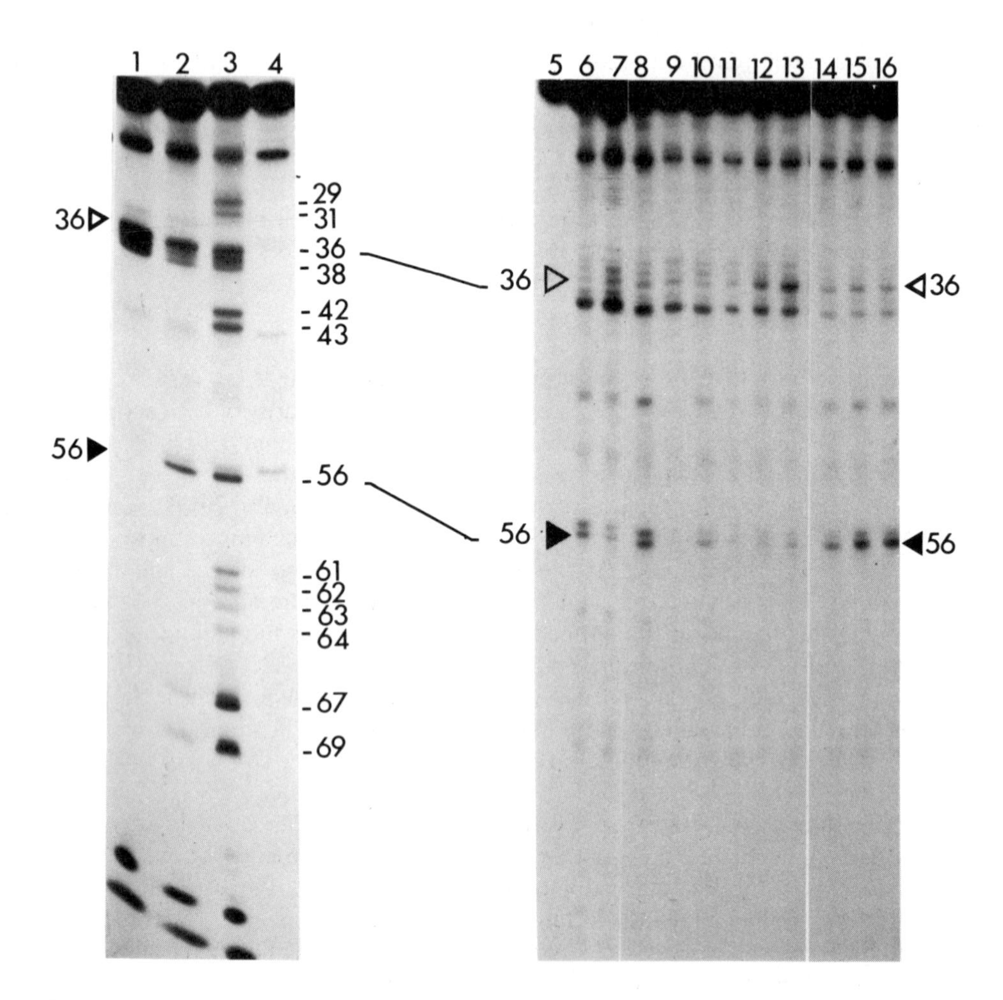

Figure 8. Structural mapping of the N3 position in cytosine residues of yeast $tRNA^{Asp}$ by dimethylsulphate. The experimental conditions are essentially those described by Peattie and Gilbert (37) and Romby et al. (38). The figure shows autoradiograms of 15% polyacrylamide gels in the presence of urea of 3′-end labelled yeast $tRNA^{Asp}$ (GU), either free (lanes 1 to 4) or mixed with cold *E. coli* $tRNA^{Val}$ (GAC) (lanes 5 to 16). Numbers on the gel correspond to C residues. Alkylation experiments were conducted under native conditions (Na cacodylate (pH 6.8) 10 mM; $MgCl_2$ 10 mM; with or without ammonium sulphate 1.6 M) (lane 1 is without salt and lanes 6, 8, 10, 12, 14 and 16 are with the salt); under semi-denaturing conditions (Na cacodylate (pH 6.8) 10 mM; EDTA 1 mM with or without ammonium sulphate 1.6M) (lane 2 is without salt and 7, 9, 11, 13 and 15 are with the salt) and under denaturing conditions (same as before, but at high temperature). Lanes 4 and 5 are incubation controls. Modifications were done at 0°C (lanes 6, 7); 15°C (lanes 8, 9); 25°C (lanes 10, 11); 30°C (lanes 1, 2, 12, 13); 50°C (lanes 14, 15); 75°C (lane 16) and 90°C (lane 3). Incubation times varied and decreased from 150 min at 0°C to 1 min at 90°C. The full triangle indicates the position of C56; the upper band corresponds to a CpA nonspecific cut (see ref. 27 for more details). The open triangle corresponds to C36 in the anticodon loop.

side the crystal structure of yeast $tRNA_f^{Met}$, solved from crystals grown in ammonium sulfate, shows striking similarities to that of $tRNA^{Phe}$ (56). In the refined structure the angle between the two limbs of the molecule is even slightly smaller than in $tRNA^{Phe}$ (57). All together these observations confirm that ammonium sulfate is important for duplex formation, but without major incidence on the conformation of the molecules. Similarly it is difficult to totally exclude packing effects but the fact that two different packings of $tRNA^{Phe}$ and another one for $tRNA_f^{Met}$ lead to similar molecules is a strong indication that these effects must be minor. One can add that to reach some of the proposed structures it is evident that more tertiary interactions have to be broken. For example to form an angle of 30° between the two limbs of $tRNA^{Phe}$, it is necessary to unstack G18. It is in fact legitimate to assume that the crystal structure gives the correct amplitude of the conformational change.

The proposed mechanism of long range transfer of conformational change could involve the base stacking, through a domino process as suggested by the variation of the twist values between $tRNA^{Phe}$ and $tRNA^{Asp}$ (5,32). In agreement with this view are laser Raman spectroscopic data obtained on $tRNA^{Phe}$ and $tRNA^{Asp}$ (58,59). Spectroscopic measurements were conducted with tRNA in the mM range which means that $tRNA^{Asp}$ was partly associated as dimers, whereas $tRNA^{Phe}$ was the free molecule. Beside the classical interpretation of most spectral lines, reflecting the base composition, the important stacking and the similarity in the ribose-phosphate backbones of the two tRNAs, it was found that for $tRNA^{Asp}$ a reversible transitions occured around 20°C (59,60). The effect is most pronounced for lines specific for bases and indicates a better stacking at low temperature. The effect is weakest for the phosphate backbone line, suggesting that the conformational transitions arise more by rearrangements in base orientations than by alterations in the backbone. Since such low temperature transitions were only found with $tRNA^{Asp}$, it is tempting to propose that they reflect the conformational rearrangements when the tRNA is going from the dimeric to the monomeric conformation and are indications of the structural changes propagating along the helical stems.

All the results and comments made about this study on anticodon-anticodon association can be generalized to codon-anticodon interactions. The present model is the closest experimentally available for structural investigations at the atomic level. Further studies, like the analysis of another duplex by crystallography or by NMR would enable to refine the mechanism of recognition. The investigation of tRNA-mRNA recognition within the ribosome is more challenging and will require more experimental data.

Acknowledgments

This research was supported by grants from the Centre National de la Recherche Scientifique (CNRS), the Ministère de la Recherche et de la Technologie (MRT), and Université Louis Pasteur (Strasbourg, France). We thank Drs. H. Grosjean, C. Houssier and J.C. Thierry for stimulating discussions.

References and Footnotes

1. Crick, F.H.C., *Biochem. Soc. Symp.* (Cambridge, England) 14-25 (1957).
2. Hoagland, M.B., Stephenson, M.L., Scott, J.F., Hecht, L.I. and Zamecwik, P.C., *J. Biol. Chem. 231,* 241-257 (1958).
3. Rich, A. and RajBhandary, U.L., *Ann. Rev. Biochem. 45,* 805-860 (1976).
4. Moras, D., Commarmond, M.C., Fischer, J., Weiss, R., Thierry, J.C., Ebel, J.P. and Giegé, R., *Nature (London) 288,* 669-674 (1980).
5. Westhof, E., Dumas, P. and Moras, D., *J. Mol. Biol. 183,* in press (1985).
6. Crick, F.H.C., *J. Mol. Biol. 19,* 548-555 (1966).
7. Ninio, J., *J. Mol. Biol. 6,* 63-82 (1971).
8. Lagerkvist, U., *Proc. Natl. Acad. Sci. U.S.A. 75,* 1759-1762 (1978).
9. Yarus, M., *Science (Washington), 218,* 646-652 (1982).
10. Eisinger, J., Feuer, B. and Yamane, T., *Nature New Biology 231,* 126-130 (1971).
11. Yoon, K., Turner, D.H. and Tinoco, I., *J. Mol. Biol. 99,* 507-518 (1975).
12. Eisinger, J. and Gross, N., *Biochemistry 14,* 4031-4041 (1975).
13. Grosjean, H. and Chantrenne, H., *Mol. Biol. Biochem. Biophys. 32,* 347-367 (1980) and ref. therein.
14. Grosjean, H., Soll, D.G. and Crothers, D.M., *J. Mol. Biol. 103,* 499-519 (1966).
15. Grosjean, H., De Henau, S. and Crothers, D., *Proc. Natl. Acad. Sci. U.S.A. 75,* 610-614 (1978).
16. Romby, P., Giegé, R., Houssier, C., and Grosjean, H., *J. Mol. Biol. 184,* in press (1985).
17. Quigley, G.J., Wang, A., Seeman, N.C., Suddath, F.L., Rich, A., Sussmann, J.L. and Kim, S.H., *Proc. Natl. Acad. Sci. U.S.A. 72,* 4866-4870 (1975).
18. Jack, A., Ladner, J.E. and Klug, A., *J. Mol. Biol. 108,* 619-649 (1976).
19. Stout, C.D., Mizuno, H., Rao, S.T., Swaminathan, P., Rubin, J., Brennan, T. and Sundaralingam, M., *Acta Crys. B54,* 1529-1544 (1978).
20. Quigley, G.J., Seeman, N.C., Wang, A.H.J., Suddath, F.L. and Rich, A., *Nucleic Acids Res. 2,* 2329-2339 (1975).
21. Dumas, P., Ebel, J.P., Giegé, R., Moras, D., Thierry, J.C. and Westhof, E., *Biochimie (Paris)* in press (1985).
22. Frauenfelder, H., Petsko, G.A. and Tsernoglou, D., *Nature (London) 280,* 558-560 (1979).
23. Artymiuk, P.J., Blake, C.C.F., Grace, D.E.P., Oatley, S.I., Phillips, D.C. and Sternberg, M.J.E., *Nature (London) 283,* 563-566 (1979).
24. Brown, D.M., in *"Basic Principles in Nucleic Acid Chemistry" (Tso, P.O.P. ed.) vol 2,* pp. 1-90, Academic Press Inc. New York and London (1974).
25. Zang, A.J., Kent, J.R. and Cech, T.R., *Science 224,* 574-577 (1984).
26. Carbon, P., Ehresmann, C., Ehresmann, B. and Ebel, J.P., *FEBS Lett. 94,* 152-156 (1978).
27. Romby, P., Moras, D., Bergdoll, M., Dumas, P., Vlassov, V.V., Westhof, E., Ebel, J.P. and Giegé, R., *J. Mol. Biol. 184,* in press (1985).
28. Riehl, N., Giegé, R., Ebel, J.P. and Ehresmann, B., *FEBS Lett. 154,* 42-46 (1983).
29. Werner, C., Krebs, B., Keith, G. and Dirheimer, G., *Biochem. Biophys. Acta 432,* 161-175 (1976).
30. Brown, R.S., Hingerty, B.E., Dewan, J.C. and Klug, A., *Nature (London) 303,* 543-546 (1983).
31. Rubin, J.R. and Sundaralingam, M., *J. Biomol. Struct. Dyn. 1,* 639-646 (1983).
32. Moras, D., Dock, A.C., Dumas, P., Westhof, E., Romby, P., Ebel, J.P. and Giegé, R., submitted.
33. Schimmel, P.R., Soll, D. and Abelson, J.N. (eds.) *Transfer RNA: Structure, Properties and Recognition, Cold Spring Harbor Monogr. Ser. 9A, New York,* 577 pp. (1979).
34. Schimmel, P.R. and Redfield, A.G., *Ann. Rev. Biophys. Bioeng. 9,* 181-221 (1980).
35. Peattie, D.A. and Gilbert, W., *Proc. Nat. Acad. Sci. U.S.A. 77,* 4679-4682 (1980).
36. Comparative mapping of yeast $tRNA^{Asp}$ and $tRNA^{Phe}$ with DMS and DEPC: paper in preparation by Romby et al. (1985).
37. Peattie, D.A., *Proc. Acad. Sci. U.S.A. 76,* 1760-1764 (1979).
38. Maxam, A.M. and Gilbert, W., *Methods in Enzymology 65,* 499-559 (1980).
39. Keith, G., Gangloff, J., Ebel, J.P. and Dirheimer, G., *Seances Acad. Sci. (Paris) 271,* 613-616 (1972).
40. Rajbhandary, U.L. and Chang, S.H., *J. Biol. Chem. 243,* 598-608 (1968).
49. Bubienko, E., Cruz, P., Thomason, J.F. and Borer, P.N., *Prog. Nucl. Acid. Res. Mol. Biol. 30,* 41-90 (1983).

50. Fuller, W. and Hodgson, A., *Nature (London) 215,* 817-821 (1967).
51. Clore, C.M., Gronenborn, A.M., Piper, E.A., McLaughlin, L.W., Graeser, E. and Van Boom, J.H., *Biochem. J. 221,* 737-751 (1984).
52. Gronenborn, A.M., Clore, G.M., MacLaughlin, L.W., Graeser, E., Lorber, B. and Giegé, R., *Eur. J. Biochem. 145,* 359-364 (1984).
53. Nilsson, L., Rigler, R. and Laggner, P., *Proc. Natl. Acad. Sci. U.S.A. 79,* 5891-5895 (1982).
54. Geerdes, H.A.M., Van Boom, J.H. and Hilbers, C.W., *J. Mol. Biol. 142,* 195-217 (1980).
55. Woese, C.R., *Nature (London) 226,* 817-820 (1970).
56. Schevitz, R.W., Podjarny, A.D., Krishnamachari, N., Hughes, J.J., Sigler, P.D. and Sussman, J.L., *Nature (London) 278,* 188-192 (1979).
57. Sussman, J.L. and Podjarny, A.D., *Acta Cryst. B39,* 495-505.
58. Chen, M.C., Giegé, R., Lord, R.C. and Rich, A., *Biochemistry 17,* 3134-3138 (1978).
59. Huong, P.U., Audrey, E., Giegé, R., Moras, D., Thierry, J.C. and Comarmond, M.B., *Biopolymers 23,* 71-81 (1984).
60. Giegé, R. and Moras, D., *Spectr. Chem. Act.,* Submitted.
61. Reprinted from the *Journal of Biomolecular Structure & Dynamics 3,* 479-493 (1985).